Penguin Books
The Blind Watchmaker

'The most brilliant contemporary preacher of Charles Darwin's theory of evolution ... Dawkins has done it again and defended modern Darwinist orthodoxy with wit and passion'

Daily Telegraph

'A lovely book, original and lively, it expounds the ins and outs of evolution with enthusiastic clarity, answering, at every point, the cavemen of creationism'

Isaac Asimov

'It succeeds quite brilliantly. Most particularly, again and again, it brings home the nature and force of the central evolutionary mechanism of natural selection in a way that I have never seen or felt previously. The closest analogy I can think of is Galileo's Dialogues which made reasonable the Copernican Revolution, and I hope I will not be thought to be pushing things to an embarrassing point if I say that Dawkins' book can be compared to Galileo's, not only in type but in standard'

Professor Michael Ruse

'He disposes of the arguments for God the Designer without diminishing our sense of the mystery and complexity of our world ... indeed he increases our sense of wonder. Dawkins's speculative attempts to account for human self-consciousness and our interest in our own doings are ingenious and daring'

A. G. Cairns-Smith, *Independent*

'I admired Richard Dawkins's *The Blind Watchmaker* – or Son of *Selfish Gene*: enough to say that it shares the grave wit and thrilling godlessness of the earlier book'

Martin Amis, Book of the Year, *Observer*

'A full-throated defence of orthodox neo-Darwinism . . . He robustly sees off creationists on the right and the spectrum of critics of his version of orthodoxy on the left . . . There is much which is good and not merely clever about this book'

Steven Rose, *New Statesman*

'One great virtue of this book is to demonstrate the incompatibility of Darwinism and theism. The author points out the inconsistency of these sophisticated theologians who wish to appear in step with modern science, while at the same time retaining a corner for God in their cosmogony'

Churchman

'Brilliant exposition, tightly argued but kept readable by plentiful recourse to analogies and examples . . . *The Blind Watchmaker* shows what a convincing scientific argument looks like; it is popular science at its best. An invigorating minor theme is provided by the sideswipes that Dawkins hands out to creationists, erring colleagues, misguided interlopers from other sciences, and the media that gleefully misreport their muddleheaded musings. Highly recommended'

The Times

'An astonishingly lucid exposition of Darwinism . . . Dawkins is a born writer with an unmatched gift for the brilliant metaphor, the inspired syntactic switch, and the relevant zoological detail. *The Blind Watchmaker* is entertaining as well as engrossing: Dawkins's most wonderful book'

Francisco J. Ayala, Professor of Genetics, University of California

'For those who like good writing, tight argument and unpulled punches, this is a satisfying book. One by one, the author takes the arguments of Darwin's critics and drives a juggernaut of logic through them'

Economist

'Dr Dawkins is determined to hunt down any theologian who tries to prove, or even suggest, God by finding gaps in neo-Darwinian biology . . . And it is surely important that those who would defend religious belief in the age of science should grasp the full force of the explanation of nature which has no need of God'

David L. Edwards, *Church Times*

'A magical guided tour through the land of the blind watchmaker . . . apparently dry-as-dust intellectual theories and concepts can, in the end, lead to the most unimaginably glorious, majestic, just plain thrilling products . . . so with Dawkins' book'

Professor Angus Martin, *Zoo News*

'He succeeds admirably in showing how natural selection allows biologists to dispense with such notions as purpose and design and he does so in a manner readily intelligible to the modern reader'

Michael T. Ghiselin, *The New York Times*

'Richard Dawkins is a superb teacher. He has a clear story to tell and it comes across with verve, wit, convincing insight, and plenty of good illustrations'

John Habgood, Archbishop of York, *Education*

Richard Dawkins was born in Nairobi in 1941. He was educated at Oxford University, and after graduation remained there to work for his doctorate with the Nobel Prize-winning ethologist Niko Tinbergen. From 1967 to 1969 he was an Assistant Professor of Zoology at the University of California at Berkeley. In 1970 he became a Lecturer in Zoology at Oxford University and a Fellow of New College. In 1995 he became the first Charles Simonyi Professor of the Public Understanding of Science at Oxford University.

Richard Dawkins's first book, *The Selfish Gene* (1976; second edition, 1989), became an immediate international bestseller and, like *The Blind Watchmaker*, was translated into all the major languages. Its sequel, *The Extended Phenotype*, followed in 1982. His other bestsellers include *The Blind Watchmaker* (1986; Penguin, 1988), *River Out of Eden* (1995), *Climbing Mount Improbable* (1996; Penguin, 1997) and *Unweaving the Rainbow* (Penguin, 1999).

Richard Dawkins won both the Royal Society of Literature Award and the *Los Angeles Times* Literary Prize in 1987 for *The Blind Watchmaker*. The television film of the book, shown in the *Horizon* series, won the Sci-Tech Prize for the Best Science Programme of 1987. He has also won the 1989 Silver Medal of the Zoological Society of London and the 1990 Royal Society Michael Faraday Award for the furtherance of the public understanding of science. In 1994 he won the Nakayama Prize for Human Science and in 1995 was awarded an Honorary D.Litt. by the University of St Andrews and by the National University, Canberra. In 1997 he was elected a Fellow of the Royal Society of Literature and won the Cosmos International Prize.

RICHARD DAWKINS

THE BLIND WATCHMAKER

PENGUIN BOOKS

PENGUIN BOOKS

Published by the Penguin Group
Penguin Books Ltd, 27 Wrights Lane, London W8 5TZ, England
Penguin Putnam Inc., 375 Hudson Street, New York, New York 10014, USA
Penguin Books Australia Ltd, Ringwood, Victoria, Australia
Penguin Books Canada Ltd, 10 Alcorn Avenue, Toronto, Ontario, Canada M4V 3B2
Penguin Books (NZ) Ltd, Private Bag 102902, NSMC, Auckland, New Zealand

Penguin Books Ltd, Registered Offices: Harmondsworth, Middlesex, England

First published by Longman 1986
Published in Penguin Books 1988
Reprinted with an Appendix 1991
13 15 17 19 20 18 16 14 12

Printed in England by Clays Ltd, St Ives plc
Filmset in Trump Medieval

To my parents

CONTENTS

PREFACE

This book is written in the conviction that our own existence once presented the greatest of all mysteries, but that it is a mystery no longer because it is solved. Darwin and Wallace solved it, though we shall continue to add footnotes to their solution for a while yet. I wrote the book because I was surprised that so many people seemed not only unaware of the elegant and beautiful solution to this deepest of problems but, incredibly, in many cases actually unaware that there was a problem in the first place!

The problem is that of complex design. The computer on which I am writing these words has an information storage capacity of about 64 kilobytes (one byte is used to hold each character of text). The computer was consciously designed and deliberately manufactured. The brain with which you are understanding my words is an array of some ten million kiloneurones. Many of these billions of nerve cells have each more than a thousand 'electric wires' connecting them to other neurones. Moreover, at the molecular genetic level, every single one of more than a trillion cells in the body contains about a thousand times as much precisely-coded digital information as my entire computer. The complexity of living organisms is matched by the elegant efficiency of their apparent design. If anyone doesn't agree that this amount of complex design cries out for an explanation, I give up. No, on second thoughts I don't give up, because one of my aims in the book is to convey something of the sheer wonder of biological complexity to those whose eyes have not been opened to it. But having built up the mystery, my other main aim is to remove it again by explaining the solution.

Explaining is a difficult art. You can explain something so that your reader understands the words; and you can explain something so that the reader feels it in the marrow of his bones. To do the latter, it sometimes isn't enough to lay the evidence before the reader in a dispassionate way. You have to become an advocate and use the tricks of the advocate's trade. This book is not a dispassionate scientific treatise. Other books on Darwinism are, and many of them are excellent and informative and should be read in conjunction with this one. Far from being dispassionate, it has to be confessed that in parts this book is written with a passion which, in a professional scientific journal, might excite comment. Certainly it seeks to inform, but it also seeks to persuade and even – one can specify *aims* without presumption – to inspire. I want to inspire the reader with a vision of our own existence as, on the face of it, a spine-chilling mystery; and simultaneously to convey the full excitement of the fact that it is a mystery with an elegant solution which is within our grasp. More, I want to persuade the reader, not just that the Darwinian world-view *happens* to be true, but that it is the only known theory that *could*, in principle, solve the mystery of our existence. This makes it a doubly satisfying theory. A good case can be made that Darwinism is true, not just on this planet but all over the universe wherever life may be found.

In one respect I plead to distance myself from professional advocates. A lawyer or a politician is paid to exercise his passion and his persuasion on behalf of a client or a cause in which he may not privately believe. I have never done this and I never shall. I may not always be right, but I care passionately about what is true and I never say anything that I do not believe to be right. I remember being shocked when visiting a university debating society to debate with creationists. At dinner after the debate, I was placed next to a young woman who had made a relatively powerful speech in favour of creationism. She clearly couldn't *be* a creationist, so I asked her to tell me honestly why she had done it. She freely admitted that she was simply practising her debating skills, and found it more challenging to advocate a position in which she did not believe. Apparently it is common practice in university debating societies for speakers simply to be *told* on which side they are to speak. Their own beliefs don't come into it. I had come a long way to perform the disagreeable task of public speaking, because I believed in the truth of the motion that I had been asked to propose. When I discovered that members of the society were using the motion as a vehicle for playing arguing games, I resolved to decline future invitations from debating societies that encourage insincere advocacy on issues where scientific truth is at stake.

For reasons that are not entirely clear to me, Darwinism seems more in need of advocacy than similarly established truths in other branches of science. Many of us have no grasp of quantum theory, or Einstein's theories of special and general relativity, but this does not in itself lead us to *oppose* these theories! Darwinism, unlike 'Einsteinism', seems to be regarded as fair game for critics with any degree of ignorance. I suppose one trouble with Darwinism is that, as Jacques Monod perceptively remarked, everybody *thinks* he understands it. It is, indeed, a remarkably simple theory; childishly so, one would have thought, in comparison with almost all of physics and mathematics. In essence, it amounts simply to the idea that non-random reproduction, where there is hereditary variation, has consequences that are far-reaching if there is time for them to be cumulative. But we have good grounds for believing that this simplicity is deceptive. Never forget that, simple as the theory may seem, nobody thought of it until Darwin and Wallace in the mid nineteenth century, nearly 200 years after Newton's *Principia*, and more than 2,000 years after Eratosthenes measured the Earth. How could such a simple idea go so long undiscovered by thinkers of the calibre of Newton, Galileo, Descartes, Leibnitz, Hume and Aristotle? Why did it have to wait for two Victorian naturalists? What was *wrong* with philosophers and mathematicians that they overlooked it? And how can such a powerful idea go still largely unabsorbed into popular consciousness?

It is almost as if the human brain were specifically designed to misunderstand Darwinism, and to find it hard to believe. Take, for instance, the issue of 'chance', often dramatized as *blind* chance. The great majority of people that attack Darwinism leap with almost unseemly eagerness to the mistaken idea that there is nothing other than random chance in it. Since living complexity embodies the very antithesis of chance, if you think that Darwinism is tantamount to chance you'll obviously find it easy to refute Darwinism! One of my tasks will be to destroy this eagerly believed myth that Darwinism is a theory of 'chance'. Another way in which we seem predisposed to disbelieve Darwinism is that our brains are built to deal with events on radically different *timescales* from those that characterize evolutionary change. We are equipped to appreciate processes that take seconds, minutes, years or, at most, decades to complete. Darwinism is a theory of cumulative processes so slow that they take between thousands and millions of decades to complete. All our intuitive judgements of what is probable turn out to be wrong by many orders of magnitude. Our well-tuned apparatus of scepticism and subjective probability-theory misfires by huge margins, because it is tuned – ironically, by evolution itself – to work within a lifetime of a few decades. It requires effort of the imagination to escape from the prison of familiar timescale, an effort that I shall try to assist.

A third respect in which our brains seem predisposed to resist Darwinism stems from our great success as creative designers. Our world is dominated by feats of engineering and works of art. We are entirely accustomed to the idea that complex elegance is an indicator of premeditated, crafted design. This is probably the most powerful reason for the belief, held by the vast majority of people that have ever lived, in some kind of supernatural deity. It took a very large leap of the imagination for Darwin and Wallace to see that, contrary to all intuition, there is another way and, once you have understood it, a far more plausible way, for complex 'design' to arise out of primeval simplicity. A leap of the imagination so large that, to this day, many people seem still unwilling to make it. It is the main purpose of this book to help the reader to make this leap.

Authors naturally hope that their books will have lasting rather than ephemeral impact. But any advocate, in addition to putting the timeless part of his case, must also respond to contemporary advocates of opposing, or apparently opposing, points of view. There is a risk that some of these arguments, however hotly they may rage today, will seem terribly dated in decades to come. The paradox has often been noted that the first edition of *The Origin of Species* makes a better case than the sixth. This is because Darwin felt obliged, in his later editions, to respond to contemporary criticisms of the first edition, criticisms which now seem so dated that the replies to them merely get in the way, and in places even mislead. Nevertheless, the temptation to ignore fashionable contemporary criticisms that one suspects of being nine days' wonders is a temptation that should not be indulged, for reasons of courtesy not just to the critics but to their otherwise confused readers. Though I have my own private ideas on which chapters of my book will eventually prove ephemeral for this reason, the reader – and time – must judge.

I am distressed to find that some women friends (fortunately not many) treat the use of the impersonal masculine pronoun as if it showed intention to exclude them. If there were any excluding to be done (happily there isn't) I think I would sooner exclude men, but when I once tentatively tried referring to my abstract reader as 'she', a feminist denounced me for patronizing condescension: I ought to say 'he-or-she', and 'his-or-her'. That is easy to do if you don't care about language, but then if you don't care about language you don't deserve readers of either sex. Here, I have returned to the normal conventions of English pronouns. I may refer to the 'reader' as 'he', but I no more think of my readers as specifically male than a French speaker thinks of a table as female. As a matter of fact I believe I do, more often than not, think of my readers as female, but that is my personal affair and I'd hate to think that such considerations impinged on how I use my native language.

Personal, too, are some of my reasons for gratitude. Those to whom I cannot do justice will understand. My publishers saw no reason to keep from me the identities of their referees (not 'reviewers' – true reviewers, *pace* many Americans under 40, criticize books only *after* they are published, when it is too late for the author to do anything about it), and I have benefited greatly from the suggestions of John Krebs (again), John Durant, Graham Cairns-Smith, Jeffrey Levinton, Michael Ruse, Anthony Hallam and David Pye. Richard Gregory kindly criticized Chapter 12, and the final version has benefited from its complete excision. Mark Ridley and Alan Grafen, now no longer even officially my students, are, together with Bill Hamilton, the leading lights of the group of colleagues with whom I discuss evolution and from whose ideas I benefit almost daily. They, Pamela Wells, Peter Atkins and John Dawkins have helpfully criticized various chapters for me. Sarah Bunney made numerous improvements, and John Gribbin corrected a major error. Alan Grafen and Will Atkinson advised on computing problems, and the Apple Macintosh Syndicate of the Zoology Department kindly allowed their laser printer to draw biomorphs.

Once again I have benefited from the relentless dynamism with which Michael Rodgers, now of Longman, carries all before him. He, and Mary Cunnane of Norton, skilfully applied the accelerator (to my morale) and the brake (to my sense of humour) when each was needed. Part of the book was written during a sabbatical leave kindly granted by the Department of Zoology and New College. Finally – a debt I should have acknowledged in both my previous books – the Oxford tutorial system and my many tutorial pupils in zoology over the years have helped me to practise what few skills I may have in the difficult art of explaining.

Richard Dawkins
Oxford, 1986

CHAPTER 1

EXPLAINING
THE VERY IMPROBABLE

We animals are the most complicated things in the known universe. The universe that we know, of course, is a tiny fragment of the actual universe. There may be yet more complicated objects than us on other planets, and some of them may already know about us. But this doesn't alter the point that I want to make. Complicated things, everywhere, deserve a very special kind of explanation. We want to know how they came into existence and why they are so complicated. The explanation, as I shall argue, is likely to be broadly the same for complicated things everywhere in the universe; the same for us, for chimpanzees, worms, oak trees and monsters from outer space. On the other hand, it will not be the same for what I shall call 'simple' things, such as rocks, clouds, rivers, galaxies and quarks. These are the stuff of physics. Chimps and dogs and bats and cockroaches and people and worms and dandelions and bacteria and galactic aliens are the stuff of biology.

The difference is one of complexity of design. Biology is the study of complicated things that give the appearance of having been designed for a purpose. Physics is the study of simple things that do not tempt us to invoke design. At first sight, man-made artefacts like computers and cars will seem to provide exceptions. They are complicated and obviously designed for a purpose, yet they are not alive, and they are made of metal and plastic rather than of flesh and blood. In this book they will be firmly treated as biological objects.

The reader's reaction to this may be to ask, 'Yes, but are they *really* biological objects?' Words are our servants, not our masters. For different purposes we find it convenient to use words in different senses. Most cookery books class lobsters as fish. Zoologists can become quite apoplectic about this, pointing out that lobsters could with greater justice call humans fish, since fish are far closer kin to humans than they are to lobsters. And, talking of justice and lobsters, I understand that a court of law recently had to decide whether lobsters were insects or 'animals' (it bore upon whether people should be allowed to boil them alive). Zoologically speaking, lobsters are certainly not insects. They are animals, but then so are insects and so are we. There is little point in getting worked up about the way different people use words (although in my nonprofessional life I am quite prepared to get worked up about people who boil lobsters alive). Cooks and lawyers need to use words in their own special ways, and so do I in this book. Never mind whether cars and computers are 'really' biological objects. The point is that if anything of that degree of complexity were found on a planet, we should have no hesitation in concluding that life existed, or had once existed, on that planet. Machines are the direct products of living objects; they derive their complexity and design from living objects, and they are diagnostic of the existence of life on a planet. The same goes for fossils, skeletons and dead bodies.

I said that physics is the study of simple things, and this, too, may seem strange at first. Physics appears to be a complicated subject, because the ideas of physics are difficult for us to understand. Our brains were designed to understand hunting and gathering, mating and child-rearing: a world of medium-sized objects moving in three dimensions at moderate speeds. We are ill-equipped to comprehend the very small and the very large; things whose duration is measured in picoseconds or gigayears; particles that don't have position; forces and fields that we cannot see or touch, which we know of only because they affect things that we can see or touch. We think that physics is complicated because it is hard for us to understand, and because physics books are full of difficult mathematics. But the objects that physicists study are still basically simple objects. They are clouds of gas or tiny particles, or lumps of uniform matter like crystals, with almost endlessly repeated atomic patterns. They do not, at least by biological standards, have intricate working parts. Even large physical objects like stars consist of a rather limited array of parts, more or less haphazardly arranged. The behaviour of physical, nonbiological objects is so simple that it is feasible to use existing mathematical language to describe it, which is why physics books are full of mathematics.

Physics *books* may be complicated, but physics books, like cars and computers, are the product of biological objects – human brains. The objects and phenomena that a physics book describes are simpler than a single cell in the body of its author. And the author consists of trillions of those cells, many of them different from each other, organized with intricate architecture and precision-engineering into a working machine capable of writing a book (my trillions are American, like all my units: one American trillion is a million millions; an American billion is a thousand millions). Our brains are no better equipped to handle extremes of complexity than extremes of size and the other difficult extremes of physics. Nobody has yet invented the mathematics for describing the total structure and behaviour of such an object as a physicist, or even of one of his cells. What we can do is understand some of the general principles of how living things work, and why they exist at all.

This was where we came in. We wanted to know why we, and all other complicated things, exist. And we can now answer that question in general terms, even without being able to comprehend the details of the complexity itself. To take an analogy, most of us don't understand in detail how an airliner works. Probably its builders don't comprehend it fully either: engine specialists don't in detail understand wings, and wing specialists understand engines only vaguely. Wing specialists don't even understand wings with full mathematical precision: they can predict how a wing will behave in turbulent conditions, only by examining a model in a wind tunnel or a computer simulation – the sort of thing a biologist might do to understand an animal. But however incompletely we understand how an airliner works, we all understand by what general process it came into existence. It was designed by humans on drawing boards. Then other humans made the bits from the drawings, then lots more humans (with the aid of other machines designed by humans) screwed, rivetted, welded or glued the bits together, each in its right place. The process by which an airliner came into existence is not fundamentally mysterious to us, because humans built it. The systematic putting together of parts to a purposeful design is something we know and understand, for we have experienced it at first hand, even if only with our childhood Meccano or Erector set.

What about our own bodies? Each one of us is a machine, like an airliner only much more complicated. Were we designed on a drawing board too, and were our parts assembled by a skilled engineer? The answer is no. It is a surprising answer, and we have known and understood it for only a century or so. When Charles Darwin first explained the matter, many people either wouldn't or couldn't grasp it. I myself flatly refused to believe Darwin's theory when I first heard about it as a child. Almost everybody throughout history, up to the second half of the nineteenth century, has firmly believed in the opposite – the

Conscious Designer theory. Many people still do, perhaps because the true, Darwinian explanation of our own existence is still, remarkably, not a routine part of the curriculum of a general education. It is certainly very widely misunderstood.

The watchmaker of my title is borrowed from a famous treatise by the eighteenth-century theologian William Paley. His *Natural Theology – or Evidences of the Existence and Attributes of the Deity Collected from the Appearances of Nature*, published in 1802, is the best-known exposition of the 'Argument from Design', always the most influential of the arguments for the existence of a God. It is a book that I greatly admire, for in his own time its author succeeded in doing what I am struggling to do now. He had a point to make, he passionately believed in it, and he spared no effort to ram it home clearly. He had a proper reverence for the complexity of the living world, and he saw that it demands a very special kind of explanation. The only thing he got wrong – admittedly quite a big thing! – was the explanation itself. He gave the traditional religious answer to the riddle, but he articulated it more clearly and convincingly than anybody had before. The true explanation is utterly different, and it had to wait for one of the most revolutionary thinkers of all time, Charles Darwin.

Paley begins *Natural Theology* with a famous passage:

> In crossing a heath, suppose I pitched my foot against a *stone*, and were asked how the stone came to be there; I might possibly answer, that, for anything I knew to the contrary, it had lain there for ever: nor would it perhaps be very easy to show the absurdity of this answer. But suppose I had found a *watch* upon the ground, and it should be inquired how the watch happened to be in that place; I should hardly think of the answer which I had before given, that for anything I knew, the watch might have always been there.

Paley here appreciates the difference between natural physical objects like stones, and designed and manufactured objects like watches. He goes on to expound the precision with which the cogs and springs of a watch are fashioned, and the intricacy with which they are put together. If we found an object such as a watch upon a heath, even if we didn't know how it had come into existence, its own precision and intricacy of design would force us to conclude

> that the watch must have had a maker: that there must have existed, at some time, and at some place or other, an artificer or artificers, who formed it for the purpose which we find it actually to answer; who comprehended its construction, and designed its use.

Nobody could reasonably dissent from this conclusion, Paley insists, yet that is just what the atheist, in effect, does when he contemplates the works of nature, for:

> every indication of contrivance, every manifestation of design, which existed in the watch, exists in the works of nature; with the difference, on the side of nature, of being greater or more, and that in a degree which exceeds all computation.

Paley drives his point home with beautiful and reverent descriptions of the dissected machinery of life, beginning with the human eye, a favourite example which Darwin was later to use and which will reappear throughout this book. Paley compares the eye with a designed instrument such as a telescope, and concludes that 'there is precisely the same proof that the eye was made for vision, as there is that the telescope was made for assisting it'. The eye must have had a designer, just as the telescope had.

Paley's argument is made with passionate sincerity and is informed by the best biological scholarship of his day, but it is wrong, gloriously and utterly wrong. The analogy between telescope and eye, between watch and living organism, is false. All appearances to the contrary, the only watchmaker in nature is the blind forces of physics, albeit deployed in a very special way. A true watchmaker has foresight: he designs his cogs and springs, and plans their interconnections, with a future purpose in his mind's eye. Natural selection, the blind, unconscious, automatic process which Darwin discovered, and which we now know is the explanation for the existence and apparently purposeful form of all life, has no purpose in mind. It has no mind and no mind's eye. It does not plan for the future. It has no vision, no foresight, no sight at all. If it can be said to play the role of watchmaker in nature, it is the *blind* watchmaker.

I shall explain all this, and much else besides. But one thing I shall not do is belittle the wonder of the living 'watches' that so inspired Paley. On the contrary, I shall try to illustrate my feeling that here Paley could have gone even further. When it comes to feeling awe over living 'watches' I yield to nobody. I feel more in common with the Reverend William Paley than I do with the distinguished modern philosopher, a well-known atheist, with whom I once discussed the matter at dinner. I said that I could not imagine being an atheist at any time before 1859, when Darwin's *Origin of Species* was published. 'What about Hume?', replied the philosopher. 'How did Hume explain the organized complexity of the living world?', I asked. 'He didn't', said the philosopher. 'Why does it need any special explanation?'

Paley knew that it needed a special explanation; Darwin knew it, and I suspect that in his heart of hearts my philosopher companion knew it too. In any case it will be my business to show it here. As for David Hume himself, it is sometimes said that that great Scottish philosopher disposed of the Argument from Design a century before Darwin. But what Hume did was criticize the logic of using apparent design in nature as *positive* evidence for the existence of a God. He did not offer any *alternative* explanation for apparent design, but left the question open. An atheist before Darwin could have said, following Hume: 'I have no explanation for complex biological design. All I know is that God isn't a good explanation, so we must wait and hope that somebody comes up with a better one.' I can't help feeling that such a position, though logically sound, would have left one feeling pretty unsatisfied, and that although atheism might have been *logically* tenable before Darwin, Darwin made it possible to be an intellectually fulfilled atheist. I like to think that Hume would agree, but some of his writings suggest that he underestimated the complexity and beauty of biological design. The boy naturalist Charles Darwin could have shown him a thing or two about that, but Hume had been dead 40 years when Darwin enrolled in Hume's university of Edinburgh.

I have talked glibly of complexity, and of apparent design, as though it were obvious what these words mean. In a sense it is obvious – most people have an intuitive idea of what complexity means. But these notions, complexity and design, are so pivotal to this book that I must try to capture a little more precisely, in words, our feeling that there is something special about complex, and apparently designed things.

So, what is a complex thing? How should we recognize it? In what sense is it true to say that a watch or an airliner or an earwig or a person is complex, but the moon is simple? The first point that might occur to us, as a necessary attribute of a complex thing, is that it has a heterogeneous structure. A pink milk pudding or blancmange is simple in the sense that, if we slice it in two, the two portions will have the same internal constitution: a blancmange is homogeneous. A car is heterogeneous: unlike a blancmange, almost any portion of the car is different from other portions. Two times half a car does not make a car. This will often amount to saying that a complex object, as opposed to a simple one, has many parts, these parts being of more than one kind.

Such heterogeneity, or 'many-partedness', may be a necessary condition, but it is not sufficient. Plenty of objects are many-parted and heterogeneous in internal structure, without being complex in the sense in which I want to use the term. Mont Blanc, for instance, consists of many different kinds of rock, all jumbled together in such a way that, if you sliced the mountain anywhere, the two portions would differ from each other in their internal constitution. Mont Blanc has a heterogeneity of structure not possessed by a blancmange, but it is still not complex in the sense in which a biologist uses the term.

Let us try another tack in our quest for a definition of complexity, and make use of the mathematical idea of probability. Suppose we try out the following definition: a complex thing is something whose constituent parts are arranged in a way that is unlikely to have arisen by chance alone. To borrow an analogy from an eminent astronomer, if you take the parts of an airliner and jumble them up at random, the likelihood that you would happen to assemble a working Boeing is vanishingly small. There are billions of possible ways of putting together the bits of an airliner, and only one, or very few, of them would actually be an airliner. There are even more ways of putting together the scrambled parts of a human.

This approach to a definition of complexity is promising, but something more is still needed. There are billions of ways of throwing together the bits of Mont Blanc, it might be said, and only one of them is Mont Blanc. So what is it that makes the airliner and the human complicated, if Mont Blanc is simple? Any old jumbled collection of parts is unique and, *with hindsight*, is as improbable as any other. The scrap-heap at an aircraft breaker's yard is unique. No two scrap-heaps are the same. If you start throwing fragments of aeroplanes into heaps, the odds of your happening to hit upon exactly the same arrangement of junk twice are just about as low as the odds of your throwing together a working airliner. So, why don't we say that a rubbish dump, or Mont Blanc, or the moon, is just as complex as an aeroplane or a dog, because in all these cases the arrangement of atoms is 'improbable'?

The combination lock on my bicycle has 4,096 different positions. Every one of these is equally 'improbable' in the sense that, if you spin the wheels at random, every one of the 4,096 positions is equally unlikely to turn up. I can spin the wheels at random, look at whatever number is displayed and exclaim with hindsight: 'How amazing. The odds against that number appearing are 4,096:1. A minor miracle!' That is equivalent to regarding the particular arrangement of rocks in a mountain, or of bits of metal in a scrap-heap, as 'complex'. But one of those 4,096 wheel positions really is interestingly unique: the combination 1207 is the only one that opens the lock. The uniqueness of 1207 has nothing to do with hindsight: it is specified in advance by the manufacturer. If you spun the wheels at random and happened to hit 1207 first time, you would be able to steal the bike, and it would seem a minor miracle. If you struck lucky on one of those multi-dialled combination locks on bank safes, it would seem a very major miracle, for the odds against it are many millions to one, and you would be able to steal a fortune.

Now, hitting upon the lucky number that opens the bank's safe is the equivalent, in our analogy, of hurling scrap metal around at random and happening to assemble a Boeing 747. Of all the millions of unique and, with hindsight equally improbable, positions of the combination lock, only one opens the lock. Similarly, of all the millions of unique and, with hindsight equally improbable, arrangements of a heap of junk, only one (or very few) will fly. The uniqueness of the arrangement that flies, or that opens the safe, is nothing to do with hindsight. It is specified in advance. The lock-manufacturer fixed the combination, and he has told the bank manager. The ability to fly is a property of an airliner that we specify in advance. If we see a plane in the air we can be sure that it was not assembled by randomly throwing scrap metal together, because we know that the odds against a random conglomeration's being able to fly are too great.

Now, if you consider all possible ways in which the rocks of Mont Blanc could have been thrown together, it is true that only one of them would make Mont Blanc as we know it. But Mont Blanc as we know it is defined with hindsight. Any one of a very large number of ways of throwing rocks together would be labelled a mountain, and might have been named Mont Blanc. There is nothing special about the particular Mont Blanc that we know, nothing specified in advance, nothing equivalent to the plane taking off, or equivalent to the safe door swinging open and the money tumbling out.

What is the equivalent of the safe door swinging open, or the plane flying, in the case of a living body? Well, sometimes it is almost literally the same. Swallows fly. As we have seen, it isn't easy to throw together a flying machine. If you took all the cells of a swallow and put them together at random, the chance that the resulting object would fly is not, for everyday purposes, different from zero. Not all living things fly, but they do other things that are just as improbable, and just as specifiable in advance. Whales don't fly, but they do swim, and swim about as efficiently as swallows fly. The chance that a random conglomeration of whale cells would swim, let alone swim as fast and efficiently as a whale actually does swim, is negligible.

At this point, some hawk-eyed philosopher (hawks have very acute eyes – you couldn't make a hawk's eye by throwing lenses and light-sensitive cells together at random) will start mumbling something about a circular argument. Swallows fly but they don't swim; and whales swim but they don't fly. It is with hindsight that we decide whether to judge the success of our random conglomeration as a swimmer or as a flyer. Suppose we agree to judge its success as an Xer, and leave open exactly what X is until we have tried throwing cells together. The random lump of cells might turn out to be an efficient burrower like a mole or an efficient climber like a monkey. It might be very good at wind-surfing, or at clutching oily rags, or at walking in ever decreasing circles until it vanished. The list could go on and on. Or could it?

If the list really *could* go on and on, my hypothetical philosopher might have a point. If, no matter how randomly you threw matter around, the resulting conglomeration could often be said, with hindsight, to be good for *something*, then it would be true to say that I cheated over the swallow and the whale. But biologists can be much more specific than that about what would constitute being 'good for something'. The minimum requirement for us to recognize an object as an animal or plant is that it should succeed in making a living *of some sort* (more precisely that it, or at least some members of its kind, should live long enough to reproduce). It is true that there are quite a number of ways of making a living – flying, swimming, swinging through the trees, and so on. But, *however many ways there may be of being alive, it is certain that there are vastly more ways of being dead*, or rather not alive. You may throw cells together at random, over and over again for a billion years, and not once will you get a conglomeration that flies or swims or burrows or runs, or does *anything*, even badly, that could remotely be construed as working to keep itself alive.

This has been quite a long, drawn-out argument, and it is time to remind ourselves of how we got into it in the first place. We were looking for a precise way to express what we mean when we refer to something as complicated. We were trying to put a finger on what it is that humans and moles and earthworms and airliners and watches have in common with each other, but not with blancmange, or Mont Blanc, or the moon. The answer we have arrived at is that complicated things have some quality, specifiable in advance, that is highly unlikely to have been acquired by random chance alone. In the case of living things, the quality that is specified in advance is, in some sense, 'proficiency'; either proficiency in a particular ability such as flying, as an aero-engineer might admire it; or proficiency in something more general, such as the ability to stave off death, or the ability to propagate genes in reproduction.

Staving off death is a thing that you have to work at. Left to itself – and that is what it is when it dies – the body tends to revert to a state of equilibrium with its environment. If you measure some quantity such as the temperature, the acidity, the water content or the electrical potential in a living body, you will typically find that it is markedly different from the corresponding measure in the surroundings. Our bodies, for instance, are usually hotter than our surroundings, and in cold climates they have to work hard to maintain the differential. When we die the work stops, the temperature differential starts to disappear, and we end up the same temperature as our surroundings. Not all animals work so hard to avoid coming into equilibrium with their surrounding temperature, but all animals do *some* comparable work. For instance, in a dry country, animals and plants work to maintain the fluid content of their cells, work against a natural tendency for water to flow from them into the dry outside world. If they fail they die. More generally, if living things didn't work actively to prevent it, they would eventually merge into their surroundings, and cease to exist as autonomous beings. That is what happens when they die.

With the exception of artificial machines, which we have already agreed to count as honorary living things, nonliving things don't work in this sense. They accept the forces that tend to bring them into equilibrium with their surroundings. Mont Blanc, to be sure, has existed for a long time, and probably will exist for a while yet, but it does not work to stay in existence. When rock comes to rest under the influence of gravity it just stays there. No work has to be done to keep it there. Mont Blanc exists, and it will go on existing until it wears away or an earthquake knocks it over. It doesn't take steps to repair wear and tear, or to right itself when it is knocked over, the way a living body does. It just obeys the ordinary laws of physics.

Is this to deny that living things obey the laws of physics? Certainly not. There is no reason to think that the laws of physics are violated in living matter. There is nothing supernatural, no 'life force' to rival the fundamental forces of physics. It is just that if you try to use the laws of physics, in a naïve way, to understand the behaviour of a *whole* living body, you will find that you don't get very far. The body is a complex thing with many constituent parts, and to understand its behaviour you must apply the laws of physics to its parts, not to the whole. The behaviour of the body as a whole will then emerge as a consequence of interactions of the parts.

Take the laws of motion, for instance. If you throw a dead bird into the air it will describe a graceful parabola, exactly as physics books say it should, then come to rest on the ground and stay there. It behaves as a solid body of a particular mass and wind resistance ought to behave.

But if you throw a live bird in the air it will not describe a parabola and come to rest on the ground. It will fly away, and may not touch land this side of the county boundary. The reason is that it has muscles which work to resist gravity and other physical forces bearing upon the whole body. The laws of physics are being obeyed within every cell of the muscles. The result is that the muscles move the wings in such a way that the bird stays aloft. The bird is not violating the law of gravity. It is constantly being pulled downwards by gravity, but its wings are performing active work – obeying laws of physics within its muscles – to keep it aloft in spite of the force of gravity. We shall think that it defies a physical law if we are naïve enough to treat it simply as a structureless lump of matter with a certain mass and wind resistance. It is only when we remember that it has many internal parts, all obeying laws of physics at their own level, that we understand the behaviour of the whole body. This is not, of course, a peculiarity of living things. It applies to all man-made machines, and potentially applies to any complex, many-parted object.

This brings me to the final topic that I want to discuss in this rather philosophical chapter, the problem of what we mean by explanation. We have seen what we are going to mean by a complex thing. But what kind of explanation will satisfy us if we wonder how a complicated machine, or living body, works? The answer is the one that we arrived at in the previous paragraph. If we wish to understand how a machine or living body works, we look to its component parts and ask how they interact with each other. If there is a complex thing that we do not yet understand, we can come to understand it in terms of simpler parts that we do already understand.

If I ask an engineer how a steam engine works, I have a pretty fair idea of the general kind of answer that would satisfy me. Like Julian Huxley I should definitely not be impressed if the engineer said it was propelled by *'force locomotif'*. And if he started boring on about the whole being greater than the sum of its parts, I would interrupt him: 'Never mind about that, tell me how it *works*.' What I would want to hear is something about how the parts of an engine interact with each other to produce the behaviour of the whole engine. I would initially be prepared to accept an explanation in terms of quite large subcomponents, whose own internal structure and behaviour might be quite complicated and, as yet, unexplained. The units of an initially satisfying explanation could have names like fire-box, boiler, cylinder, piston, steam governor. The engineer would assert, without explanation initially, what each of these units does. I would accept this for the moment, without asking how each unit does its own particular thing. *Given* that the units each do their particular thing, I can then understand how they interact to make the whole engine move.

Of course, I am then at liberty to ask how each part works. Having previously accepted the *fact* that the steam governor regulates the flow of steam, and having used this fact in my understanding of the behaviour of the whole engine, I now turn my curiosity on the steam governor itself. I now want to understand how it achieves its own behaviour, in terms of its own internal parts. There is a hierarchy of subcomponents within components. We explain the behaviour of a component at any given level, in terms of interactions between subcomponents whose own internal organization, for the moment, is taken for granted. We peel our way down the hierarchy, until we reach units so simple that, for everyday purposes, we no longer feel the need to ask questions about them. Rightly or wrongly for instance, most of us are happy about the properties of rigid rods of iron, and we are prepared to use them as units of explanation of more complex machines that contain them.

Physicists, of course, do not take iron rods for granted. They ask why they are rigid, and they continue the hierarchical peeling for several more layers yet, down to fundamental particles and quarks. But life is too short for most of us to follow them. For any given level of complex organization, satisfying explanations may normally be attained if we peel the hierarchy down one or two layers from our starting layer, but not more. The behaviour of a motor car is explained in terms of cylinders, carburettors and sparking plugs. It is true that each one of these components rests atop a pyramid of explanations at lower levels. But if you asked me how a motor car worked you would think me somewhat pompous if I answered in terms of Newton's laws and the laws of thermodynamics, and downright obscurantist if I answered in terms of fundamental particles. It is doubtless true that at bottom the behaviour of a motor car is to be explained in terms of interactions between fundamental particles. But it is much more useful to explain it in terms of interactions between pistons, cylinders and sparking plugs.

The behaviour of a computer can be explained in terms of interactions between semiconductor electronic gates, and the behaviour of these, in turn, is explained by physicists at yet lower levels. But, for most purposes, you would in practice be wasting your time if you tried to understand the behaviour of the whole computer at either of those levels. There are too many electronic gates and too many interconnections between them. A satisfying explanation has to be in terms of a manageably small number of interactions. This is why, if we want to understand the workings of computers, we prefer a preliminary explanation in terms of about half a dozen major subcomponents – memory, processing mill, backing store, control unit, input–output handler, etc. Having grasped the interactions between the half-dozen major components, we then may wish to ask questions about the internal organization of these major components. Only specialist engineers are likely to go down to the level of AND gates and NOR gates, and only physicists will go down further, to the level of how electrons behave in a semiconducting medium.

For those that like '-ism' sorts of names, the aptest name for my approach to understanding how things work is probably 'hierarchical reductionism'. If you read trendy intellectual magazines, you may have noticed that 'reductionism' is one of those things, like sin, that is only mentioned by people who are against it. To call oneself a reductionist will sound, in some circles, a bit like admitting to eating babies. But, just as nobody actually eats babies, so nobody is really a reductionist in any sense worth being against. The nonexistent reductionist – the sort that everybody is against, but who exists only in their imaginations – tries to explain complicated things *directly* in terms of the *smallest* parts, even, in some extreme versions of the myth, as the *sum* of the parts! The hierarchical reductionist, on the other hand, explains a complex entity at any particular level in the hierarchy of organization, in terms of entities only one level down the hierarchy; entities which, themselves, are likely to be complex enough to need further reducing to their own component parts; and so on. It goes without saying – though the mythical, baby-eating reductionist is reputed to deny this – that the kinds of explanations which are suitable at high levels in the hierarchy are quite different from the kinds of explanations which are suitable at lower levels. This was the point of explaining cars in terms of carburettors rather than quarks. But the hierarchical reductionist believes that carburettors are explained in terms of smaller units . . ., which are explained in terms of smaller units . . . , which are ultimately explained in terms of the smallest of fundamental particles. Reductionism, in this sense, is just another name for an honest desire to understand how things work.

We began this section by asking what kind of explanation for complicated things would satisfy us. We have just considered the question from the point of view of mechanism: how does it work? We concluded that the behaviour of a complicated thing should be explained in terms of interactions between its component parts, considered as successive layers of an orderly hierarchy. But another kind of question is how the complicated thing came into existence in the first place. This is the question that this whole book is particularly concerned with, so I won't say much more about it here. I shall just mention that the same general principle applies as for understanding mechanism. A complicated thing is one whose existence we do not feel inclined to take for granted, because it is too 'improbable'. It could not have come into existence in a single act of chance. We shall explain its coming into existence as a consequence of gradual, cumulative, step-by-step transformations from simpler things, from primordial objects sufficiently simple to have come into being by chance. Just as 'big-step reductionism' cannot work as an explanation of mechanism, and must be replaced by a series of small step-by-step peelings down through the hierarchy, so we can't explain a complex thing as *originating* in a single step. We must again resort to a series of small steps, this time arranged sequentially in time.

In his beautifully written book, *The Creation*, the Oxford physical chemist Peter Atkins begins:

> I shall take your mind on a journey. It is a journey of comprehension, taking us to the edge of space, time, and understanding. On it I shall argue that there is nothing that cannot be understood, that there is nothing that cannot be explained, and that everything is extraordinarily simple . . . A great deal of the universe does not need any explanation. Elephants, for instance. Once molecules have learnt to compete and to create other molecules in their own image, elephants, and things resembling elephants, will in due course be found roaming through the countryside.

Atkins assumes the evolution of complex things – the subject matter of this book – to be inevitable once the appropriate physical conditions have been set up. He asks what the minimum necessary physical conditions are, what is the minimum amount of design work that a very lazy Creator would have to do, in order to see to it that the universe and, later, elephants and other complex things, would one day come into existence. The answer, from his point of view as a physical scientist, is that the Creator could be infinitely lazy. The fundamental original units that we need to postulate, in order to understand the coming into existence of everything, either consist of literally nothing (according to some physicists), or (according to other physicists) they are units of the utmost simplicity, far too simple to need anything so grand as deliberate Creation.

Atkins says that elephants and complex things do not need any explanation. But that is because he is a physical scientist, who takes for granted the biologists' theory of evolution. He doesn't really mean that elephants don't need an explanation; rather that he is satisfied that biologists can explain elephants, provided they are allowed to take certain facts of physics for granted. His task as a physical scientist, therefore, is to justify our taking those facts for granted. This he succeeds in doing. My position is complementary. I am a biologist. I take the facts of physics, the facts of the world of simplicity, for granted. If physicists still don't agree over whether those simple facts are yet understood, that is not my problem. My task is to explain elephants, and the world of complex things, in terms of the simple things that physicists either understand, or are working on. The physicist's problem is the problem of ultimate origins and ultimate natural laws. The biologist's problem is the problem of complexity. The biologist tries to explain the workings, and the coming into existence, of complex things, in terms of simpler things. He can regard his task as done when he has arrived at entities so simple that they can safely be handed over to physicists.

I am aware that my characterization of a complex object – statistically improbable in a direction that is specified not with hindsight – may seem idiosyncratic. So, too, may seem my characterization of physics as the study of simplicity. If you prefer some other way of defining complexity, I don't care and I would be happy to go along with your definition for the sake of discussion. But what I do care about is that, whatever we choose to *call* the quality of being statistically-improbable-in-a-direction-specified-without-hindsight, it is an important quality that needs a special effort of explanation. It is the quality that characterizes biological objects as opposed to the objects of physics. The kind of explanation we come up with must not contradict the laws of physics. Indeed it will make use of the laws of physics, and nothing more than the laws of physics. But it will deploy the laws of physics in a special way that is not ordinarily discussed in physics textbooks. That special way is Darwin's way. I shall introduce its fundamental essence in Chapter 3 under the title of *cumulative selection*.

Meanwhile I want to follow Paley in emphasizing the magnitude of the problem that our explanation faces, the sheer hugeness of biological complexity and the beauty and elegance of biological design. Chapter 2 is an extended discussion of a particular example, 'radar' in bats, discovered long after Paley's time. And here, in this chapter, I have placed an illustration (Figure 1) – how Paley would have loved the electron microscope! – of an eye together with two successive 'zoomings in' on detailed portions. At the top of the figure is a section through an eye itself. This level of magnification shows the eye as an optical instrument. The resemblance to a camera is obvious. The iris diaphragm is responsible for constantly varying the aperture, the *f* stop.

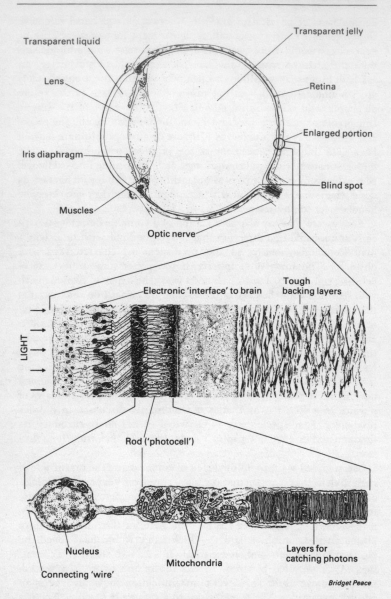

Figure 1

The lens, which is really only part of a compound lens system, is responsible for the variable part of the focusing. Focus is changed by squeezing the lens with muscles (or in chameleons by moving the lens forwards or backwards, as in a man-made camera). The image falls on the retina at the back, where it excites photocells.

The middle part of Figure 1 shows a small section of the retina enlarged. Light comes from the left. The light-sensitive cells ('photocells') are not the first thing the light hits, but they are buried inside and facing away from the light. This odd feature is mentioned again later. The first thing the light hits is, in fact, the layer of ganglion cells which constitute the 'electronic interface' between the photocells and the brain. Actually the ganglion cells are responsible for preprocessing the information in sophisticated ways before relaying it to the brain, and in some ways the word 'interface' doesn't do justice to this. 'Satellite computer' might be a fairer name. Wires from the ganglion cells run along the surface of the retina to the 'blind spot', where they dive through the retina to form the main trunk cable to the brain, the optic nerve. There are about three million ganglion cells in the 'electronic interface', gathering data from about 125 million photocells.

At the bottom of the figure is one enlarged photocell, a rod. As you look at the fine architecture of this cell, keep in mind the fact that all that complexity is repeated 125 million times in each retina. And comparable complexity is repeated trillions of times elsewhere in the body as a whole. The figure of 125 million photocells is about 5,000 times the number of separately resolvable points in a good-quality magazine photograph. The folded membranes on the right of the illustrated photocell are the actual light-gathering structures. Their layered form increases the photocell's efficiency in capturing photons, the fundamental particles of which light is made. If a photon is not caught by the first membrane, it may be caught by the second, and so on. As a result of this, some eyes are capable of detecting a single photon. The fastest and most sensitive film emulsions available to photographers need about 25 times as many photons in order to detect a point of light. The lozenge-shaped objects in the middle section of the cell are mostly mitochondria. Mitochondria are found not just in photocells, but in most other cells. Each one can be thought of as a chemical factory which, in the course of delivering its primary product of usable energy, processes more than 700 different chemical substances, in long, interweaving assembly-lines strung out along the surface of its intricately folded internal membranes. The round globule at the left of Figure 1 is the nucleus. Again, this is characteristic of all animal and plant cells. Each nucleus, as we shall see in Chapter 5, contains a digitally coded database larger, in information content, than all 30 volumes of the *Encyclopaedia Britannica* put together. And this figure is for *each* cell, not all the cells of a body put together.

The rod at the base of the picture is one single cell. The total number of cells in the body (of a human) is about 10 trillion. When you eat a steak, you are shredding the equivalent of more than 100 billion copies of the *Encyclopaedia Britannica*.

CHAPTER 2

GOOD DESIGN

Natural selection is the blind watchmaker, blind because it does not see ahead, does not plan consequences, has no purpose in view. Yet the living results of natural selection overwhelmingly impress us with the appearance of design as if by a master watchmaker, impress us with the illusion of design and planning. The purpose of this book is to resolve this paradox to the satisfaction of the reader, and the purpose of this chapter is further to impress the reader with the power of the illusion of design. We shall look at a particular example and shall conclude that, when it comes to complexity and beauty of design, Paley hardly even began to state the case.

We may say that a living body or organ is well designed if it has attributes that an intelligent and knowledgeable engineer might have built into it in order to achieve some sensible purpose, such as flying, swimming, seeing, eating, reproducing, or more generally promoting the survival and replication of the organism's genes. It is not necessary to suppose that the design of a body or organ is the *best* that an engineer could conceive of. Often the best that one engineer can do is, in any case, exceeded by the best that another engineer can do, especially another who lives later in the history of technology. But any engineer can recognize an object that has been designed, even poorly designed, for a purpose, and he can usually work out what that purpose is just by looking at the structure of the object. In Chapter 1 we bothered ourselves mostly with philosophical aspects. In this chapter, I shall develop a particular factual example that I believe would impress any engineer, namely sonar ('radar') in bats. In explaining each point, I shall begin by posing a problem that the living machine faces; then I shall consider possible solutions to the problem that a sensible engineer might consider; I shall finally come to the solution that nature has actually adopted. This one example is, of course, just for illustration. If an engineer is impressed by bats, he will be impressed by countless other examples of living design.

Bats have a problem: how to find their way around in the dark. They hunt at night, and cannot use light to help them find prey and avoid obstacles. You might say that if this is a problem it is a problem of their own making, a problem that they could avoid simply by changing their habits and hunting by day. But the daytime economy is already heavily exploited by other creatures such as birds. Given that there is a living to be made at night, and given that alternative daytime trades are thoroughly occupied, natural selection has favoured bats that make a go of the night-hunting trade. It is probable, by the way, that the nocturnal trades go way back in the ancestry of all us mammals. In the time when the dinosaurs dominated the daytime economy, our mammalian ancestors probably only managed to survive at all because they found ways of scraping a living at night. Only after the mysterious mass extinction of the dinosaurs about 65 million years ago were our ancestors able to emerge into the daylight in any substantial numbers.

Returning to bats, they have an engineering problem: how to find their way and find their prey in the absence of light. Bats are not the only creatures to face this difficulty today. Obviously the night-flying insects that they prey on must find their way about somehow. Deep-sea fish and whales have little or no light by day or by night, because the sun's rays cannot penetrate far below the surface. Fish and dolphins that live in extremely muddy water cannot see because, although there is light, it is obstructed and scattered by the dirt in the water. Plenty of other modern animals make their living in conditions where seeing is difficult or impossible.

Given the question of how to manoeuvre in the dark, what solutions might an engineer consider? The first one that might occur to him is to manufacture light, to use a lantern or a searchlight. Fireflies and some fish (usually with the help of bacteria) have the power to manufacture their own light, but the process seems to consume a large amount of energy. Fireflies use their light for attracting mates. This doesn't require prohibitively much energy: a male's tiny pinprick can be seen by a female from some distance on a dark night, since her eyes are exposed directly to the light source itself. Using light to find one's own way around requires vastly more energy, since the eyes have to detect the tiny fraction of the light that bounces off each part of the scene. The light source must therefore be immensely brighter if it is to be used as a headlight to illuminate the path, than if it is to be used as a signal to others. Anyway, whether or not the reason is the energy expense, it seems to be the case that, with the possible exception of some weird deep-sea fish, no animal apart from man uses manufactured light to find its way about.

What else might the engineer think of? Well, blind humans sometimes seem to have an uncanny sense of obstacles in their path. It has been given the name 'facial vision', because blind people have reported that it feels a bit like the sense of touch, on the face. One report tells of a totally blind boy who could ride his tricycle at a good speed round the block near his home, using 'facial vision'. Experiments showed that, in fact, 'facial vision' is nothing to do with touch or the front of the face, although the sensation may be *referred* to the front of the face, like the referred pain in a phantom (severed) limb. The sensation of 'facial vision', it turns out, really goes in through the ears. The blind people, without even being aware of the fact, are actually using *echoes*, of their own footsteps and other sounds, to sense the presence of obstacles. Before this was discovered, engineers had already built instruments to exploit the principle, for example to measure the depth of the sea under a ship. After this technique had been invented, it was only a matter of time before weapons designers adapted it for the detection of submarines. Both sides in the Second World War relied heavily on these devices, under such code names as Asdic (British) and Sonar (American), as well as the similar technology of Radar (American) or RDF (British), which uses radio echoes rather than sound echoes.

The Sonar and Radar pioneers didn't know it then, but all the world now knows that bats, or rather natural selection working on bats, had perfected the system tens of millions of years earlier, and their 'radar' achieves feats of detection and navigation that would strike an engineer dumb with admiration. It is technically incorrect to talk about bat 'radar', since they do not use radio waves. It is sonar. But the underlying mathematical theories of radar and sonar are very similar, and much of our scientific understanding of the details of what bats are doing has come from applying radar theory to them. The American zoologist Donald Griffin, who was largely responsible for the discovery of sonar in bats, coined the term 'echolocation' to cover both sonar and radar, whether used by animals or by human instruments. In practice, the word seems to be used mostly to refer to animal sonar.

It is misleading to speak of bats as though they were all the same. It is as though we were to speak of dogs, lions, weasels, bears, hyenas, pandas and otters all in one breath, just because they are all carnivores. Different groups of bats use sonar in radically different ways, and they seem to have 'invented' it separately and independently, just as the British, Germans and Americans all independently developed radar. Not all bats use echolocation. The Old World tropical fruit bats have good vision, and most of them use only their eyes for finding their way around. One or two species of fruit bats, however, for instance *Rousettus*, are capable of finding their way around in total darkness where eyes, however good, must be powerless. They are using sonar, but it is a cruder kind of sonar than is used by the smaller bats with which we, in temperate regions, are familiar. *Rousettus* clicks its tongue loudly and rhythmically as it flies, and navigates by measuring the time interval between each click and its echo. A good proportion of *Rousettus*'s clicks are clearly audible to us (which by definition makes them sound rather than ultrasound: ultrasound is just the same·as sound except that it is too high for humans to hear).

In theory, the higher the pitch of a sound, the better it is for accurate sonar. This is because low-pitched sounds have long wavelengths which cannot resolve the difference between closely spaced objects. All other things being equal therefore, a missile that used echoes for its guidance system would ideally produce very high-pitched sounds. Most bats do, indeed, use extremely high-pitched sounds, far too high for humans to hear – ultrasound. Unlike *Rousettus*, which can see very well and which uses unmodified relatively low-pitched sounds to do a modest amount of echolocation to supplement its good vision, the smaller bats appear to be technically highly advanced echo-machines. They have tiny eyes which, in most cases, probably can't see much. They live in a world of echoes, and probably their brains can use echoes to do something akin to 'seeing' images, although it is next to impossible for us to 'visualize' what those images might be like. The noises that they produce are not just slightly too high for humans to hear, like a kind of super dog whistle. In many cases they are vastly higher than the highest note anybody has heard or can imagine. It is fortunate that we can't hear them, incidentally, for they are immensely powerful and would be deafeningly loud if we could hear them, and impossible to sleep through.

These bats are like miniature spy planes, bristling with sophisticated instrumentation. Their brains are delicately tuned packages of miniaturized electronic wizardry, programmed with the elaborate software necessary to decode a world of echoes in real time. Their faces are often distorted into gargoyle shapes that appear hideous to us until we see them for what they are, exquisitely fashioned instruments for beaming ultrasound in desired directions.

Although we can't hear the ultrasound pulses of these bats directly, we can get some idea of what is going on by means of a translating machine or 'bat-detector'. This receives the pulses through a special ultrasonic microphone, and turns each pulse into an audible click or tone which we can hear through headphones. If we take such a 'bat-detector' out to a clearing where a bat is feeding, we shall hear *when* each bat pulse is emitted, although we cannot hear what the pulses really 'sound' like. If our bat is *Myotis*, one of the common little brown bats, we shall hear a chuntering of clicks at a rate of about 10 per second as the bat cruises about on a routine mission. This is about the rate of a standard teleprinter, or a Bren machine gun.

Presumably the bat's image of the world in which it is cruising is being updated 10 times per second. Our own visual image appears to be continuously updated as long as our eyes are open. We can see what it might be like to have an intermittently updated world image, by using a stroboscope at night. This is sometimes done at discotheques, and it produces some dramatic effects. A dancing person appears as a succession of frozen statuesque attitudes. Obviously, the faster we set the strobe, the more the image corresponds to normal 'continuous' vision. Stroboscopic vision 'sampling' at the bat's cruising rate of about 10 samples per second would be nearly as good as normal 'continuous' vision for some ordinary purposes, though not for catching a ball or an insect.

This is just the sampling rate of a bat on a routine cruising flight. When a little brown bat detects an insect and starts to move in on an interception course, its click rate goes up. Faster than a machine gun, it can reach peak rates of 200 pulses per second as the bat finally closes in on the moving target. To mimic this, we should have to speed up our stroboscope so that its flashes came twice as fast as the cycles of mains electricity, which are not noticed in a fluorescent strip light. Obviously we have no trouble in performing all our normal visual functions, even playing squash or ping-pong, in a visual world 'pulsed' at such a high frequency. If we may imagine bat brains as building up an image of the world analogous to our visual images, the pulse rate alone seems to suggest that the bat's echo image might be at least as detailed and 'continuous' as our visual image. Of course, there may be other reasons why it is not so detailed as our visual image.

If bats are capable of boosting their sampling rates to 200 pulses per second, why don't they keep this up all the time? Since they evidently have a rate control 'knob' on their 'stroboscope', why don't they turn it permanently to maximum, thereby keeping their perception of the world at its most acute, all the time, to meet any emergency? One reason is that these high rates are suitable only for near targets. If a pulse follows too hard on the heels of its predecessor it gets mixed up with the echo of its predecessor returning from a distant target. Even if this weren't so, there would probably be good economic reasons for not keeping up the maximum pulse rate all the time. It must be costly producing loud ultrasonic pulses, costly in energy, costly in wear and tear on voice and ears, perhaps costly in computer time. A brain that is processing 200 distinct echoes per second might not find surplus capacity for thinking about anything else. Even the ticking-over rate of about 10 pulses per second is probably quite costly, but much less so than the maximum rate of 200 per second. An individual bat that boosted its tickover rate would pay an additional price in energy, etc., which would not be justified by the increased sonar acuity. When the only moving object in the immediate vicinity is the bat itself, the apparent world is sufficiently similar in successive tenths of seconds that it need not be sampled more frequently than this. When the salient vicinity includes another moving object, particularly a flying insect twisting and turning and diving in a desperate attempt to shake off its pursuer, the extra benefit to the bat of increasing its sample rate more than justifies the increased cost. Of course, the considerations of cost and benefit in this paragraph are all surmise, but something like this almost certainly must be going on.

The engineer who sets about designing an efficient sonar or radar device soon comes up against a problem resulting from the need to make the pulses extremely loud. They have to be loud because when a sound is broadcast its wavefront advances as an ever-expanding sphere. The intensity of the sound is distributed and, in a sense, 'diluted' over the whole surface of the sphere. The surface area of any sphere is proportional to the radius squared. The intensity of the sound at any particular point on the sphere therefore decreases, not in proportion to the distance (the radius) but in proportion to the square of the distance from the sound source, as the wavefront advances and the sphere swells. This means that the sound gets quieter pretty fast, as it travels away from its source, in this case the bat.

When this diluted sound hits an object, say a fly, it bounces off the fly. This reflected sound now, in its turn, radiates away from the fly in an expanding spherical wavefront. For the same reason as in the case of the original sound, it decays as the square of the distance from the fly. By the time the echo reaches the bat again, the decay in its intensity is proportional, not to the distance of the fly from the bat, not even to the square of that distance, but to something more like the square of the square – the fourth power, of the distance. This means that it is very very quiet indeed. The problem can be partially overcome if the bat beams the sound by means of the equivalent of a megaphone, but only if it already knows the direction of the target. In any case, if the bat is to receive any reasonable echo at all from a distant target, the outgoing squeak as it leaves the bat must be very loud indeed, and the instrument that detects the echo, the ear, must be highly sensitive to very quiet sounds – the echoes. Bat cries, as we have seen, are indeed often very loud, and their ears are very sensitive.

Now here is the problem that would strike the engineer trying to design a bat-like machine. If the microphone, or ear, is as sensitive as all that, it is in grave danger of being seriously damaged by its own enormously loud outgoing pulse of sound. It is no good trying to combat the problem by making the sounds quieter, for then the echoes would be too quiet to hear. And it is no good trying to combat *that* by making the microphone ('ear') more sensitive, since this would only make it more vulnerable to being damaged by the, albeit now slightly quieter, outgoing sounds! It is a dilemma inherent in the dramatic difference in intensity between outgoing sound and returning echo, a difference that is inexorably imposed by the laws of physics.

What other solution might occur to the engineer? When an analogous problem struck the designers of radar in the Second World War, they hit upon a solution which they called 'send/receive' radar. The radar signals were sent out in necessarily very powerful pulses, which might have damaged the highly sensitive aerials (American 'antennas') waiting for the faint returning echoes. The 'send/receive' circuit temporarily disconnected the receiving aerial just before the outgoing pulse was about to be emitted, then switched the aerial on again in time to receive the echo.

Bats developed 'send/receive' switching technology long long ago, probably millions of years before our ancestors came down from the trees. It works as follows. In bat ears, as in ours, sound is transmitted from the eardrum to the microphonic, sound-sensitive cells by means of a bridge of three tiny bones known (in Latin) as the hammer, the anvil and the stirrup, because of their shape. The mounting and hinging of these three bones, by the way, is exactly as a hi-fi engineer might have designed it to serve a necessary 'impedance-matching' function, but that is another story. What matters here is that some bats have well-developed muscles attached to the stirrup and to the hammer. When these muscles are contracted the bones don't transmit sound so efficiently – it is as though you muted a microphone by jamming your thumb against the vibrating diaphragm. The bat is able to use these muscles to switch its ears off temporarily. The muscles contract immediately before the bat emits each outgoing pulse, thereby switching the ears off so that they are not damaged by the loud pulse. Then they relax so that the ear returns to maximal sensitivity just in time for the returning echo. This send/receive switching system works only if split-second accuracy in timing is maintained. The bat called *Tadarida* is capable of alternately contracting and relaxing its switching muscles 50 times per second, keeping in perfect synchrony with the machine gun-like pulses of ultrasound. It is a formidable feat of timing, comparable to a clever trick that was used in some fighter planes during the First World War. Their machine guns fired 'through' the propeller, the timing being carefully synchronized with the rotation of the propeller so that the bullets always passed between the blades and never shot them off.

The next problem that might occur to our engineer is the following. If the sonar device is measuring the distance of targets by measuring the duration of silence between the emission of a sound and its re-turning echo – the method which *Rousettus*, indeed, seems to be using – the sounds would seem to have to be very brief, staccato pulses. A long drawn-out sound would still be going on when the echo returned, and, even if partially muffled by send/receive muscles, would get in the way of detecting the echo. Ideally, it would seem, bat pulses should be very brief indeed. But the briefer a sound is, the more difficult it is to make it energetic enough to produce a decent echo. We seem to have another unfortunate trade-off imposed by the laws of physics. Two solutions might occur to ingenious engineers, indeed did occur to them when they encountered the same problem, again in the analogous case of radar. Which of the two solutions is preferable depends on whether it is more important to measure range (how far away an object is from the instrument) or velocity (how fast the object is moving relative to the instrument). The first solution is that known to radar engineers as 'chirp radar'.

We can think of radar signals as a series of pulses, but each pulse has a so-called carrier frequency. This is analogous to the 'pitch' of a pulse of sound or ultrasound. Bat cries, as we have seen, have a pulse-repetition rate in the tens or hundreds per second. Each one of those pulses has a carrier frequency of tens of thousands to hundreds of thousands of cycles per second. Each pulse, in other words, is a high-pitched shriek. Similarly, each pulse of radar is a 'shriek' of radio waves, with a high carrier frequency. The special feature of chirp radar is that it does not have a fixed carrier frequency during each shriek. Rather, the carrier frequency swoops up or down about an octave. If you think of it as its sound equivalent, each radar emission can be thought of as a swooping wolf-whistle. The advantage of chirp radar, as opposed to the fixed pitch pulse, is the following. It doesn't matter if the original chirp is still going on when the echo returns. They won't be confused with each other. This is because the echo being detected at any given moment will be a reflection of an earlier part of the chirp, and will therefore have a different pitch.

Human radar designers have made good use of this ingenious technique. Is there any evidence that bats have 'discovered' it too, just as they did the send/receive system? Well, as a matter of fact, numerous species of bats do produce cries that sweep down, usually through about an octave, during each cry. These wolf-whistle cries are known as frequency modulated (FM). They appear to be just what would be required to exploit the 'chirp radar' technique. However, the evidence so far suggests that bats are using the technique, not to distinguish an echo from the original sound that produced it, but for the more subtle task of distinguishing echoes from other echoes. A bat lives in a world of echoes from near objects, distant objects and objects at all intermediate distances. It has to sort these echoes out from each other. If it gives downward-swooping, wolf-whistle chirps, the sorting is neatly done by pitch. When an echo from a distant object finally arrives back at the bat, it will be an 'older' echo than an echo that is simultaneously arriving back from a near object. It will therefore be of higher pitch. When the bat is faced with clashing echoes from several objects, it can apply the rule of thumb: higher pitch means farther away.

The second clever idea that might occur to the engineer, especially one interested in measuring the speed of a moving target, is to exploit what physicists call the Doppler Shift. This may be called the 'ambulance effect' because its most familiar manifestation is the sudden drop in pitch of an ambulance's siren as it speeds past the listener. The Doppler Shift occurs whenever a source of sound (or light or any other kind of wave) and a receiver of that sound move relative to one another. It is easiest to think of the sound source as motionless and the listener as moving. Assume that the siren on a factory roof is wailing continuously, all on one note. The sound is broadcast outwards as a series of waves. The waves can't be seen, because they are waves of air pressure. If they could be seen they would resemble the concentric circles spreading outwards when we throw pebbles into the middle of a still pond. Imagine that a series of pebbles is being dropped in quick succession into the middle of a pond, so that waves are continuously radiating out from the middle. If we moor a tiny toy boat at some fixed point in the pond, the boat will bob up and down rhythmically as the waves pass under it. The frequency with which the boat bobs is analogous to the pitch of a sound. Now suppose that the boat, instead of being moored, is steaming across the pond, in the general direction of the centre from which the wave circles are originating. It will still bob up and down as it hits the successive wavefronts. But now the frequency with which it hits waves will be higher, since it is travelling towards the source of the waves. It will bob up and down at a higher rate. On the other hand, when it has passed the source of the waves and is travelling away the other side, the frequency with which it bobs up and down will obviously go down.

For the same reason, if we ride fast on a (preferably quiet) motorbike past a wailing factory siren, when we are approaching the factory the pitch will be raised: our ears are, in effect, gobbling up the waves at a faster rate than they would if we just sat still. By the same kind of argument, when our motorbike has passed the factory and is moving away from it, the pitch will be lowered. If we stop moving we shall hear the pitch of the siren as it actually is, intermediate between the two Doppler-shifted pitches. It follows that if we know the exact pitch of the siren, it is theoretically possible to work out how fast we are moving towards or away from it simply by listening to the apparent pitch and comparing it with the known 'true' pitch.

The same principle works when the sound source is moving and the listener is still. That is why it works for ambulances. It is rather implausibly said that Christian Doppler himself demonstrated his effect by hiring a brass band to play on an open railway truck as it rushed past his amazed audience. It is relative motion that matters, and as far as the Doppler Effect is concerned it doesn't matter whether we consider the sound source to be moving past the ear, or the ear moving past the sound source. If two trains pass in opposite directions, each travelling at 125 m.p.h., a passenger in one train will hear the whistle of the other train swoop down through a particularly dramatic Doppler Shift, since the relative velocity is 250 m.p.h.

The Doppler Effect is used in police radar speed-traps for motorists. A static instrument beams radar signals down a road. The radar waves bounce back off the cars that approach, and are registered by the receiving apparatus. The faster a car is moving, the higher is the Doppler shift in frequency. By comparing the outgoing frequency with the frequency of the returning echo the police, or rather their automatic instrument, can calculate the speed of each car. If the police can exploit the technique for measuring the speed of road hogs, dare we hope to find that bats use it for measuring the speed of insect prey?

The answer is yes. The small bats known as horseshoe bats have long been known to emit long, fixed-pitch hoots rather than staccato clicks or descending wolf-whistles. When I say long, I mean long by bat standards. The 'hoots' are still less than a tenth of a second long. And there is often a 'wolf-whistle' tacked onto the end of each hoot, as we shall see. Imagine, first, a horseshoe bat giving out a continuous hum of ultrasound as it flies fast towards a still object, like a tree. The wavefronts will hit the tree at an accelerated rate because of the movement of the bat towards the tree. If a microphone were concealed in the tree, it would 'hear' the sound Doppler-shifted upwards in pitch because of the movement of the bat. There isn't a microphone in the tree, but the echo reflected back from the tree will be Doppler-shifted upwards in pitch in this way. Now, as the echo wavefronts stream back from the tree towards the approaching bat, the bat is still moving fast towards them. Therefore there is a further Doppler shift upwards in the bat's perception of the pitch of the echo. The movement of the bat leads to a kind of double Doppler shift, whose magnitude is a precise indication of the velocity of the bat relative to the tree. By comparing the pitch of its cry with the pitch of the returning echo, therefore, the bat (or rather its on-board computer in the brain) could, in theory, calculate how fast it was moving towards the tree. This wouldn't tell the bat how far away the tree was, but it might still be very useful information, nevertheless.

If the object reflecting the echoes were not a static tree but a moving insect, the Doppler consequences would be more complicated, but the bat could still calculate the velocity of relative motion between itself and its target, obviously just the kind of information a sophisticated guided missile like a hunting bat needs. Actually some bats play a trick that is more interesting than simply emitting hoots of constant pitch and measuring the pitch of the returning echoes. They carefully adjust the pitch of the outgoing hoots, in such a way as to keep the pitch of the echo constant after it has been Doppler-shifted. As they speed towards a moving insect, the pitch of their cries is constantly changing, continuously hunting for just the pitch needed to keep the returning echoes at a fixed pitch. This ingenious trick keeps the echo at the pitch to which their ears are maximally sensitive – important since the echoes are so faint. They can then obtain the necessary information for their Doppler calculations, by monitoring the pitch at which they are obliged to hoot in order to achieve the fixed-pitch echo. I don't know whether man-made devices, either sonar or radar, use this subtle trick. But on the principle that most clever ideas in this field seem to have been developed first by bats, I don't mind betting that the answer is yes.

It is only to be expected that these two rather different techniques, the Doppler shift technique and the 'chirp radar' technique, would be useful for different special purposes. Some groups of bats specialize in one of them, some in the other. Some groups seem to try to get the best of both worlds, tacking an FM 'wolf-whistle' onto the end (or sometimes the beginning) of a long, constant-frequency 'hoot'. Another curious trick of horseshoe bats concerns movements of their outer ear flaps. Unlike other bats, horseshoe bats move their outer ear flaps in fast alternating forward and backward sweeps. It is conceivable that this additional rapid movement of the listening surface relative to the target causes useful modulations in the Doppler shift, modulations that supply additional information. When the ear is flapping towards the target, the apparent velocity of movement towards the target goes up. When it is flapping away from the target, the reverse happens. The bat's brain 'knows' the direction of flapping of each ear, and in principle could make the necessary calculations to exploit the information.

Possibly the most difficult problem of all that bats face is the danger of inadvertent 'jamming' by the cries of other bats. Human experimenters have found it surprisingly difficult to put bats off their stride by playing loud artificial ultrasound at them. With hindsight one might have predicted this. Bats must have come to terms with the jamming-avoidance problem long ago. Many species of bats roost in enormous aggregations, in caves that must be a deafening babel of ultrasound and echoes, yet the bats can still fly rapidly about the cave, avoiding the walls and each other in total darkness. How does a bat keep track of its own echoes, and avoid being misled by the echoes of others? The first solution that might occur to an engineer is some sort of frequency coding: each bat might have its own private frequency, just like separate radio stations. To some extent this may happen, but it is by no means the whole story.

How bats avoid being jammed by other bats is not well understood, but an interesting clue comes from experiments on trying to put bats off. It turns out that you can actively deceive some bats if you play back to them their *own* cries with an artificial *delay*. Give them, in other words, false echoes of their own cries. It is even possible, by carefully controlling the electronic apparatus delaying the false echo, to make the bats attempt to land on a 'phantom' ledge. I suppose it is the bat equivalent of looking at the world through a lens.

It seems that bats may be using something that we could call a 'strangeness filter'. Each successive echo from a bat's own cries produces a picture of the world that makes sense in terms of the previous picture of the world built up with earlier echoes. If the bat's brain hears an echo from another bat's cry, and attempts to incorporate it into the picture of the world that it has previously built up, it will make no sense. It will appear as though objects in the world have suddenly jumped in various random directions. Objects in the real world do not behave in such a crazy way, so the brain can safely filter out the apparent echo as background noise. If a human experimenter feeds the bat artificially delayed or accelerated 'echoes' of its own cries, the false echoes *will* make sense in terms of the world picture that the bat has previously built up. The false echoes are accepted by the strangeness filter because they are plausible in the context of the previous echoes. They cause objects to seem to shift in position by only a small amount, which is what objects plausibly can be expected to do in the real world. The bat's brain relies upon the assumption that the world portrayed by any one echo pulse will be either the same as the world portrayed by previous pulses, or only slightly different: the insect being tracked may have moved a little, for instance.

There is a well-known paper by the philosopher Thomas Nagel called 'What is it like to be a bat?'. The paper is not so much about bats as about the philosophical problem of imagining what it is 'like' to be anything that we are not. The reason a bat is a particularly telling example for a philosopher, however, is that the experiences of an echolocating bat are assumed to be peculiarly alien and different from our own. If you want to share a bat's experience, it is almost certainly grossly misleading to go into a cave, shout or bang two spoons together, consciously time the delay before you hear the echo, and calculate from this how far the wall must be.

That is no more what it is like to be a bat than the following is a good picture of what it is like to see colour: use an instrument to measure the wavelength of the light that is entering your eye: if it is long, you are seeing red, if it is short you are seeing violet or blue. It happens to be a physical fact that the light that we call red has a longer wavelength than the light that we call blue. Different wavelengths switch on the red-sensitive and the blue-sensitive photocells in our retinas. But there is no trace of the concept of wavelength in our subjective sensation of the colours. Nothing about 'what it is like' to see blue or red tells us which light has the longer wavelength. If it matters (it usually doesn't), we just have to remember it, or (what I always do) look it up in a book. Similarly, a bat perceives the position of an insect using what we call echoes. But the bat surely no more thinks in terms of delays of echoes when it perceives an insect, than we think in terms of wavelengths when we perceive blue or red.

Indeed, if I were forced to try the impossible, to imagine what it is like to be a bat, I would guess that echolocating, for them, might be rather like seeing for us. We are such thoroughly visual animals that we hardly realize what a complicated business seeing is. Objects are 'out there', and we think that we 'see' them out there. But I suspect that really our percept is an elaborate computer model in the brain, constructed on the basis of information coming from out there, but transformed in the head into a form in which that information can be *used*. Wavelength differences in the light out there become coded as 'colour' differences in the computer model in the head. Shape and other attributes are encoded in the same kind of way, encoded into a form that is convenient to handle. The sensation of seeing is, for us, very different from the sensation of hearing, but this cannot be directly due to the physical differences between light and sound. Both light and sound are, after all, translated by the respective sense organs into the same kind of nerve impulses. It is impossible to tell, from the physical attributes of a nerve impulse, whether it is conveying information about light, about sound or about smell. The reason the sensation of seeing is so different from the sensation of hearing and the sensation of smelling is that the brain finds it convenient to use different kinds of internal model of the visual world, the world of sound and the world of smell. It is because we *internally use* our visual information and our sound information in different ways and for different purposes that the sensations of seeing and hearing are so different. It is not directly because of the physical differences between light and sound.

But a bat uses its *sound* information for very much the same kind of purpose as we use our *visual* information. It uses sound to perceive, and continuously update its perception of, the position of objects in three-dimensional space, just as we use light. The type of internal computer model that it needs, therefore, is one suitable for the internal representation of the changing positions of objects in three-dimensional space. My point is that the form that an animal's subjective experience takes will be a property of the internal computer model. That model will be designed, in evolution, for its suitability for useful internal representation, irrespective of the physical stimuli that come to it from outside. Bats and we *need* the same kind of internal model for representing the position of objects in three-dimensional space. The fact that bats construct their internal model with the aid of echoes, while we construct ours with the aid of light, is irrelevant. That outside information is, in any case, translated into the same kind of nerve impulses on its way to the brain.

My conjecture, therefore, is that bats 'see' in much the same way as we do, even though the physical medium by which the world 'out there' is translated into nerve impulses is so different – ultrasound rather than light. Bats may even use the sensations that we call colour for their own purposes, to represent differences in the world out there that have nothing to do with the physics of wavelength, but which play a functional role, for the bat, similar to the role that colours play to us. Perhaps male bats have body surfaces that are subtly textured so that the echoes that bounce off them are perceived by females as gorgeously coloured, the sound equivalent of the nuptial plumage of a bird of paradise. I don't mean this just as some vague metaphor. It is possible that the subjective sensation experienced by a female bat when she perceives a male really is, say, bright red: the same sensation as I experience when I see a flamingo. Or, at least, the bat's sensation of her mate may be no more different from my visual sensation of a flamingo, than my visual sensation of a flamingo is different from a flamingo's visual sensation of a flamingo.

Donald Griffin tells a story of what happened when he and his colleague Robert Galambos first reported to an astonished conference of zoologists in 1940 their new discovery of the facts of bat echolocation. One distinguished scientist was so indignantly incredulous that

> he seized Galambos by the shoulders and shook him while complaining that we could not possibly mean such an outrageous suggestion. Radar and sonar were still highly classified developments in military technology, and the notion that bats might do anything even remotely analogous to the latest triumphs of electronic engineering struck most people as not only implausible but emotionally repugnant.

It is easy to sympathize with the distinguished sceptic. There is something very human in his reluctance to believe. And that, really, says it: human is precisely what it is. It is precisely because our own human senses are *not* capable of doing what bats do that we find it hard to believe. Because we can only understand it at a level of artificial instrumentation, and mathematical calculations on paper, we find it hard to imagine a little animal doing it in its head. Yet the mathematical calculations that would be necessary to explain the principles of vision are just as complex and difficult, and nobody has ever had any difficulty in believing that little animals can see. The reason for this double standard in our scepticism is, quite simply, that we can see and we can't echolocate.

I can imagine some other world in which a conference of learned, and totally blind, bat-like creatures is flabbergasted to be told of animals called humans that are actually capable of using the newly discovered inaudible rays called 'light', still the subject of top-secret military development, for finding their way about. These otherwise humble humans are almost totally deaf (well, they can hear after a fashion and even utter a few ponderously slow, deep drawling growls, but they only use these sounds for rudimentary purposes like communicating with each other; they don't seem capable of using them to detect even the most massive objects). They have, instead, highly specialized organs called 'eyes' for exploiting 'light' rays. The sun is the main source of light rays, and humans, remarkably, manage to exploit the complex echoes that bounce off objects when light rays from the sun hit them. They have an ingenious device called a 'lens', whose shape appears to be mathematically calculated so that it bends these silent rays in such a way that there is an exact one-to-one mapping between objects in the world and an 'image' on a sheet of cells called the 'retina'. These retinal cells are capable, in some mysterious way, of rendering the light 'audible' (one might say), and they relay their information to the brain. Our mathematicians have shown that it is theoretically possible, by doing the right highly complex calculations, to navigate safely through the world using these light rays, just as effectively as one can in the ordinary way using ultrasound – in some respects even *more* effectively! But who would have thought that a humble human could do these calculations?

Echo-sounding by bats is just one of the thousands of examples that I could have chosen to make the point about good design. Animals give the appearance of having been designed by a theoretically sophisticated and practically ingenious physicist or engineer, but there is no suggestion that the bats themselves know or understand the theory in the same sense as a physicist understands it. The bat should be thought of as analogous to the police radar trapping *instrument*, not to the person who designed that instrument. The designer of the police radar speed-meter understood the theory of the Doppler Effect, and expressed this understanding in mathematical equations, explicitly written out on paper. The designer's understanding is embodied in the design of the instrument, but the instrument itself does not understand how it works. The instrument contains electronic components, which are wired up so that they automatically compare two radar frequencies and convert the result into convenient units – miles per hour. The computation involved is complicated, but well within the powers of a small box of modern electronic components wired up in the proper way. Of course, a sophisticated conscious brain did the wiring up (or at least designed the wiring diagram), but no conscious brain is involved in the moment-to-moment working of the box.

Our experience of electronic technology prepares us to accept the idea that unconscious machinery can behave as if it understands complex mathematical ideas. This idea is directly transferable to the workings of living machinery. A bat is a machine, whose internal electronics are so wired up that its wing muscles cause it to home in on insects, as an unconscious guided missile homes in on an aeroplane. So far our intuition, derived from technology, is correct. But our experience of technology also prepares us to see the mind of a conscious and purposeful designer in the genesis of sophisticated machinery. It is this second intuition that is wrong in the case of living machinery. In the case of living machinery, the 'designer' is unconscious natural selection, the blind watchmaker.

I hope that the reader is as awestruck as I am, and as William Paley would have been, by these bat stories. My aim has been in one respect identical to Paley's aim. I do not want the reader to underestimate the prodigious works of nature and the problems we face in explaining them. Echolocation in bats, although unknown in Paley's time, would have served his purpose just as well as any of his examples. Paley rammed home his argument by multiplying up his examples. He went right through the body, from head to toe, showing how every part, every last detail, was like the interior of a beautifully fashioned watch. In many ways I should like to do the same, for there are wonderful stories to be told, and I love storytelling. But there is really no need to multiply examples. One or two will do. The hypothesis that can explain bat navigation is a good candidate for explaining anything in the world of life, and if Paley's explanation for any one of his examples was wrong we can't make it right by multiplying up examples. His hypothesis was that living watches were literally designed and built by a master watchmaker. Our modern hypothesis is that the job was done in gradual evolutionary stages by natural selection.

Nowadays theologians aren't quite so straightforward as Paley. They don't point to complex living mechanisms and say that they are self-evidently designed by a creator, just like a watch. But there is a tendency to point to them and say 'It is impossible to believe' that such complexity, or such perfection, could have evolved by natural selection. Whenever I read such a remark, I always feel like writing 'Speak for yourself' in the margin. There are numerous examples (I counted 35 in one chapter) in a recent book called *The Probability of God* by the Bishop of Birmingham, Hugh Montefiore. I shall use this book for all my examples in the rest of this chapter, because it is a sincere and honest attempt, by a reputable and educated writer, to bring natural theology up to date. When I say honest, I mean honest. Unlike some of his theological colleagues, Bishop Montefiore is not afraid to state that the question of whether God exists is a definite question of fact. He has no truck with shifty evasions such as 'Christianity is a way of life. The question of God's *existence* is eliminated: it is a mirage created by the illusions of realism'. Parts of his book are about physics and cosmology, and I am not competent to comment on those except to note that he seems to have used genuine physicists as his authorities. Would that he had done the same in the biological parts. Unfortunately, he preferred here to consult the works of Arthur Koestler, Fred Hoyle, Gordon Rattray-Taylor and Karl Popper! The Bishop believes in evolution, but cannot believe that natural selection is an adequate explanation for the course that evolution has taken (partly because, like many others, he sadly misunderstands natural selection to be 'random' and 'meaningless').

He makes heavy use of what may be called the Argument from Personal Incredulity. In the course of one chapter we find the following phrases, in this order:

> . . . there seems no explanation on Darwinian grounds . . . It is no easier to explain . . . It is hard to understand . . . It is not easy to understand . . . It is equally difficult to explain . . . I do not find it easy to comprehend . . . I do not find it easy to see . . . I find it hard to understand . . . it does not seem feasible to explain . . . I cannot see how . . . neo-Darwinism seems inadequate to explain many of the complexities of animal behaviour . . . it is not easy to comprehend how such behaviour could have evolved solely through natural selection . . . It is impossible . . . How could an organ so complex evolve? . . . It is not easy to see . . . It is difficult to see . . .

The Argument from Personal Incredulity is an extremely weak argument, as Darwin himself noted. In some cases it is based upon simple ignorance. For instance, one of the facts that the Bishop finds it difficult to understand is the white colour of polar bears.

> As for camouflage, this is not always easily explicable on neo-Darwinian premises. If polar bears are dominant in the Arctic, then there would seem to have been no need for them to evolve a white-coloured form of camouflage.

This should be translated:

> I personally, off the top of my head sitting in my study, never having visited the Arctic, never having seen a polar bear in the wild, and having been educated in classical literature and theology, have not so far managed to think of a reason why polar bears might benefit from being white.

In this particular case, the assumption being made is that only animals that are preyed upon need camouflage. What is overlooked is that predators also benefit from being concealed from their prey. Polar bears stalk seals resting on the ice. If the seal sees the bear coming from far-enough away, it can escape. I suspect that, if he imagines a dark grizzly bear trying to stalk seals over the snow, the Bishop will immediately see the answer to his problem.

The polar bear argument turned out to be almost too easy to demolish but, in an important sense, this is not the point. Even if the foremost authority in the world can't explain some remarkable biological phenomenon, this doesn't mean that it is inexplicable. Plenty of mysteries have lasted for centuries and finally yielded to explanation. For what it is worth, most modern biologists wouldn't find it difficult to explain every one of the Bishop's 35 examples in terms of the theory of natural selection, although not all of them are quite as easy as the polar bears. But we aren't testing human ingenuity. Even if we found one example that we *couldn't* explain, we should hesitate to draw any grandiose conclusions from the fact of our own inability. Darwin himself was very clear on this point.

There are more serious versions of the argument from personal incredulity, versions which do not rest simply upon ignorance or lack of ingenuity. One form of the argument makes direct use of the extreme sense of wonder which we all feel when confronted with highly complicated machinery, like the detailed perfection of the echolocation equipment of bats. The implication is that it is somehow self-evident that anything so wonderful as this could not possibly have evolved by natural selection. The Bishop quotes, with approval, G. Bennett on spider webs:

> It is impossible for one who has watched the work for many hours to have any doubt that neither the present spiders of this species nor their ancestors were ever the architects of the web or that it could conceivably have been produced step by step through random variation; it would be as absurd to suppose that the intricate and exact proportions of the Parthenon were produced by piling together bits of marble.

It is not impossible at all. That is exactly what I firmly believe, and I have some experience of spiders and their webs.

The Bishop goes on to the human eye, asking rhetorically, and with the implication that there is no answer, 'How could an organ so complex evolve?' This is not an argument, it is simply an affirmation of incredulity. The underlying basis for the intuitive incredulity that we all are tempted to feel about what Darwin called organs of extreme perfection and complication is, I think, twofold. First we have no intuitive grasp of the immensities of time available for evolutionary change. Most sceptics about natural selection are prepared to accept that it can bring about minor changes like the dark coloration that has evolved in various species of moth since the industrial revolution. But, having accepted this, they then point out how small a change this is. As the Bishop underlines, the dark moth is not a *new species*. I agree that this is a small change, no match for the evolution of the eye, or of echolocation. But equally, the moths only took a hundred years to make their change. One hundred years seems like a long time to us, because it is longer than our lifetime. But to a geologist it is about a thousand times shorter than he can ordinarily measure!

Eyes don't fossilize, so we don't know how long our type of eye took to evolve its present complexity and perfection from nothing, but the time available is several hundred million years. Think, by way of comparison, of the change that man has wrought in a much shorter time by genetic selection of dogs. In a few hundreds, or at most thousands, of years we have gone from wolf to Pekinese, Bulldog, Chihuahua and Saint Bernard. Ah, but they are still *dogs* aren't they? They haven't turned into a different '*kind*' of animal? Yes, if it comforts you to play with words like that, you can call them all dogs. But just think about the time involved. Let's represent the total time it took to evolve all these breeds of dog from a wolf, by one ordinary walking pace. Then, on the same scale, how far would you have to walk, in order to get back to Lucy and her kind, the earliest human fossils that unequivocally walked upright? The answer is about 2 miles. And how far would you have to walk, in order to get back to the start of evolution on Earth? The answer is that you would have to slog it out all the way from London to Baghdad. Think of the total quantity of change involved in going from wolf to Chihuahua, and then multiply it up by the number of walking paces between London and Baghdad. This will give some intuitive idea of the amount of change that we can expect in real natural evolution.

The second basis for our natural incredulity about the evolution of very complex organs like human eyes and bat ears is an intuitive application of probability theory. Bishop Montefiore quotes C. E. Raven on cuckoos. These lay their eggs in the nests of other birds, which then act as unwitting foster parents. Like so many biological adaptations, that of the cuckoo is not single but multiple. Several different facts about cuckoos fit them to their parasitic way of life. For instance, the mother has the habit of laying in other birds' nests, and the baby has the habit of throwing the host's own chicks out of the nest. Both habits help the cuckoo succeed in its parasitic life. Raven goes on:

It will be seen that each one of this sequence of conditions is essential for the success of the whole. Yet each by itself is useless. The whole *opus perfectum* must have been achieved simultaneously. The odds against the random occurrence of such a series of coincidences are, as we have already stated, astronomical.

Arguments such as this are in principle more respectable than the argument based on sheer, naked incredulity. Measuring the statistical improbability of a suggestion is the right way to go about assessing its believability. Indeed, it is a method that we shall use in this book several times. But you have to do it right! There are two things wrong with the argument put by Raven. First, there is the familiar, and I have to say rather irritating, confusion of natural selection with 'randomness'. Mutation is random; natural selection is the very opposite of random. Second, it just isn't *true* that 'each by itself is useless'. It isn't true that the whole perfect work must have been achieved simultaneously. It isn't true that each part is essential for the success of the whole. A simple, rudimentary, half-cocked eye/ear/echolocation system/cuckoo parasitism system, etc., is better than none at all. Without an eye you are totally blind. With half an eye you may at least be able to detect the general direction of a predator's movement, even if you can't focus a clear image. And this may make all the difference between life and death. These matters will be taken up again in more detail in the next two chapters.

CHAPTER 3

ACCUMULATING
SMALL CHANGE

We have seen that living things are too improbable and too beautifully 'designed' to have come into existence by chance. How, then, did they come into existence? The answer, Darwin's answer, is by gradual, step-by-step transformations from simple beginnings, from primordial entities sufficiently simple to have come into existence by chance. Each successive change in the gradual evolutionary process was simple enough, *relative to its predecessor*, to have arisen by chance. But the whole sequence of cumulative steps constitutes anything but a chance process, when you consider the complexity of the final end-product relative to the original starting point. The cumulative process is directed by nonrandom survival. The purpose of this chapter is to demonstrate the power of this *cumulative selection* as a fundamentally nonrandom process.

If you walk up and down a pebbly beach, you will notice that the pebbles are not arranged at random. The smaller pebbles typically tend to be found in segregated zones running along the length of the beach, the larger ones in different zones or stripes. The pebbles have been sorted, arranged, selected. A tribe living near the shore might wonder at this evidence of sorting or arrangement in the world, and might develop a myth to account for it, perhaps attributing it to a Great Spirit in the sky with a tidy mind and a sense of order. We might give a superior smile at such a superstitious notion, and explain that the arranging was really done by the blind forces of physics, in this case the action of waves. The waves have no purposes and no intentions, no tidy mind, no mind at all. They just energetically throw the pebbles around, and big pebbles and small pebbles respond differently to this treatment so they end up at different levels of the beach. A small amount of order has come out of disorder, and no mind planned it.

The waves and the pebbles together constitute a simple example of a system that automatically generates non-randomness. The world is full of such systems. The simplest example I can think of is a hole. Only objects smaller than the hole can pass through it. This means that if you start with a random collection of objects above the hole, and some force shakes and jostles them about at random, after a while the objects above and below the hole will come to be nonrandomly sorted. The space below the hole will tend to contain objects smaller than the hole, and the space above will tend to contain objects larger than the hole. Mankind has, of course, long exploited this simple principle for generating non-randomness, in the useful device known as the sieve.

The Solar System is a stable arrangement of planets, comets and debris orbiting the sun, and it is presumably one of many such orbiting systems in the universe. The nearer a satellite is to its sun, the faster it has to travel if it is to counter the sun's gravity and remain in stable orbit. For any given orbit, there is only one speed at which a satellite can travel and remain in that orbit. If it were travelling at any other velocity, it would either move out into deep space, or crash into the Sun, or move into another orbit. And if we look at the planets of our solar system, lo and behold, every single one of them is travelling at exactly the right velocity to keep it in its stable orbit around the Sun. A blessed miracle of provident design? No, just another natural 'sieve'. Obviously all the planets that we see orbiting the sun must be travelling at exactly the right speed to keep them in their orbits, or we wouldn't see them there because they wouldn't be there! But equally obviously this is not evidence for conscious design. It is just another kind of sieve.

Sieving of this order of simplicity is not, on its own, enough to account for the massive amounts of nonrandom order that we see in living things. Nowhere near enough. Remember the analogy of the combination lock. The kind of non-randomness that can be generated by simple sieving is roughly equivalent to opening a combination lock with only one dial: it is easy to open it by sheer luck. The kind of non-randomness that we see in living systems, on the other hand, is equivalent to a gigantic combination lock with an almost uncountable number of dials. To generate a biological molecule like haemoglobin, the red pigment in blood, by simple sieving would be equivalent to taking all the amino-acid building blocks of haemoglobin, jumbling them up at random, and hoping that the haemoglobin molecule would reconstitute itself by sheer luck. The amount of luck that would be required for this feat is unthinkable, and has been used as a telling mind-boggler by Isaac Asimov and others.

A hacmoglobin molecule consists of four chains of amino acids twisted together. Let us think about just one of these four chains. It consists of 146 amino acids. There are 20 different kinds of amino acids commonly found in living things. The number of possible ways of arranging 20 kinds of thing in chains 146 links long is an inconceivably large number, which Asimov calls the 'haemoglobin number'. It is easy to calculate, but impossible to visualize the answer. The first link in the 146-long chain could be any one of the 20 possible amino acids. The second link could also be any one of the 20, so the number of possible 2-link chains is 20 × 20, or 400. The number of possible 3-link chains is 20 × 20 × 20, or 8,000. The number of possible 146-link chains is 20 times itself 146 times. This is a staggeringly large number. A million is a 1 with 6 noughts after it. A billion (1,000 million) is a 1 with 9 noughts after it. The number we seek, the 'haemoglobin number', is (near enough) a 1 with 190 noughts after it! This is the chance against happening to hit upon haemoglobin by luck. And a haemoglobin molecule has only a minute fraction of the complexity of a living body. Simple sieving, on its own, is obviously nowhere near capable of generating the amount of order in a living thing. Sieving is an essential ingredient in the generation of living order, but it is very far from being the whole story. Something else is needed. To explain the point, I shall need to make a distinction between 'single-step' selection and 'cumulative' selection. The simple sieves we have been considering so far in this chapter are all examples of single-step selection. Living organization is the product of cumulative selection.

The essential difference between single-step selection and cumulative selection is this. In single-step selection the entities selected or sorted, pebbles or whatever they are, are sorted once and for all. In cumulative selection, on the other hand, they 'reproduce'; or in some other way the results of one sieving process are fed into a subsequent sieving, which is fed into . . . , and so on. The entities are subjected to selection or sorting over many 'generations' in succession. The end-product of one generation of selection is the starting point for the next generation of selection, and so on for many generations. It is natural to borrow such words as 'reproduce' and 'generation', which have associations with living things, because living things are the main examples we know of things that participate in cumulative selection. They may in practice be the only things that do. But for the moment I don't want to beg that question by saying so outright.

Sometimes clouds, through the random kneading and carving of the winds, come to look like familiar objects. There is a much published photograph, taken by the pilot of a small aeroplane, of what looks a bit like the face of Jesus, staring out of the sky. We have all seen clouds that reminded us of something – a sea horse, say, or a smiling face. These resemblances come about by single-step selection, that is to say by a single coincidence. They are, consequently, not very impressive. The resemblance of the signs of the zodiac to the animals after which they are named, Scorpio, Leo, and so on, is as unimpressive as the predictions of astrologers. We don't feel overwhelmed by the resemblance, as we are by biological adaptations – the products of cumulative selection. We describe as weird, uncanny or spectacular, the resemblance of, say, a leaf insect to a leaf or a praying mantis to a cluster of pink flowers. The resemblance of a cloud to a weasel is only mildly diverting, barely worth calling to the attention of our companion. Moreover, we are quite likely to change our mind about exactly what the cloud most resembles.

Hamlet.	Do you see yonder cloud that's almost in shape of a camel?
Polonius.	By the mass, and 'tis like a camel, indeed.
Hamlet.	Methinks it is like a weasel.
Polonius.	It is backed like a weasel.
Hamlet.	Or like a whale?
Polonius.	Very like a whale.

I don't know who it was first pointed out that, given enough time, a monkey bashing away at random on a typewriter could produce all the works of Shakespeare. The operative phrase is, of course, given enough time. Let us limit the task facing our monkey somewhat. Suppose that he has to produce, not the complete works of Shakespeare but just the short sentence 'Methinks it is like a weasel', and we shall make it relatively easy by giving him a typewriter with a restricted keyboard, one with just the 26 (capital) letters, and a space bar. How long will he take to write this one little sentence?

The sentence has 28 characters in it, so let us assume that the monkey has a series of discrete 'tries', each consisting of 28 bashes at the keyboard. If he types the phrase correctly, that is the end of the experiment. If not, we allow him another 'try' of 28 characters. I don't know any monkeys, but fortunately my 11-month old daughter is an experienced randomizing device, and she proved only too eager to step into the role of monkey typist. Here is what she typed on the computer:

```
UMMK JK CDZZ F ZD DSDSKSM
S SS FMCV PU I DDRGLKDXRRDO
RDTE QDWFDVIOY UDSKZWDCCVYT
H CHVY NMGNBAYTDFCCVD D
RCDFYYYRM N DFSKD LD K WDWK
JJKAUIZMZI UXDKIDISFUMDKUDXI
```

She has other important calls on her time, so I was obliged to program the computer to simulate a randomly typing baby or monkey:

```
WDLDMNLT DTJBKWIRZREZLMQCO P
Y YVMQKZPGJXWVHGLAWFVCHQYOPY
MWR SWTNUXMLCDLEUBXTQHNZVJQF
FU OVAODVYKDGXDEKYVMOGGS VT
HZQZDSFZIHIVPHZPETPWVOVPMZGF
GEWRGZRPBCTPGQMCKHFDBGW ZCCF
```

And so on and on. It isn't difficult to calculate how long we should reasonably expect to wait for the random computer (or baby or monkey) to type METHINKS IT IS LIKE A WEASEL. Think about the total number of *possible* phrases of the right length that the monkey or baby or random computer *could* type. It is the same kind of calculation as we did for haemoglobin, and it produces a similarly large result. There are 27 possible letters (counting 'space' as one letter) in the first position. The chance of the monkey happening to get the first letter – M – right is therefore 1 in 27. The chance of it getting the first two letters – ME – right is the chance of it getting the second letter – E – right (1 in 27) *given that* it has also got the first letter – M – right, therefore $1/27 \times 1/27$, which equals $1/729$. The chance of it getting the first word – METHINKS – right is $1/27$ for each of the 8 letters, therefore $(1/27) \times (1/27) \times (1/27) \times (1/27) \ldots$, etc. 8 times, or $(1/27)$ to the power 8. The chance of it getting the entire phrase of 28 characters right is $(1/27)$ to the power 28, i.e. $(1/27)$ multiplied by itself 28 times. These are very small odds, about 1 in 10,000 million million million million million million. To put it mildly, the phrase we seek would be a long time coming, to say nothing of the complete works of Shakespeare.

So much for single-step selection of random variation. What about cumulative selection; how much more effective should this be? Very very much more effective, perhaps more so than we at first realize, although it is almost obvious when we reflect further. We again use our computer monkey, but with a crucial difference in its program. It again begins by choosing a random sequence of 28 letters, just as before:

```
WDLMNLT DTJBKWIRZREZLMQCO P
```

It now 'breeds from' this random phrase. It duplicates it repeatedly, but with a certain chance of random error – 'mutation' – in the copying. The computer examines the mutant nonsense phrases, the 'progeny' of the original phrase, and chooses the one which, *however slightly*, most resembles the target phrase, METHINKS IT IS LIKE A

WEASEL. In this instance the winning phrase of the next 'generation' happened to be:

WDLTMNLT DTJBSWIRZREZLMQCO P

Not an obvious improvement! But the procedure is repeated, again mutant 'progeny' are 'bred from' the phrase, and a new 'winner' is chosen. This goes on, generation after generation. After 10 generations, the phrase chosen for 'breeding' was:

MDLDMNLS ITJISWHRZREZ MECS P

After 20 generations it was:

MELDINLS IT ISWPRKE Z WECSEL

By now, the eye of faith fancies that it can see a resemblance to the target phrase. By 30 generations there can be no doubt:

METHINGS IT ISWLIKE B WECSEL

Generation 40 takes us to within one letter of the target:

METHINKS IT IS LIKE I WEASEL

And the target was finally reached in generation 43. A second run of the computer began with the phrase:

Y   YVMQKZPFJXWVHGLAWFVCHQXYOPY,

passed through (again reporting only every tenth generation):

Y YVMQKSPFTXWSHLIKEFV HQYSPY
YETHINKSPITXISHLIKEFA WQYSEY
METHINKS IT ISSLIKE A WEFSEY
METHINKS IT ISBLIKE A WEASES
METHINKS IT ISJLIKE A WEASEO
METHINKS IT IS LIKE A WEASEP

and reached the target phrase in generation 64. In a third run the computer started with:

GEWRGZRPBCTPGQMCKHFDBGW ZCCF

and reached METHINKS IT IS LIKE A WEASEL in 41 generations of selective 'breeding'.

The exact time taken by the computer to reach the target doesn't matter. If you want to know, it completed the whole exercise for me, the first time, while I was out to lunch. It took about half an hour. (Computer enthusiasts may think this unduly slow. The reason is that the program was written in BASIC, a sort of computer baby-talk. When I rewrote it in Pascal, it took 11 seconds.) Computers are a bit faster at this kind of thing than monkeys, but the difference really isn't significant. What matters is the difference between the time taken by *cumulative* selection, and the time which the same computer, working flat out at the same rate, would take to reach the target phrase if it were forced to use the other procedure of *single-step selection*: about a million million million million million years. This is more than a million million million times as long as the universe has so far existed. Actually it would be fairer just to say that, in comparison with the time it would take either a monkey or a randomly programmed computer to type our target phrase, the total age of the universe so far is a negligibly small quantity, so small as to be well within the margin of error for this sort of back-of-an-envelope calculation. Whereas the time taken for a computer working randomly but with the constraint of *cumulative selection* to perform the same task is of the same order as humans ordinarily can understand, between 11 seconds and the time it takes to have lunch.

There is a big difference, then, between cumulative selection (in which each improvement, however slight, is used as a basis for future building), and single-step selection (in which each new 'try' is a fresh one). If evolutionary progress had had to rely on single-step selection, it would never have got anywhere. If, however, there was any way in which the necessary conditions for *cumulative* selection could have been set up by the blind forces of nature, strange and wonderful might have been the consequences. As a matter of fact that is exactly what happened on this planet, and we ourselves are among the most recent, if not the strangest and most wonderful, of those consequences.

It is amazing that you can still read calculations like my haemoglobin calculation, used as though they constituted arguments *against* Darwin's theory. The people who do this, often expert in their own field, astronomy or whatever it may be, seem sincerely to believe that Darwinism explains living organization in terms of chance – 'single- step selection' – alone. This belief, that Darwinian evolution is 'random', is not merely false. It is the exact opposite of the truth. Chance is a minor ingredient in the Darwinian recipe, but the most important ingredient is cumulative selection which is quintessentially *non*random.

Clouds are not capable of entering into cumulative selection. There is no mechanism whereby clouds of particular shapes can spawn daughter clouds resembling themselves. If there were such a mechanism, if a cloud resembling a weasel or a camel could give rise to a lineage of other clouds of roughly the same shape, cumulative selection would have the opportunity to get going. Of course, clouds do break up and form 'daughter' clouds sometimes, but this isn't enough for cumulative selection. It is also necessary that the 'progeny' of any given cloud should resemble its 'parent' *more* than it resembles any old 'parent' in the 'population'. This vitally important point is apparently misunderstood by some of the philosophers who have, in recent years, taken an interest in the theory of natural selection. It is further necessary that the chances of a given cloud's surviving and spawning copies should depend upon its shape. Maybe in some distant galaxy these conditions did arise, and the result, if enough millions of years have gone by, is an ethereal, wispy form of life. This might make a good science fiction story – *The White Cloud*, it could be called – but for our purposes a computer model like the monkey/Shakespeare model is easier to grasp.

Although the monkey/Shakespeare model is useful for explaining the distinction between single-step selection and cumulative selection, it is misleading in important ways. One of these is that, in each generation of selective 'breeding', the mutant 'progeny' phrases were judged according to the criterion of resemblance to a *distant ideal* target, the phrase METHINKS IT IS LIKE A WEASEL. Life isn't like that. Evolution has no long-term goal. There is no long-distance target, no final perfection to serve as a criterion for selection, although human vanity cherishes the absurd notion that our species is the final goal of evolution. In real life, the criterion for selection is always short-term, either simple survival or, more generally, reproductive success. If, after the aeons, what looks like progress towards some distant goal seems, with hindsight, to have been achieved, this is always an incidental consequence of many generations of short-term selection. The 'watchmaker' that is cumulative natural selection is blind to the future and has no long-term goal.

We can change our computer model to take account of this point. We can also make it more realistic in other respects. Letters and words are peculiarly human manifestations, so let's make the computer draw pictures instead. Maybe we shall even see animal-like shapes evolving in the computer, by cumulative selection of mutant forms. We shan't prejudge the issue by building-in specific animal pictures to start with. We want them to emerge solely as a result of cumulative selection of random mutations.

In real life, the form of each individual animal is produced by embryonic development. Evolution occurs because, in successive generations, there are slight differences in embryonic development.

These differences come about because of changes (mutations – this is the small random element in the process that I spoke of) in the genes controlling development. In our computer model, therefore, we must have something equivalent to embryonic development, and something equivalent to genes that can mutate. There are many ways in which we could meet these specifications in a computer model. I chose one and wrote a program that embodied it. I shall now describe this computer model, because I think it is revealing. If you don't know anything about computers, just remember that they are machines that do exactly what you tell them but often surprise you in the result. A list of instructions for a computer is called a program (this is standard American spelling, and it is also recommended by the Oxford Dictionary: the alternative, 'programme', commonly used in Britain, appears to be a Frenchified affectation).

Embryonic development is far too elaborate a process to simulate realistically on a small computer. We must represent it by some simplified analogue. We must find a simple picture-drawing rule that the computer can easily obey, and which can then be made to vary under the influence of 'genes'. What drawing rule shall we choose? Textbooks of computer science often illustrate the power of what they call 'recursive' programming with a simple *tree-growing* procedure. The computer starts by drawing a single vertical line. Then the line branches into two. Then each of the branches splits into two sub-branches. Then each of the sub-branches splits into sub-sub-branches, and so on. It is 'recursive' because the same rule (in this case a branching rule) is applied locally all over the growing tree. No matter how big the tree may grow, the same branching rule goes on being applied at the tips of all its twigs.

The 'depth' of recursion, means the number of sub-sub- . . . branches that are allowed to grow, before the process is brought to a halt. Figure 2 shows what happens when you tell the computer to obey exactly the same drawing rule, but going on to various depths of recursion. At high levels of recursion the pattern becomes quite elaborate, but you can easily see in Figure 2 that it is still produced by the same very simple branching rule. This is, of course, just what happens in a real tree. The branching pattern of an oak tree or an apple tree looks complex, but it really isn't. The basic branching rule is very simple. It is because it is applied recursively at the growing tips all over the tree – branches make sub-branches, then each sub-branch makes sub-sub-branches, and so on – that the whole tree ends up large and bushy.

Recursive branching is also a good metaphor for the embryonic development of plants and animals generally. I don't mean that animal

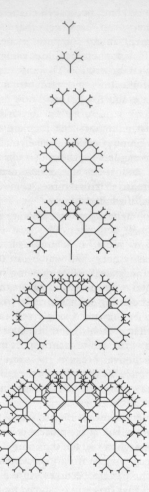

Figure 2

embryos look like branching trees. They don't. But all embryos grow by cell division. Cells always split into two daughter cells. And genes always exert their final effects on bodies by means of *local* influences on cells, and on the two-way branching patterns of cell division. An animal's genes are never a grand design, a blueprint for the whole body. The genes, as we shall see, are more like a recipe than like a blueprint; and a recipe, moreover, that is obeyed *not* by the developing embryo as a whole, but by each cell or each local cluster of dividing cells. I'm not denying that the embryo, and later the adult, *has* a large-scale form. But this large-scale form *emerges* because of lots of little local cellular effects all over the developing body, and these local effects consist primarily of two-way branchings, in the form of two-way cell splittings. It is by influencing these local events that genes ultimately exert influences on the adult body.

The simple branching rule for drawing trees, then, looks like a promising analogue for embryonic development. Accordingly, we wrap it up in a little computer procedure, label it DEVELOPMENT, and prepare to embed it in a larger program labelled EVOLUTION. As a first step towards writing this larger program, we now turn our attention to genes. How shall we represent 'genes' in our computer model? Genes in real life do two things. They influence development, and they get passed on to future generations. In real animals and plants there are tens of thousands of genes, but we shall modestly limit our computer model to nine. Each of the nine genes is simply represented by a number in the computer, which will be called its *value*. The value of a particular gene might be, say 4, or −7.

How shall we make these genes influence development? There are lots of things they could do. The basic idea is that they should exert some minor quantitative influence on the drawing rule that is DE-VELOPMENT. For instance, one gene might influence the angle of branching, another might influence the length of some particular branch. Another obvious thing for a gene to do is to influence the depth of the recursion, the number of successive branchings. I made Gene 9 have this effect. You can regard Figure 2, therefore, as a picture of seven related organisms, identical to each other except with respect to Gene 9. I shan't spell out in detail what each one of the other eight genes does. You can get a general idea of the *kinds* of things they do by studying Figure 3. In the middle of the picture is the basic tree, one of the ones from Figure 2. Encircling this central tree are eight others. All are the same as the central tree, except that one gene, a different gene in each of the eight, has been changed − 'mutated'. For instance, the picture to the right of the central tree shows what happens when Gene 5 mutates by having +1 added to its value. If there'd been room, I'd have liked to print a ring of 18 mutants around the central tree. The reason for wanting 18 is that there are nine genes, and each one can mutate in an 'upward' direction (1 is added to its value) or in a 'down-ward' direction (1 is subtracted from its value). So a ring of 18 trees would be enough to represent all *possible* single-step mutants that you can derive from the one central tree.

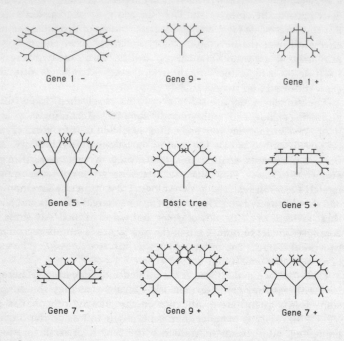

Figure 3

Each of these trees has its own, unique 'genetic formula', the numerical values of its nine genes. I haven't written the genetic formulae down, because they wouldn't mean anything to you, in themselves. That is true of real genes too. Genes only start to mean something when they are translated, via protein synthesis, into growing-rules for a developing embryo. And in the computer model too, the numerical values of the nine genes only mean something when they are translated into growing rules for the branching tree pattern. But you can get an idea of what each gene does by *comparing* the bodies of two organisms known to differ with respect to a certain gene. Compare, for instance, the basic tree in the middle of the picture with the two trees on either side, and you'll get some idea of what Gene 5 does.

This, too, is exactly what real-life geneticists do. Geneticists normally don't know how genes exert their effects on embryos. Nor do they know the complete genetic formula of any animal. But by comparing the bodies of two adult animals that are known to *differ* according to a single gene, they can see what effects that single gene has. It is more complicated than that, because the effects of genes interact with each other in ways that are more complicated than simple addition. Exactly the same is true of the computer trees. Very much so, as later pictures will show.

You will notice that all the shapes are symmetrical about a left/right axis. This is a constraint that I imposed on the DEVELOPMENT procedure. I did it partly for aesthetic reasons; partly to economize on the number of genes necessary (if genes didn't exert mirror-image effects on the two sides of the tree, we'd need separate genes for the left and the right sides); and partly because I was hoping to evolve animal-like shapes, and most animal bodies are pretty symmetrical. For the same reason, from now on I shall stop calling these creatures 'trees', and shall call them 'bodies' or 'biomorphs'. Biomorph is the name coined by Desmond Morris for the vaguely animal-like shapes in his surrealist paintings. These paintings have a special place in my affections, because one of them was reproduced on the cover of my first book. Desmond Morris claims that his biomorphs 'evolve' in his mind, and that their evolution can be traced through successive paintings.

Back to the computer biomorphs, and the ring of 18 possible mutants, of which a representative eight are drawn in Figure 3. Since each member of the ring is only one mutational step away from the central biomorph, it is easy for us to see them as *children* of the central parent. We have our analogue of REPRODUCTION, which, like DE-VELOPMENT, we can wrap up in another small computer program, ready to embed in our big program called EVOLUTION. Note two things about REPRODUCTION. First, there is no sex; reproduction is asexual. I think of the biomorphs as female, therefore, because asexual animals like greenfly are nearly always basically female in form. Second, my mutations are all constrained to occur one at a time. A child differs from its parent at only one of the nine genes; moreover, all mutation occurs by +1 or −1 being added to the value of the corresponding parental gene. These are just arbitrary conventions: they could have been different and still remained biologically realistic.

The same is not true of the following feature of the model, which embodies a fundamental principle of biology. The shape of each child is not derived directly from the shape of the parent. Each child gets its shape from the values of its own nine genes (influencing angles, distances, and so on). And each child gets its nine genes from its parent's nine genes. This is just what happens in real life. Bodies don't get passed down the generations; genes do. Genes influence embryonic development of the body in which they are sitting. Then those same genes either get passed on to the next generation or they don't. The nature of the genes is unaffected by their participation in bodily development, but their likelihood of being passed on may be affected by the success of the body that they helped to create. This is why, in the computer model, it is important that the two procedures called DEVELOPMENT and REPRODUCTION are written as two watertight compartments. They are watertight except that REPRODUCTION passes gene values across to DEVELOPMENT, where they influence the growing rules. DEVELOPMENT most emphatically does not pass gene values back to REPRODUCTION – that would be tantamount to 'Lamarckism' (see Chapter 11).

We have assembled our two program modules, then, labelled DEVELOPMENT and REPRODUCTION. REPRODUCTION passes genes down the generations, with the possibility of mutation. DEVELOPMENT takes the genes provided by REPRODUCTION in any given generation, and translates those genes into drawing action, and hence into a picture of a body on the computer screen. The time has come to bring the two modules together in the big program called EVOLUTION.

EVOLUTION basically consists of endless repetition of REPRODUCTION. In every generation, REPRODUCTION takes the genes that are supplied to it by the previous generation, and hands them on to the next generation but with minor random errors – mutations. A mutation simply consists in +1 or −1 being added to the value of a randomly chosen gene. This means that, as the generations go by, the total amount of genetic difference from the original ancestor can become very large, cumulatively, one small step at a time. But although the mutations are random, the cumulative change over the generations is not random. The progeny in any one generation are different from their parent in random directions. But which of those progeny is selected to go forward into the next generation is not random. This is where Darwinian selection comes in. The criterion for selection is not the genes themselves, but the bodies whose shape the genes influence through DEVELOPMENT.

In addition to being REPRODUCED, the genes in each generation are also handed to DEVELOPMENT, which grows the appropriate body on the screen, following its own strictly laid-down rules. In every generation, a whole 'litter' of 'children' (i.e. individuals of the next generation) is displayed. All these children are mutant children of the same parent, differing from their parent with respect to one gene each. This very high mutation rate is a distinctly unbiological feature of the computer model. In real life, the probability that a gene will mutate is often less than one in a million. The reason for building a high mutation rate into the model is that the whole performance on the computer screen is for the benefit of human eyes, and humans haven't the patience to wait a million generations for a mutation!

The human eye has an active role to play in the story. It is the selecting agent. It surveys the litter of progeny and chooses one for breeding. The chosen one then becomes the parent of the next generation, and a litter of *its* mutant children are displayed simultaneously on the screen. The human eye is here doing exactly what it does in the breeding of pedigree dogs or prize roses. Our model, in other words, is strictly a model of artificial selection, not natural selection. The criterion for 'success' is not the direct criterion of survival, as it is in true natural selection. In true natural selection, if a body has what it takes to survive, its genes automatically survive because they are inside it. So the genes that survive tend to be, automatically, those genes that confer on bodies the qualities that assist them to survive. In the computer model, on the other hand, the selection criterion is not survival, but the ability to appeal to human whim. Not necessarily idle, casual whim, for we can resolve to select consistently for some quality such as 'resemblance to a weeping willow'. In my experience, however, the human selector is more often capricious and opportunistic. This, too, is not unlike certain kinds of natural selection.

The human tells the computer which one of the current litter of progeny to breed from. The genes of the chosen one are passed across to REPRODUCTION, and a new generation begins. This process, like real-life evolution, goes on indefinitely. Each generation of biomorphs is only a single mutational step away from its predecessor and its successor. But after 100 generations of EVOLUTION, the biomorphs can be anything up to 100 mutational steps away from their original ancestor. And in 100 mutational steps, much can happen.

I never dreamed *how* much, when I first started to play with my newly written EVOLUTION program. The main thing that surprised me was that the biomorphs can pretty quickly cease to look like trees. The basic two-way branching structure is always there, but it is easily smothered as lines cross and recross one another, making solid masses of colour (only black and white in the printed pictures). Figure 4 shows one particular evolutionary history consisting of no more than 29 generations. The ancestor is a tiny creature, a single dot. Although the ancestor's body is a dot, like a bacterium in the primeval slime, hidden inside it is the potential for branching in exactly the pattern of the

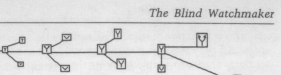

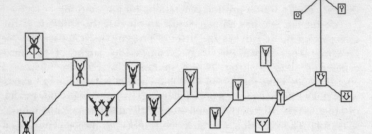

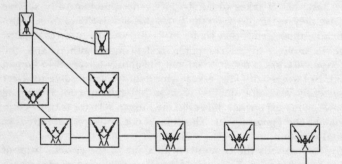

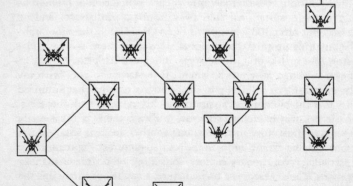

Figure 4

central tree of Figure 3: it is just that its Gene 9 tells it to branch zero times! All the creatures pictured on the page are descended from the dot but, in order to avoid cluttering the page, I haven't printed all the descendants that I actually saw. I've printed only the successful child of each generation (i.e. the parent of the next generation) and one or two of its unsuccessful sisters. So, the picture basically shows just the one main line of evolution, guided by my aesthetic selection. All the stages in the main line are shown.

Let's briefly go through the first few generations of the main line of evolution in Figure 4. The dot becomes a Y in generation 2. In the next two generations, the Y becomes larger. Then the branches become slightly curved, like a well-made catapult. In generation 7, the curve is accentuated, so that the two branches almost meet. The curved branches get bigger, and each acquires a couple of small appendages in generation 8. In generation 9 these appendages are lost again, and the stem of the catapult becomes longer. Generation 10 looks like a section through a flower; the curved side-branches resemble petals cupping a central appendage or 'stigma'. In generation 11, the same 'flower' shape has become bigger and slightly more complicated.

I won't pursue the narrative. The picture speaks for itself, on through the 29 generations. Notice how each generation is just a little different from its parent and from its sisters. Since each is a little different from its parent, it is only to be expected that each will be slightly *more* different from its grandparents (and its grandchildren), and even more different still from its great grandparents (and great grandchildren). This is what *cumulative* evolution is all about, although, because of our high mutation rate, we have speeded it up here to unrealistic rates. Because of this, Figure 4 looks more like a pedigree of *species* than a pedigree of individuals, but the principle is the same.

When I wrote the program, I never thought that it would evolve anything more than a variety of tree-like shapes. I had hoped for weeping willows, cedars of Lebanon, Lombardy poplars, seaweeds, perhaps deer antlers. Nothing in my biologist's intuition, nothing in my 20 years' experience of programming computers, and nothing in my wildest dreams, prepared me for what actually emerged on the screen. I can't remember exactly when in the sequence it first began to dawn on me that an evolved resemblance to something like an insect was possible. With a wild surmise, I began to breed, generation after generation, from whichever child looked most like an insect. My incredulity grew in parallel with the evolving resemblance. You see the eventual results at the bottom of Figure 4. Admittedly they have eight legs like a spider, instead of six like an insect, but even so! I still cannot conceal from you my feeling of exultation as I first watched these exquisite creatures emerging before my eyes. I distinctly heard the triumphal opening chords of *Also sprach Zarathustra* (the '2001 theme') in my mind. I couldn't eat, and that night 'my' insects swarmed behind my eyelids as I tried to sleep.

There are computer games on the market in which the player has the illusion that he is wandering about in an underground labyrinth, which has a definite if complex geography and in which he encounters dragons, minotaurs or other mythic adversaries. In these games the monsters are rather few in number. They are all designed by a human programmer, and so is the geography of the labyrinth. In the evolution game, whether the computer version or the real thing, the player (or observer) obtains the same feeling of wandering metaphorically through a labyrinth of branching passages, but the number of possible pathways is all but infinite, and the monsters that one encounters are undesigned and unpredictable. On my wanderings through the backwaters of Biomorph Land, I have encountered fairy shrimps, Aztec temples, Gothic church windows, aboriginal drawings of kangaroos, and, on one memorable but unrecapturable occasion, a passable caricature of the Wykeham Professor of Logic. Figure 5 is another little collection from my trophy room, all of which developed in the same kind of way. I want to emphasize that these shapes are not artists' impressions. They have not been touched-up or doctored in any way whatever. They are exactly as the computer drew them when they evolved inside it. The role of the human eye was limited to *selecting*, among randomly mutated progeny over many generations of cumulative evolution.

We now have a much more realistic model of evolution than the monkeys typing Shakespeare gave us. But the biomorph model is still deficient. It shows us the power of cumulative selection to generate an almost endless variety of quasi-biological form, but it uses artificial selection, not natural selection. The human eye does the selecting. Could we dispense with the human eye, and make the computer itself do the selecting, on the basis of some biologically realistic criterion? This is more difficult than it may seem. It is worth spending a little time explaining why.

It is trivially easy to select for a particular genetic formula, so long as you can read the genes of all the animals. But natural selection doesn't choose genes directly, it chooses the *effects* that genes have on bodies, technically called phenotypic effects. The human eye is good at choosing phenotypic effects, as is shown by the numerous breeds of

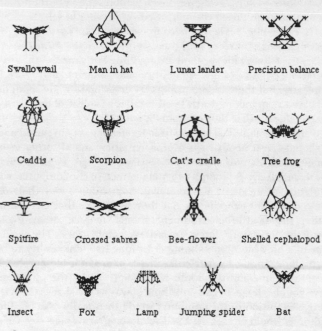

Swallowtail Man in hat Lunar lander Precision balance

Caddis Scorpion Cat's cradle Tree frog

Spitfire Crossed sabres Bee-flower Shelled cephalopod

Insect Fox Lamp Jumping spider Bat

Figure 5

dogs, cattle and pigeons, and also, if I may say so, as is shown by Figure 5. To make the computer choose phenotypic effects directly, we should have to write a very sophisticated pattern-recognition program. Pattern-recognizing programs exist. They are used to read print and even handwriting. But they are difficult, 'state of the art' programs, needing very large and fast computers. Even if such a pattern-recognition program were not beyond my programming capabilities, and beyond the capacity of my little 64-kilobyte computer, I wouldn't bother with it. This is a task that is better done by the human eye, together with – and this is more to the point – the 10-giganeurone computer inside the skull.

It wouldn't be too difficult to make the computer select for vague general features like, say, tall-thinness, short-fatness, perhaps curvaceousness, spikiness, even rococo ornamentation. One method would be to program the computer to remember the *kinds* of qualities that humans have favoured in the past, and to exert continued selection of the same general kind in the future. But this isn't getting us any closer to simulating *natural* selection. The important point is that nature doesn't need computing power in order to select, except in special cases like peahens choosing peacocks. In nature, the usual selecting agent is direct, stark and simple. It is the grim reaper. Of course, the *reasons* for survival are anything but simple – that is why natural selection can build up animals and plants of such formidable complexity. But there is something very crude and simple about death itself. And nonrandom death is all it takes to select phenotypes, and hence the genes that they contain, in nature.

To simulate natural selection in an interesting way in the computer, we should forget about rococo ornamentation and all other visually defined qualities. We should concentrate, instead, upon simulating nonrandom death. Biomorphs should interact, in the computer, with a simulation of a hostile environment. Something about their shape should determine whether or not they survive in that environment. Ideally, the hostile environment should include other evolving biomorphs: 'predators', 'prey', 'parasites', 'competitors'. The particular shape of a prey biomorph should determine its vulnerability to being caught, for example, by particular shapes of predator biomorphs. Such criteria of vulnerability should not be built in by the programmer. They should *emerge*, in the same kind of way as the shapes themselves emerge. Evolution in the computer would then really take off, for the conditions would be met for a self-reinforcing 'arms race' (see Chapter 7), and I dare not speculate where it would all end. Unfortunately, I think it may be beyond my powers as a programmer to set up such a counterfeit world.

If anybody is clever enough to do it, it would be the programmers who develop those noisy and vulgar arcade games – Space Invaders' derivatives. In these programs a counterfeit world is simulated. It has a geography, often in three dimensions, and it has a fast-moving time dimension. Entities zoom around in simulated three-dimensional space, colliding with each other, shooting each other down, swallowing each other amid revolting noises. So good can the simulation be that the player handling the joystick receives a powerful illusion that he himself is part of the counterfeit world. I imagine that the summit of this kind of programming is achieved in the chambers used to train aeroplane and spacecraft pilots. But even these programs are small-fry compared to the program that would have to be written to simulate an emerging arms race between predators and prey, embedded in a complete, counterfeit ecosystem. It certainly could be done, however. If there is a professional programmer out there who feels like collaborating on the challenge, I should like to hear from him or her.

Meanwhile, there is something else that is much easier, and which I intend trying when summer comes. I shall put the computer in a shady place in the garden. The screen can display in colour. I already have a version of the program which uses a few more 'genes' to control colour, in the same kind of way as the other nine genes control shape. I shall begin with any more-or-less compact and brightly coloured biomorph. The computer will simultaneously display a range of mutant progeny of the biomorph, differing from it in shape and/or colour pattern. I believe that bees, butterflies and other insects will visit the screen, and 'choose' by bumping into a particular spot on the screen. When a certain number of choices have been logged, the computer will wipe the screen clean, 'breed' from the preferred biomorph, and display the next generation of mutant progeny.

I have high hopes that, over a large number of generations, the wild insects will actually cause the evolution, in the computer, of flowers. If they do, the computer flowers will have evolved under exactly the same selection pressure as caused real flowers to evolve in the wild. I am encouraged in my hope by the fact that insects frequently visit bright blobs of colour on women's dresses (and also by more systematic experiments that have been published). An alternative possibility, which I would find even more exciting, is that the wild insects might cause the evolution of insect-like shapes. The precedent for this – and hence the reason for hope – is that bees in the past caused the evolution of bee-orchids. Male bees, over many generations of cumulative orchid evolution, have built up the bee-like shape through trying to copulate with flowers, and hence carrying pollen. Imagine the 'bee-flower' of Figure 5 in colour. Wouldn't you fancy it if you were a bee?

My main reason for pessimism is that insect vision works in a very different way from ours. Video-screens are designed for human eyes not bee eyes. This could easily mean that, although both we and bees see bee-orchids, in our very different ways, as bee-like, bees might not see video-screen images at all. Bees might see nothing but 625 scanning lines! Still, it is worth a try. By the time this book is published, I shall know the answer.

There is a popular cliché, usually uttered in the tones Stephen Potter would have called 'plonking', which says that you cannot get out of computers any more than you put in. Other versions are that computers only do exactly what you tell them to, and that therefore computers are never creative. The cliché is true only in a crashingly trivial sense, the same sense in which Shakespeare never wrote anything except what his first schoolteacher taught him to write – words. I programmed EVOLUTION into the computer, but I did not plan 'my' insects, nor the scorpion, nor the spitfire, nor the lunar lander. I had not the slightest inkling that they would emerge, which is why 'emerge' is the right word. True, my eyes did the selecting that guided their evolution, but at every stage I was limited to a small clutch of progeny offered up by random mutation, and my selection 'strategy', such as it was, was opportunistic, capricious and short-term. I was not aiming for any distant target, and nor does natural selection.

I can dramatize this by discussing the one time when I *did* try to aim for a distant target. First I must make a confession. You will have guessed it anyway. The evolutionary history of Figure 4 is a reconstruction. It was not the first time I had seen 'my' insects. When they originally emerged to the sound of trumpets, I had no means of recording their genes. There they were, sitting on the computer screen, and I couldn't get at them, couldn't decipher their genes. I delayed switching the computer off while I racked my brain trying to think of some way of saving them, but there was none. The genes were too deeply buried, just as they are in real life. I could print out pictures of the insects' bodies, but I had lost their genes. I immediately modified the program so that in future it would keep accessible records of genetic formulae, but it was too late. I had lost my insects.

I set about trying to 'find' them again. They had evolved once, so it seemed that it must be possible to evolve them again. Like the lost chord, they haunted me. I wandered through Biomorph Land, moving through an endless landscape of strange creatures and things, but I couldn't find my insects. I knew that they must be lurking there somewhere. I knew the genes from which the original evolution had started. I had a picture of my insects' bodies. I even had a picture of the evolutionary sequence of bodies leading up to my insects by slow degrees from a dot ancestor. But I didn't know their genetic formula.

You might think that it would have been easy enough to reconstruct the evolutionary pathway, but it wasn't. The reason, which I shall come back to, is the astronomical number of *possible* biomorphs that a sufficiently long evolutionary pathway can offer, even when there are only nine genes varying. Several times on my pilgrimage through Biomorph Land I seemed to come close to a precursor of my insects, but, then, in spite of my best efforts as a selecting agent, evolution went off on what proved to be a false trail. Eventually, during my evolutionary wanderings through Biomorph Land – the sense of triumph was scarcely less than on the first occasion – I finally cornered them again. I didn't know (still don't) if these insects were exactly the same as my original, 'lost chords of Zarathustra' insects, or whether they were superficially 'convergent' (see next chapter), but it was good enough. This time there was no mistake: I wrote down the genetic formula, and now I can 'evolve' insects whenever I want.

Yes I am piling on the drama a bit, but there is a serious point being made. The point of the story is that even though it was I that programmed the computer, telling it in great detail what to do, nevertheless I didn't plan the animals that evolved, and I was totally surprised by them when I first saw their precursors. So powerless was I to control the evolution that, even when I very much wanted to retrace a particular evolutionary pathway it proved all but impossible to do so. I don't believe I would ever have found my insects again if I hadn't had a printed picture of the *complete set* of their evolutionary precursors, and even then it was difficult and tedious. Does the powerlessness of the programmer to control or predict the course of evolution in the computer seem paradoxical? Does it mean that something mysterious, even mystical was going on inside the computer? Of course not. Nor is there anything mystical going on in the evolution of real animals and plants. We can use the computer model to resolve the paradox, and learn something about real evolution in the process.

To anticipate, the basis of the resolution of the paradox will turn out to be as follows. There is a definite set of biomorphs, each permanently sitting in its own unique place in a mathematical space. It is permanently sitting there in the sense that, if only you knew its genetic formula, you could instantly find it; moreover, its neighbours in this special kind of space are the biomorphs that differ from it by only one gene. Now that I know the genetic formula of my insects, I can reproduce them at will, and I can tell the computer to 'evolve' towards them from any arbitrary starting point. When you first evolve a new creature by artificial selection in the computer model, it feels like a creative process. So it is, indeed. But what you are really doing is *finding* the creature, for it is, in a mathematical sense, already sitting in its own place in the genetic space of Biomorph Land. The reason it is a truly creative process is that finding any particular creature is extremely difficult, simply and purely because Biomorph Land is very very large, and the total number of creatures sitting there is all but infinite. It isn't feasible just to search aimlessly and at random. You have to adopt some more efficient – creative – searching procedure.

Some people fondly believe that chess-playing computers work by internally trying out all possible combinations of chess moves. They find this belief comforting when a computer beats them, but their belief is utterly false. There are far too many possible chess moves: the search-space is billions of times too large to allow blind stumbling to succeed. The art of writing a good chess program is thinking of efficient short cuts through the search-space. Cumulative selection, whether artificial selection as in the computer model or natural selection out there in the real world, is an efficient searching procedure, and its consequences look very like creative intelligence. That, after all, is what William Paley's Argument from Design was all about. Technically, all that we are doing, when we play the computer biomorph game, is *finding* animals that, in a mathematical sense, are waiting to be found. What it feels like is a process of artistic creation. Searching a small space, with only a few entities in it, doesn't ordinarily feel like a creative process. A child's game of hunt the thimble doesn't feel creative. Turning things over at random and hoping to stumble on the sought object usually works when the space to be searched is small. As the search-space gets larger, more and more sophisticated searching procedures become necessary. Effective searching procedures become, when the search-space is *sufficiently* large, indistinguishable from true creativity.

The computer biomorph models make these points well, and they constitute an instructive bridge between human creative processes, such as planning a winning strategy at chess, and the evolutionary creativity of natural selection, the blind watchmaker. To see this, we must develop the idea of Biomorph Land as a mathematical 'space', an endless but orderly vista of morphological variety, but one in which every creature is sitting in its correct place, waiting to be discovered. The 17 creatures of Figure 5 are arranged in no special order on the page. But in Biomorph Land itself each occupies its own unique position, determined by its genetic formula, surrounded by its own particular neighbours. All the creatures in Biomorph Land have a definite spatial relationship one to another. What does that mean? What meaning can we attach to spatial position?

The space we are talking about is genetic space. Each animal has its own position in genetic space. Near neighbours in genetic space are animals that differ from one another by only a single mutation. In Figure 3, the basic tree in the centre is surrounded by 8 of its 18 immediate neighbours in genetic space. The 18 neighbours of an animal are the 18 different kinds of children that it can give rise to, and the 18 different kinds of parent from which it could have come, given the rules of our computer model. At one remove, each animal has 324 (18 × 18, ignoring back-mutations for simplicity) neighbours, the set of its possible grandchildren, grandparents, aunts or nieces. At one remove again, each animal has 5,832 (18 × 18 × 18) neighbours, the set of possible great grandchildren, great grandparents, first cousins, etc.

What is the point of thinking in terms of genetic space? Where does it get us? The answer is that it provides us with a way to understand evolution as a gradual, cumulative process. In any one generation, according to the rules of the computer model, it is possible to move only a single step through genetic space. In 29 generations it isn't possible to move farther than 29 steps, in genetic space, away from the starting ancestor. Every evolutionary history consists of a particular pathway, or trajectory, through genetic space. For instance, the evolutionary history recorded in Figure 4 is a particular winding trajectory through genetic space, connecting a dot to an insect, and passing through 28 intermediate stages. It is this that I mean when I talk metaphorically about 'wandering' through Biomorph Land.

I wanted to try to represent this genetic space in the form of a picture. The trouble is, pictures are two-dimensional. The genetic space in which the biomorphs sit is not two-dimensional space. It isn't even three-dimensional space. It is nine-dimensional space! (The important thing to remember about mathematics is not to be frightened. It isn't as difficult as the mathematical priesthood sometimes pretends. Whenever I feel intimidated, I always remember Silvanus Thompson's dictum in *Calculus Made Easy*: 'What one fool can do, another can'.) If only we could draw in nine dimensions we could make each dimension correspond to one of the nine genes. The position of a particular animal, say the scorpion or the bat or the insect, is fixed in genetic space by the numerical value of its nine genes. Evolutionary change consists of a step by step walk through nine-dimensional space. The amount of genetic difference between one animal and another, and hence the time taken to evolve, and the difficulty of evolving from one to the other, is measured as the *distance* in nine-dimensional space from one to the other.

Alas, we can't draw in nine dimensions. I sought a way of fudging it, of drawing a two-dimensional picture that conveyed something of what it feels like to move from point to point in the nine-dimensional genetic space of Biomorph Land. There are various possible ways in which this could be done, and I chose one that I call the triangle trick. Look at Figure 6. At the three corners of the triangle are three arbitrarily chosen biomorphs. The one at the top is the basic tree, the one on the left is one of 'my' insects, and the one on the right has no name but I thought it looked pretty. Like all biomorphs, each of these three has its own genetic formula, which determines its unique position in nine-dimensional genetic space.

The triangle lies on a flat two-dimensional 'plane' that cuts through the nine-dimensional hypervolume (what one fool can do, another

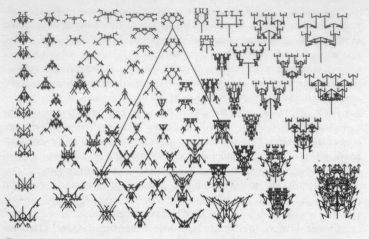

Figure 6

can). The plane is like a flat piece of glass stuck through a jelly. On the glass is drawn the triangle, and also some of the biomorphs whose genetic formulae entitle them to sit on that particular flat plane. What is it that entitles them? This is where the three biomorphs at the corners of the triangle come in. They are called the anchor biomorphs.

Remember that the whole idea of 'distance' in genetic 'space' is that genetically similar biomorphs are near neighbours, genetically different biomorphs are distant neighbours. On this particular plane, the distances are all calculated with reference to the three anchor biomorphs. For any given point on the sheet of glass, whether inside the triangle or outside it, the appropriate genetic formula for that point is calculated as a 'weighted average' of the genetic formulae of the three anchor biomorphs. You will already have guessed how the weighting is done. It is done by the distances on the page, more precisely the *nearnesses*, from the point in question to the three anchor biomorphs. So, the nearer you are to the insect on the plane, the more insect-like are the local biomorphs. As you move along the glass towards the tree, the 'insects' gradually become less insect-like and more tree-like. If you walk into the centre of the triangle the animals that you find there, for instance the spider with a Jewish seven-branched candelabra on its head, will be various 'genetic compromises' between the three anchor biomorphs.

But this account gives altogether too much prominence to the three anchor biomorphs. Admittedly the computer did use them to calculate the appropriate genetic formula for every point on the picture. But actually any three anchor points on the plane would have done the trick just as well, and would have given identical results. For this reason, in Figure 7 I haven't actually drawn the triangle. Figure 7 is exactly the same kind of picture as Figure 6. It just shows a different plane. The same insect is one of the three anchor points, this time the right-hand one. The other anchor points, in this case, are the spitfire and the bee-flower, both as seen in Figure 5. On this plane, too, you will notice that neighbouring biomorphs resemble each other more than distant biomorphs. The spitfire, for instance, is part of a squadron of similar aircraft, flying in formation. Because the insect is on both sheets of glass, you can think of the two sheets as passing, at an angle, through each other. Relative to Figure 6, the plane of Figure 7 is said to be 'rotated about' the insect.

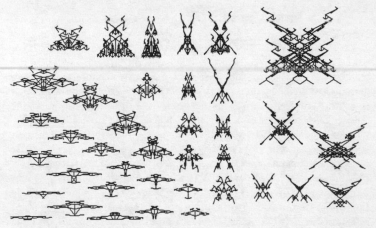

Figure 7

The elimination of the triangle is an improvement to our method, because it was a distraction. It gave undue prominence to three particular points in the plane. We still have one further improvement to make. In Figures 6 and 7, spatial distance represents genetic distance, but the *scaling* is all distorted. One inch upwards is not necessarily equivalent to one inch across. To remedy this, we must choose our three anchor biomorphs carefully, so that their genetic distances, one from the other, are all the same. Figure 8 does just this. Again the triangle is not actually drawn. The three anchors are the scorpion from Figure 5, the insect again (we have yet another 'rotation about' the insect), and the rather nondescript biomorph at the top. These three biomorphs are all 30 mutations distant from each other. This means that it is equally easy to evolve from any one to any other one. In all three cases, a minimum of 30 genetic steps must be taken. The little blips along the lower margin of Figure 8 represent units of distance measured in genes. You can think of it as a genetic ruler. The ruler doesn't only work in the horizontal direction. You can tilt it in any direction, and measure the genetic distance, and hence the minimum evolution time, between any point on the plane and any other (annoyingly, that is not quite true on the page, because the computer's printer distorts proportions, but this effect is too trivial to make a fuss about, although it does mean that you will get slightly the wrong answer if you simply count blips on the scale).

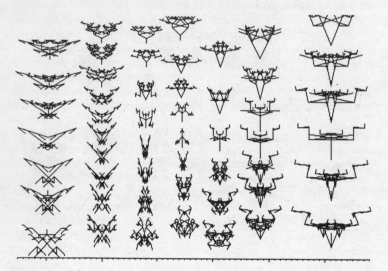

Figure 8

These two-dimensional planes cutting through nine-dimensional genetic space give some feeling for what it means to walk through Biomorph Land. To improve that feeling, you have to remember that evolution is not restricted to one flat plane. On a true evolutionary walk you could 'drop through', at any time, to another plane, for instance from the plane of Figure 6 to the plane of Figure 7 (in the vicinity of the insect, where the two planes come close to each other).

I said that the 'genetic ruler' of Figure 8 enables us to calculate the minimum time it would take to evolve from one point to another. So it does, given the restrictions of the original model, but the emphasis is on the word *minimum*. Since the insect and the scorpion are 30 genetic units distant from one another, it takes only 30 generations to evolve from one to the other *if you never take a wrong turning*; if, that is, you know exactly what genetic formula you are heading towards, and how to steer towards it. In real-life evolution there is nothing that corresponds to steering towards some distant genetic target.

Let's now use the biomorphs to return to the point made by the monkeys typing Hamlet, the importance of gradual, step-by-step change in evolution, as opposed to pure chance. Begin by relabelling the graticules along the bottom of Figure 8, but in different units. Instead of measuring distance as 'number of genes that have to change in evolution', we are going to measure distance as 'odds of happening to jump the distance, by sheer luck, in a single hop'. To think about this, we now have to relax one of the restrictions that I built into the computer game: we shall end by seeing why I built that restriction in in the first place. The restriction was that children were only 'allowed' to be one mutation distant from their parents. In other words, only one gene was allowed to mutate at a time, and that gene was allowed to change its 'value' only by +1 or −1. By relaxing the restriction, we are now allowing any number of genes to mutate simultaneously, and they can add any number, positive or negative, to their current value. Actually, that is *too* great a relaxation, since it allows genetic values to range from minus infinity to plus infinity. The point is adequately made if we restrict gene values to single figures, that is if we allow them to range from −9 to +9.

So, within these wide limits, we are theoretically allowing mutation, at a stroke, in a single generation, to change any combination of the nine genes. Moreover, the value of each gene can change any amount, so long as it doesn't stray into double figures. What does this mean? It means that, theoretically, evolution can jump, in a single generation, from any point in Biomorph Land to any other. Not just any point on one plane, but any point in the entire nine-dimensional hypervolume. If, for instance, you should want to jump in one fell swoop from the insect to the fox in Figure 5, here is the recipe. Add the following numbers to the values of Genes 1 to 9, respectively: −2,2,2,−2,2,0,−4,−1,1. But since we are talking about random jumps, *all* points in Biomorph Land are equally likely as destinations for one of these jumps. So, the odds against jumping to any *particular* destination, say the fox, by sheer luck, are easy to calculate. They are simply the total number of biomorphs in the space. As you can see, we are embarking on another of those astronomical calculations. There are nine genes, and each of them can take any of 19 values. So the total number of biomorphs that we *could* jump to in a single step is 19 times itself 9 times over: 19 to the power 9. This works out as about half a trillion biomorphs. Paltry compared with Asimov's 'haemoglobin number', but still what I would call a large number. If you started from the insect, and jumped like a demented flea half a trillion times, you could expect to arrive at the fox once.

What is all this telling us about real evolution? Once again, it is ramming home the importance of *gradual*, step-by-step change. There have been evolutionists who have denied that gradualism of this kind is necessary in evolution. Our biomorph calculation shows us *exactly* one reason why gradual, step-by-step change is important. When I say that you can expect evolution to jump from the insect to one of its immediate neighbours, but *not* to jump from the insect directly to the fox or the scorpion, what I exactly mean is the following. If genuinely random jumps really occurred, then a jump from insect to scorpion would be perfectly possible. Indeed it would be just *as* probable as a jump from insect to one of its immediate neighbours. But it would also be just as probable as a jump to any other biomorph in the land. And there's the rub. For the number of biomorphs in the land is half a trillion, and if no one of them is any more probable as a destination than any other, the odds of jumping to any *particular* one are small enough to ignore.

Notice that it doesn't help us to assume that there is a powerful nonrandom 'selection pressure'. It wouldn't matter if you'd been promised a king's ransom if you achieved a lucky jump to the scorpion. The odds against your doing so are still half a trillion to one. But if, instead of jumping you *walked*, one step at a time, and were given one small coin as a reward every time you happened to take a step in the right direction, you would reach the scorpion in a very short time. Not necessarily in the fastest possible time of 30 generations, but very fast, nevertheless. Jumping could *theoretically* get you the prize faster – in a single hop. But because of the astronomical odds against success, a series of small steps, each one building on the accumulated success of previous steps, is the only feasible way.

The tone of my previous paragraphs is open to a misunderstanding which I must dispel. It sounds, once again, as though evolution deals in distant targets, homing in on things like scorpions. As we have seen, it never does. But if we think of our target as *anything that would improve survival chances*, the argument still works. If an animal is a parent, it must be good enough to survive at least to adulthood. It is possible that a mutant child of that parent might be even better at surviving. But if a child mutates in a big way, so that it has moved a long distance away from its parent in genetic space, what are the odds of its being better than its parent? The answer is that the odds against are very large indeed. And the reason is the one we have just seen with our biomorph model. If the mutational jump we are considering is a very large one, the number of *possible* destinations of that jump is astronomically large. And because, as we saw in Chapter 1, the number of different ways of being dead is so much greater than the number of different ways of being alive, the chances are very high that a big random jump in genetic space will end in death. Even a small random jump in genetic space is pretty likely to end in death. But the smaller the jump the less likely death is, and the more likely is it that the jump will result in improvement. We shall return to this theme in a later chapter.

That is as far as I want to go in drawing morals from Biomorph Land. I hope that you didn't find it too abstract. There is another mathematical space filled, not with nine-gened biomorphs but with flesh and blood animals made of billions of cells, each containing tens of thousands of genes. This is not biomorph space but real genetic space. The actual animals that have ever lived on Earth are a tiny subset of the theoretical animals that *could* exist. These real animals are the products of a very small number of evolutionary trajectories through genetic space. The vast majority of theoretical trajectories through animal space give rise to impossible monsters. Real animals are dotted around here and there among the hypothetical monsters, each perched in its own unique place in genetic hyperspace. Each real animal is surrounded by a little cluster of neighbours, most of whom have never existed, but a few of whom are its ancestors, its descendants and its cousins.

Sitting somewhere in this huge mathematical space are humans and hyenas, amoebas and aardvarks, flatworms and squids, dodos and dinosaurs. In theory, if we were skilled enough at genetic engineering, we could move from any point in animal space to any other point. From any starting point we could move through the maze in such a way as to recreate the dodo, the tyrannosaur and trilobites. If only we knew which genes to tinker with, which bits of chromosome to duplicate, invert or delete. I doubt if we shall ever know enough to do it, but these dear dead creatures are lurking there forever in their private corners of that huge genetic hypervolume, waiting to be *found* if we but had the knowledge to navigate the right course through the maze. We might even be able to *evolve* an exact reconstruction of a dodo by selectively breeding pigeons, though we'd have to live a million years in order to complete the experiment. But when we are prevented from making a journey in reality, the imagination is not a bad substitute. For those, like me, who are not mathematicians, the computer can be a powerful friend to the imagination. Like mathematics, it doesn't only stretch the imagination. It also disciplines and controls it.

CHAPTER 4

MAKING TRACKS THROUGH ANIMAL SPACE

As we saw in Chapter 2, many people find it hard to believe that something like the eye, Paley's favourite example, so complex and well designed, with so many interlocking working parts, could have arisen from small beginnings by a gradual series of step-by-step changes. Let's return to the problem in the light of such new intuitions as the biomorphs may have given us. Answer the following two questions:

1. Could the human eye have arisen directly from no eye at all, in a single step?

2. Could the human eye have arisen directly from something slightly different from itself, something that we may call X?

The answer to Question 1 is clearly a decisive *no*. The odds against a 'yes' answer for questions like Question 1 are many billions of times greater than the number of atoms in the universe. It would need a gigantic and vanishingly improbable leap across genetic hyperspace. The answer to Question 2 is equally clearly *yes*, provided only that the difference between the modern eye and its immediate predecessor X is sufficiently small. Provided, in other words, that they are sufficiently close to one another in the space of all possible structures. If the answer to Question 2 for any particular degree of difference is no, all we have to do is repeat the question for a smaller degree of difference. Carry on doing this until we find a degree of difference sufficiently small to give us a 'yes' answer to Question 2.

X is *defined* as something very like a human eye, sufficiently similar that the human eye could plausibly have arisen by a single alteration in X. If you have a mental picture of X and you find it implausible that the human eye could have arisen directly from it, this simply means that you have chosen the wrong X. Make your mental picture of X progressively more like a human eye, until you find an X that you *do* find plausible as an immediate predecessor to the human eye. There has to be one for you, even if your idea of what is plausible may be more, or less, cautious than mine!

Now, having found an X such that the answer to Question 2 is yes, we apply the same question to X itself. By the same reasoning we must conclude that X could plausibly have arisen, directly by a single change, from something slightly different again, which we may call X'. Obviously we can then trace X' back to something else slightly different from it, X", and so on. By interposing a large enough series of Xs, we can derive the human eye from something not slightly different from itself but *very* different from itself. We can 'walk' a large distance across 'animal space', and our move will be plausible provided we take small-enough steps. We are now in a position to answer a third question.

3. Is there a continuous series of Xs connecting the modern human eye to a state with no eye at all?

It seems to me clear that the answer has to be yes, provided only that we allow ourselves a *sufficiently large* series of Xs. You might feel that 1,000 Xs is ample, but if you need more steps to make the total transition plausible in your mind, simply allow yourself to assume 10,000 Xs. And if 10,000 is not enough for you, allow yourself 100,000, and so on. Obviously the available time imposes an upper ceiling on this game, for there can be only one X per generation. In practice the question therefore resolves itself into: Has there been enough time for enough successive generations? We can't give a precise answer to the number of generations that would be necessary. What we do know is that geological time is awfully long. Just to give you an idea of the order of magnitude we are talking about, the number of generations that separate us from our earliest ancestors is certainly measured in the thousands of millions. Given, say, a hundred million Xs, we should be able to construct a plausible series of tiny gradations linking a human eye to just about anything!

So far, by a process of more-or-less abstract reasoning, we have concluded that there is a series of imaginable Xs, each sufficiently similar to its neighbours that it could plausibly turn into one of its neighbours, the whole series linking the human eye back to no eye at all. But we still haven't demonstrated that it is plausible that this series of Xs actually existed. We have two more questions to answer.

4. Considering each member of the series of hypothetical Xs connecting the human eye to no eye at all, is it plausible that every one of them was made available by random mutation of its predecessor?

This is really a question about embryology, not genetics; and it is an entirely separate question from the one that worried the Bishop of Birmingham and others. Mutation has to work by modifying the existing processes of embryonic development. It is arguable that certain kinds of embryonic process are highly amenable to variation in certain directions, recalcitrant to variation in others. I shall return to this matter in Chapter 11, so here I'll just stress again the difference between small change and large. The smaller the change you postulate, the smaller the difference between X″ and X′, the more embryologically plausible is the mutation concerned. In the previous chapter we saw, on purely statistical grounds, that any *particular* large mutation is inherently less probable than any particular small mutation. Whatever problems may be raised by Question 4, then, we can at least see that the smaller we make the difference between any given X′ and X″, the smaller will be the problems. My feeling is that, provided the difference between neighbouring intermediates in our series leading to the eye is *sufficiently small*, the necessary mutations are almost bound to be forthcoming. We are, after all, always talking about minor quantitative changes in an existing embryonic process. Remember that, however complicated the embryological status quo may be in any given generation, each mutational *change* in the status quo can be very small and simple.

We have one final question to answer:

5. Considering each member of the series of Xs connecting the human eye to no eye at all, is it plausible that every one of them worked sufficiently well that it assisted the survival and reproduction of the animals concerned?

Rather oddly, some people have thought that the answer to this question is a self-evident 'no'. For instance, I quote from Francis Hitching's book of 1982 called *The Neck of the Giraffe or Where Darwin Went Wrong*. I could have quoted basically the same words from almost any Jehovah's Witness tract, but I choose this book because a reputable publisher (Pan Books Ltd) saw fit to publish it, despite a very large number of errors which would quickly have been spotted if an unemployed biology graduate, or indeed undergraduate, had been asked to glance through the manuscript. (My favourites, if you'll indulge me just two in-jokes, are the conferring of a knighthood

on Professor John Maynard Smith, and the description of Professor
Ernst Mayr, that eloquent and most unmathematical arch-critic of
mathematical genetics, as 'the high priest' of mathematical genetics.)

> For the eye to work the following minimum perfectly coordinated steps
> have to take place (there are many others happening simultaneously, but
> even a grossly simplified description is enough to point up the problems for
> Darwinian theory). The eye must be clean and moist, maintained in this
> state by the interaction of the tear gland and movable eyelids, whose
> eyelashes also act as a crude filter against the sun. The light then passes
> through a small transparent section of the protective outer coating (the
> *cornea*), and continues via a *lens* which focuses it on the back of the *retina*.
> Here 130 million light-sensitive rods and cones cause photochemical
> reactions which transform the light into electrical impulses. Some 1,000
> million of these are transmitted every second, by means that are not
> properly understood, to a brain which then takes appropriate action.
>
> Now it is quite evident that if the slightest thing goes wrong *en route* – if
> the cornea is fuzzy, or the pupil fails to dilate, or the lens becomes opaque,
> or the focussing goes wrong – then a recognizable image is not formed. The
> eye either functions as a whole, or not at all. So how did it come to evolve
> by slow, steady, infinitesimally small Darwinian improvements? Is it really
> plausible that thousands upon thousands of lucky chance mutations
> happened coincidentally so that the lens and the retina, which cannot work
> without each other, evolved in synchrony? What survival value can there be
> in an eye that doesn't see?

This remarkable argument is very frequently made, presumably
because people *want* to believe its conclusion. Consider the statement
that 'if the slightest thing goes wrong . . . if the focusing goes wrong . . .
a recognizable image is not formed'. The odds cannot be far from 50/50
that you are reading these words through glass lenses. Take them off
and look around. Would you agree that 'a recognizable image is not
formed'? If you are male, the odds are about 1 in 12 that you are
colourblind. You may well be astigmatic. It is not unlikely that, without glasses, your vision is a misty blur. One of today's most distinguished (though not yet knighted) evolutionary theorists so seldom
cleans his glasses that his vision is probably a misty blur anyway, but
he seems to get along pretty well and, by his own account, he used to
play a mean game of monocular squash. If you have lost your glasses, it
may be that you upset your friends by failing to recognize them in the
street. But you yourself would be even more upset if somebody said to
you: 'Since your vision is now not absolutely perfect, you might as
well go around with your eyes tight shut until you find your glasses
again.' Yet that is essentially what the author of the passage I have
quoted is suggesting.

He also states, as though it were obvious, that the lens and the retina cannot work without each other. On what authority? Someone close to me has had a cataract operation in both eyes. She has no lenses in her eyes at all. Without glasses she couldn't even begin to play lawn tennis or aim a rifle. But she assures me that you are far better off with a lensless eye than with no eye at all. You can tell if you are about to walk into a wall or another person. If you were a wild creature, you could certainly use your lensless eye to detect the looming shape of a predator, and the direction from which it was approaching. In a primitive world where some creatures had no eyes at all and others had lensless eyes, the ones with lensless eyes would have all sorts of advantages. And there is a continuous series of Xs, such that each tiny improvement in sharpness of image, from swimming blur to perfect human vision, plausibly increases the organism's chances of surviving.

The book goes on to quote Stephen Jay Gould, the noted Harvard palaeontologist, as saying:

> We avoid the excellent question, What good is 5 percent of an eye? by arguing that the possessor of such an incipient structure did not use it for sight.

An ancient animal with 5 per cent of an eye might indeed have used it for something other than sight, but it seems to me at least as likely that it used it for 5 per cent vision. And actually I don't think it is an excellent question. Vision that is 5 per cent as good as yours or mine is very much worth having in comparison with no vision at all. So is 1 per cent vision better than total blindness. And 6 per cent is better than 5, 7 per cent better than 6, and so on up the gradual, continuous series.

This kind of problem has worried some people interested in animals that gain protection from predators by 'mimicry'. Stick insects look like sticks and so are saved from being eaten by birds. Leaf insects look like leaves. Many edible species of butterfly gain protection by resembling noxious or poisonous species. These resemblances are far more impressive than the resemblance of clouds to weasels. In many cases they are more impressive than the resemblance of 'my' insects to real insects. Real insects, after all, have six legs, not eight! Real natural selection has had a least a million times as many generations as I had, in which to perfect the resemblance.

We use the word 'mimicry' for these cases, not because we think that the animals consciously imitate other things, but because natural selection has favoured those individuals whose bodies were mistaken for other things. To put it another way, ancestors of stick insects that did not resemble sticks did not leave descendants. The German–American geneticist Richard Goldschmidt is the most distinguished of those who

have argued that the *early* evolution of such resemblances could not
have been favoured by natural selection. As Gould, an admirer of
Goldschmidt, said of dung-mimicking insects: 'can there be any edge
in looking 5 per cent like a turd?' Largely through Gould's influence, it
has recently become fashionable to say that Goldschmidt was under-
rated in his own lifetime, and that he really has much to teach us. Here
is a sample of his reasoning.

> Ford speaks . . . of any mutation which chances to give a 'remote
> resemblance' to a more protected species, from which some advantage,
> however slight, might accrue. We must ask how remote the resemblance
> can be to have selective value. Can we really assume that the birds and
> monkeys and also mantids are such wonderful observers (or that some very
> clever ones among them are) to notice a 'remote' resemblance and be
> repelled by it? I think that this is asking too much.

Such sarcasm ill becomes anybody on the shaky ground that
Goldschmidt here treads. *Wonderful* observers? Very *clever* ones
among them? Anybody would think the birds and monkeys *benefited*
from being fooled by the remote resemblance! Goldschmidt might
rather have said: 'Can we really assume that the birds, etc. are such
poor observers (or that some very stupid ones among them are)?'
Nevertheless, there is a real dilemma here. The initial resemblance of
the ancestral stick insect to a stick must have been very remote. A bird
would need extremely *poor* vision to be fooled by it. Yet the resemblance of a modern stick insect to a stick is marvellously good,
down to the last fine details of fake buds and leaf-scars. The birds
whose selective predation put the finishing touches to their evolution
must, at least collectively, have had excellently *good* vision. They
must have been extremely hard to fool, otherwise the insects would
not have evolved to become as perfect mimics as they are: they would
have remained relatively imperfect mimics. How can we resolve this
apparent contradiction?

One kind of answer suggests that bird vision has been improving
over the same evolutionary timespan as insect camouflage. Perhaps, to
be a little facetious, an ancestral insect that looked only 5 per cent like
a turd would have fooled an ancestral bird with only 5 per cent vision.
But that is not the kind of answer I want to give. I suspect, indeed, that
the whole process of evolution, from remote resemblance to near
perfect mimicry, has gone on, rather rapidly, many times over in
different insect groups, during the whole long period that bird vision
has been just about as good as it is today.

Another kind of answer that has been offered to the dilemma is the
following. Perhaps each species of bird or monkey has poor vision and

latches onto just one limited aspect of an insect. Maybe one predator species notices only the colour, another only the shape, another only the texture, and so on. Then an insect that resembles a stick in only one limited respect will fool one kind of predator, even though it is eaten by all other kinds of predators. As evolution progresses, more and more features of resemblance are added to the repertoire of the insects. The final multifaceted perfection of mimicry has been put together by the summed natural selection provided by many different species of predators. No one predator sees the whole perfection of mimicry, only we do that.

This seems to imply that only we are 'clever' enough to see the mimicry in all its glory. Not only because of this human snobbishness, I prefer yet another explanation. This is that, no matter how good any one predator's vision may be under some conditions, it can be exceedingly poor under other conditions. We can easily, in fact, appreciate from our own familiar experience the whole spectrum from exceedingly poor vision to excellent vision. If I am looking directly at a stick insect, 8 inches in front of my nose and in strong daylight, I shall not be fooled by it. I shall notice the long legs hugging the line of the trunk. I may spot the unnatural symmetry which a real stick would not have. But if I, with the very same eyes and brain, am walking through a forest at dusk, I may well fail to distinguish almost any dull-coloured insect from the twigs that abound everywhere. The image of the insect may pass over the edge of my retina rather than the more acute central region. The insect may be 50 yards away, and so make only a tiny image on my retina. The light may be so poor that I can hardly see anything at all anyway.

In fact, it doesn't matter *how* remote, how poor is the resemblance of an insect to a stick, there must be *some* level of twilight, or some degree of distance away from the eye, or some degree of distraction of the predator's attention, such that even a very good eye will be fooled by the remote resemblance. If you don't find that plausible for some particular example that you have imagined, just turn down the imaginary light a bit, or move a bit further away from the imaginary object! The point is that many an insect was saved by an exceedingly slight resemblance to a twig or a leaf or a fall of dung, on occasions when it was far away from a predator, or on occasions when the predator was looking at it at dusk, or looking at it through a fog, or looking at it while distracted by a receptive female. And many an insect was saved, perhaps from the very same predator, by an uncannily close resemblance to a twig, on occasions when the predator happened to be seeing it at relatively close range and in a good light.

The important thing about light intensity, distance of insect from predator, distance of image from centre of retina, and similar variables, is that they are all *continuous* variables. They vary by insensible degrees all the way from the extreme of invisibility to the extreme of visibility. Such continuous variables foster continuous and gradual evolution.

Richard Goldschmidt's problem – which was one of a set that made him resort, for most of his professional life, to the extreme belief that evolution takes great leaps rather than small steps – turns out to be no problem at all. And incidentally, we have also demonstrated to ourselves, yet again, that 5 per cent vision is better than no vision at all. The quality of my vision right at the edge of my retina is probably even poorer than 5 per cent of the quality at the centre of my retina, however you care to measure quality. Yet I can still detect the presence of a large lorry or bus out of the extreme corner of my eye. Since I ride a bicycle to work every day this fact has quite probably saved my life. I notice the difference on those occasions when it is raining and I wear a hat. The quality of our vision on a dark night must be far poorer than 5 per cent of what it is at midday. Yet many an ancestor was probably saved through seeing something that really mattered, a sabre-tooth 'tiger' perhaps, or a precipice, in the middle of the night.

Every one of us knows from personal experience, for example on dark nights, that there is an insensibly graded continuous series running all the way from total blindness up to perfect vision, and that every step along this series confers significant benefits. By looking at the world through progressively defocused and focused binoculars, we can quickly convince ourselves that there is a graded series of focusing quality, each step in the series being an improvement over the previous one. By progressively turning the colour-balance knob of a colour television set, we can convince ourselves that there is a graded series of progressive improvement from black and white to full colour vision. The iris diaphragm that opens and shuts the pupil prevents us from being dazzled in bright light, while allowing us to see in dim light. We all experience what it is like not to have an iris diaphragm, when we are momentarily dazzled by oncoming car headlights. Unpleasant, and even dangerous, as this dazzling can be, it still doesn't mean that the whole eye ceases to work! The claim that 'The eye either functions as a whole, or not at all' turns out to be, not merely false but self-evidently false to anybody who thinks for 2 seconds about his own familiar experience.

Let us return to our Question 5. Considering each member of the series of Xs connecting the human eye to no eye at all, is it plausible that every one of them worked sufficiently well that it assisted the survival and reproduction of the animals concerned? We have now seen the silliness of the anti-evolutionist's assumption that the answer is an obvious no. But is the answer yes? It is less obvious, but I think that it is. Not only is it clear that part of an eye is better than no eye at all. We also can find a plausible series of intermediates among modern animals. This doesn't mean, of course, that these modern intermediates really represent ancestral types. But it does show that intermediate designs are capable of working.

Some single-celled animals have a light-sensitive spot with a little pigment screen behind it. The screen shields it from light coming from one direction, which gives it some 'idea' of where the light is coming from. Among many-celled animals, various types of worm and some shellfish have a similar arrangement, but the pigment-backed light-sensitive cells are set in a little cup. This gives slightly better direction-finding capability, since each cell is selectively shielded from light rays coming into the cup from its own side. In a continuous series from flat sheet of light-sensitive cells, through shallow cup to deep cup, each step in the series, however small (or large) the step, would be an optical improvement. Now, if you make a cup very deep and turn the sides over, you eventually make a lensless pinhole camera. There is a continuously graded series from shallow cup to pinhole camera (see, for illustration, the first seven generations of the evolutionary series in Figure 4).

A pinhole camera forms a definite image, the smaller the pinhole the sharper (but dimmer) the image, the larger the pinhole the brighter (but fuzzier) the image. The swimming mollusc *Nautilus*, a rather strange squid-like creature that lives in a shell like the extinct ammonites (see the 'shelled cephalopod' of Figure 5), has a pair of pinhole cameras for eyes. The eye is basically the same shape as ours, but there is no lens and the pupil is just a hole that lets the seawater into the hollow interior of the eye. Actually, *Nautilus* is a bit of a puzzle in its own right. Why, in all the hundreds of millions of years since its ancestors first evolved a pinhole eye, did it never discover the principle of the lens? The advantage of a lens is that it allows the image to be both sharp *and* bright. What is worrying about *Nautilus* is that the quality of its retina suggests that it would really benefit, greatly and immediately, from a lens. It is like a hi-fi system with an excellent amplifier fed by a gramophone with a blunt needle. The system is crying out for a particular simple change. In genetic hyperspace, *Nautilus* appears to be sitting right next door to an obvious and immediate improvement, yet it doesn't take the small step necessary. Why not?

Michael Land of Sussex University, our foremost authority on invertebrate eyes, is worried, and so am I. Is it that the necessary mutations cannot arise, given the way *Nautilus* embryos develop? I don't want to believe it, but I don't have a better explanation. At least *Nautilus* dramatizes the point that a lensless eye is better than no eye at all.

When you have a cup for an eye, almost any vaguely convex, vaguely transparent or even translucent material over its opening will constitute an improvement, because of its slight lens-like properties. It collects light over its area and concentrates it on a smaller area of retina. Once such a crude proto-lens is there, there is a continuously graded series of improvements, thickening it and making it more transparent and less distorting, the trend culminating in what we would all recognize as a true lens. *Nautilus*'s relatives, the squids and octopuses, have a true lens, very like ours although their ancestors certainly evolved the whole camera-eye principle completely independently of ours. Incidentally, Michael Land reckons that there are nine basic principles for image-forming that eyes use, and that most of them have evolved many times independently. For instance, the curved dish-reflector principle is radically different from our own camera-eye (we use it in radiotelescopes, and also in our largest optical telescopes because it is easier to make a large mirror than a large lens), and it has been independently 'invented' by various molluscs and crustaceans. Other crustaceans have a compound eye like insects (really a bank of lots of tiny eyes), while other molluscs, as we have seen, have a lensed camera-eye like ours, or a pinhole camera-eye. For each of these types of eye, stages corresponding to evolutionary intermediates exist as working eyes among other modern animals.

Anti-evolution propaganda is full of alleged examples of complex systems that 'could not possibly' have passed through a gradual series of intermediates. This is often just another case of the rather pathetic 'Argument from Personal Incredulity' that we met in Chapter 2. Immediately after the section on the eye, for example, *The Neck of the Giraffe* goes on to discuss the bombardier beetle, which

> squirts a lethal mixture of hydroquinone and hydrogen peroxide into the
> face of its enemy. These two chemicals, when mixed together, literally
> explode. So in order to store them inside its body, the Bombardier Beetle has
> evolved a chemical inhibitor to make them harmless. At the moment the
> beetle squirts the liquid out of its tail, an anti-inhibitor is added to make the
> mixture explosive once again. The chain of events that could have led to the
> evolution of such a complex, coordinated and subtle process is beyond
> biological explanation on a simple step-by-step basis. The slightest

alteration in the chemical balance would result immediately in a race of
exploded beetles.

A biochemist colleague has kindly provided me with a bottle of
hydrogen peroxide, and enough hydroquinone for 50 bombardier
beetles. I am now about to mix the two together. According to the
above, they will explode in my face. Here goes . . .

Well, I'm still here. I poured the hydrogen peroxide into the
hydroquinone, and absolutely nothing happened. It didn't even get
warm. Of course I knew it wouldn't: I'm not that foolhardy! The
statement that 'these two chemicals, when mixed together, literally
explode', is, quite simply, false, although it is regularly repeated
throughout creationist literature. If you are curious about the
bombardier beetle, by the way, what actually happens is as follows. It
is true that it squirts a scaldingly hot mixture of hydrogen peroxide and
hydroquinone at enemies. But hydrogen peroxide and hydroquinone
don't react violently together unless a catalyst is *added*. This is what
the bombardier beetle does. As for the evolutionary precursors of the
system, both hydrogen peroxide and various kinds of quinones are used
for other purposes in body chemistry. The bombardier beetle's
ancestors simply pressed into different service chemicals that already
happened to be around. That's often how evolution works.

On the same page of the book as the bombardier beetle passage is the
question: 'What use would be . . . half a lung? Natural selection would
surely eliminate creatures with such oddities, not preserve them.' In a
healthy adult human, each of the two lungs is divided into about 300
million tiny chambers, at the tips of a branching system of tubes. The
architecture of these tubes resembles the biomorph tree at the bottom
of Figure 2 in the previous chapter. In that tree, the number of successive branchings, determined by 'Gene 9', is eight, and the number of
twig tips is 2 to the power 8, or 256. As you go down the page in Figure
2, the number of twig tips successively doubles. In order to provide 300
million twig tips, only 29 successive doublings would be required.
Note that there is a continuous gradation from a single chamber to 300
million tiny chambers, each step in the gradation being provided by
another two-way branching. This transition can be accomplished in 29
branchings, which we may naively think of as a stately walk of 29
steps across genetic space.

In the lungs, the result of all this branching is that the surface area
inside each lung is rather more than 70 square yards. Area is the
important variable for a lung, for it is area that determines the rate at
which oxygen can be taken in, and waste carbon dioxide pushed out.
Now, the thing about area is that it is a *continuous* variable. Area is

not one of those things that you either have or you don't. It is a thing that you can have a little bit more of, or a little bit less of. More than most things, lung area lends itself to *gradual*, step-by-step change, all the way from 0 square yards up to 70 square yards.

There are plenty of surgical patients walking around with only one lung, and some of them are down to a third of normal lung area. They may be walking, but they aren't walking very far, nor very fast. That is the point. The effect of gradually reducing lung area is not an absolute, all-or-none effect on survival. It is a gradual, continuously varying effect on how far you can walk, and how fast. A gradual, continuously varying effect, indeed, on how long you can expect to live. Death doesn't suddenly arrive below a particular threshold lung area! It becomes gradually more probable as lung area decreases below an optimum (and as it increases above the same optimum, for different reasons connected with economic waste).

The first of our ancestors to develop lungs almost certainly lived in water. We can get an idea of how they might have breathed by looking at modern fish. Most modern fish breathe in water with gills, but many species living in foul, swampy water supplement this by gulping air at the surface. They use the internal chamber of the mouth as a kind of crude proto-lung, and this cavity is sometimes enlarged into a breathing pocket rich in blood vessels. As we've seen, there is no problem in imagining a continuous series of Xs connecting a single pocket to a branching set of 300 million pockets as in a modern human lung.

Interestingly, many modern fish have kept their pocket single, and use it for a completely different purpose. Although it probably began as a lung, over the course of evolution it has become the swimbladder, an ingenious device with which the fish maintains itself as a hydrostat in permanent equilibrium. An animal without an air bladder inside it is normally slightly heavier than water, so sinks to the bottom. This is why sharks have to swim continuously to stop themselves sinking. An animal with large air pockets inside it, like us with our great lungs, tends to rise to the surface. Somewhere in the middle of this continuum, an animal with an air bladder of exactly the right size neither sinks nor rises, but floats steadily in effortless equilibrium. This is the trick that modern fish, other than sharks, have perfected. Unlike sharks, they don't waste energy preventing themselves from sinking. Their fins and tail are freed for guidance and rapid propulsion. They no longer rely on outside air to fill the bladder, but have special glands for manufacturing gas. Using these glands and other means, they accurately regulate the volume of gas in the bladder, and hence keep themselves in precise hydrostatic equilibrium.

Several species of modern fish can leave the water. An extreme is the Indian climbing perch, which hardly ever goes into the water. It has independently evolved a quite different kind of lung from that of our ancestors – an air chamber surrounding the gills. Other fish live basically in water but make brief forays out of it. This is probably what our ancestors did. The thing about forays is that their duration can vary continuously, all the way down to zero. If you are a fish who basically lives and breathes in water, but who occasionally ventures on land, perhaps to cross from one mud puddle to another thereby surviving a drought, you might benefit not just from half a lung but from one-hundredth of a lung. It doesn't matter *how* small your primordial lung is, there must be *some* time out of water that you can just endure with the lung, which is a little bit longer than you could have endured without the lung. Time is a continuous variable. There is no hard-and-fast divide between water-breathing and air-breathing animals. Different animals may spend 99 per cent of their time in water, 98 per cent, 97 per cent, and so on all the way to 0 per cent. At every step of the way, some fractional increase in lung area will be an advantage. There is continuity, gradualism, all the way.

What use is half a wing? How did wings get their start? Many animals leap from bough to bough, and sometimes fall to the ground. Especially in a small animal, the whole body surface catches the air and assists the leap, or breaks the fall, by acting as a crude aerofoil. Any tendency to increase the ratio of surface area to weight would help, for example flaps of skin growing out in the angles of joints. From here, there is a continuous series of gradations to gliding wings, and hence to flapping wings. Obviously there are distances that could not have been jumped by the earliest animals with proto-wings. Equally obviously, for *any* degree of smallness or crudeness of ancestral air-catching surfaces, there must be *some* distance, however short, which can be jumped with the flap and which cannot be jumped without the flap.

Or, if prototype wingflaps worked to break the animal's fall, you cannot say 'Below a certain size the flaps would have been of no use at all'. Once again, it doesn't matter *how* small and un-winglike the first wingflaps were. There must be some height, call it *h*, such that an animal would just break its neck if it fell from that height, but would just survive if it fell from a slightly lower height. In this critical zone, any improvement in the body surface's ability to catch the air and break the fall, however slight that improvement, can make the difference between life and death. Natural selection will then favour slight, prototype wingflaps. When these small wingflaps have become the norm, the critical height *h* will become slightly greater. Now a slight further increase in the wingflaps will make the difference between life and death. And so on, until we have proper wings.

There are animals alive today that beautifully illustrate every stage in the continuum. There are frogs that glide with big webs between their toes, tree-snakes with flattened bodies that catch the air, lizards with flaps along their bodies; and several different kinds of mammals that glide with membranes stretched between their limbs, showing us the kind of way bats must have got their start. Contrary to the creationist literature, not only are animals with 'half a wing' common, so are animals with a quarter of a wing, three quarters of a wing, and so on. The idea of a flying continuum becomes even more persuasive when we remember that very small animals tend to float gently in air, whatever their shape. The reason this is persuasive is that there is an infinitesimally graded continuum from small to large.

The idea of tiny changes cumulated over many steps is an immensely powerful idea, capable of explaining an enormous range of things that would be otherwise inexplicable. How did snake venom get its start? Many animals bite, and any animal's spit contains proteins which, if they get into a wound, may cause an allergic reaction. Even so-called non-venomous snakes can give bites that cause a painful reaction in some people. There is a continuous, graded series from ordinary spit to deadly venom.

How did ears get their start? Any piece of skin can detect vibrations if they come in contact with vibrating objects. This is a natural outgrowth of the sense of touch. Natural selection could easily have enhanced this faculty by gradual degrees until it was sensitive enough to pick up very *slight* contact vibrations. At this point it would automatically have been sensitive enough to pick up *airborne* vibrations of sufficient loudness and/or sufficient nearness of origin. Natural selection would then favour the evolution of special organs – ears – for picking up airborne vibrations originating from steadily increasing distances. It is easy to see that there would have been a continuous trajectory of step-by-step improvement, all the way. How did echolocation get its start? Any animal that can hear at all may hear echoes. Blind humans frequently learn to make use of these echoes. A rudimentary version of such a skill in ancestral mammals would have provided ample raw material for natural selection to build upon, leading up by gradual degrees to the high perfection of bats.

Five per cent vision is better than no vision at all. Five per cent hearing is better than no hearing at all. Five per cent flight efficiency is better than no flight at all. It is thoroughly believable that every organ or apparatus that we actually see is the product of a smooth trajectory through animal space, a trajectory in which every intermediate stage assisted survival and reproduction. Wherever we have an X in a real live animal, where X is some organ too complex to have arisen by chance in a single step, then according to the theory of evolution by natural selection it must be the case that a fraction of an X is better than no X at all; and two fractions of an X must be better than one; and a whole X must be better than nine-tenths of an X. I have no trouble at all in accepting that these statements are true of eyes, ears including bat ears, wings, camouflaged and mimicking insects, snake jaws, stings, cuckoo habits and all the other examples trotted out in anti-evolution propaganda. No doubt there are plenty of *conceivable* Xs for which these statements would *not* be true, plenty of conceivable evolutionary pathways for which the intermediates would *not* be improvements on their predecessors. But those Xs are not found in the real world.

Darwin wrote (in *The Origin of Species*):

> If it could be demonstrated that any complex organ existed which could not possibly have been formed by numerous, successive, slight modifications, my theory would absolutely break down.

One hundred and twenty five years on, we know a lot more about animals and plants than Darwin did, and still not a single case is known to me of a complex organ that could not have been formed by numerous successive slight modifications. I do not believe that such a case will ever be found. If it is – it'll have to be a *really* complex organ, and, as we'll see in later chapters, you have to be sophisticated about what you mean by 'slight' – I shall cease to believe in Darwinism.

Sometimes the history of gradual, intermediate stages is clearly written into the shape of modern animals, even taking the form of outright imperfections in the final design. Stephen Gould, in his excellent essay on *The Panda's Thumb*, has made the point that evolution can be more strongly supported by evidence of telling imperfections than by evidence of perfection. I shall give just two examples.

Fish living on the sea bottom benefit by being flat and hugging the contours. There are two very different kinds of flat fish living on the sea bottom, and they have evolved their flatness in quite different ways. The skates and rays, relatives of sharks, have become flat in what might be called the obvious way. Their bodies have grown out sideways to form great 'wings'. They are like sharks that have passed under a steam roller, but they remain symmetrical and 'the right way up'. Plaice, sole, halibut and their relatives have become flat in a different way. They are bony fish (with swimbladders) related to herrings, trout, etc., and are nothing to do with sharks. Unlike sharks, bony fish as a rule have a marked tendency to be flattened in a vertical direction. A herring, for instance, is much 'taller' than it is wide. It uses its whole, vertically flattened body as a swimming surface, which undulates through the water as it swims. It was natural, therefore, that when the ancestors of plaice and sole took to the sea bottom, they should have lain on one *side* rather than on the belly like the ancestors of skates and rays. But this raised the problem that one eye was always looking down into the sand and was effectively useless. In evolution this problem was solved by the lower eye 'moving' round to the upper side.

We see this process of moving round re-enacted in the development of every young bony flatfish. A young flatfish starts life swimming near the surface, and it is symmetrical and vertically flattened just like a herring. But then the skull starts to grow in a strange, asymmetrical, twisted fashion, so that one eye, for instance the left, moves over the top of the head to finish up on the other side. The young fish settles on the bottom, with both its eyes looking upwards, a strange Picasso-like vision. Incidentally, some species of flatfish settle on the right side, others on the left, and others on either side.

The whole skull of a bony flatfish retains the twisted and distorted evidence of its origins. Its very imperfection is powerful testimony of its ancient history, a history of step-by-step change rather than of deliberate design. No sensible designer would have conceived such a monstrosity if given a free hand to create a flatfish on a clean drawing board. I suspect that most sensible designers would think in terms of something more like a skate. But evolution never starts from a clean drawing board. It has to start from what is already there. In the case of the ancestors of skates this was free-swimming sharks. Sharks in general aren't flattened from side to side as free-swimming bony fish like herrings are. If anything, sharks are already slightly flattened from back to belly. This meant that when some ancient sharks first took to the sea bottom, there was an easy smooth progression to the skate shape, with each intermediate being a slight improvement, given bottom conditions, over its slightly less flattened predecessor.

On the other hand, when the free-swimming ancestor of plaice and halibut, being, like a herring, vertically flattened from side to side, took to the bottom, it was better off lying on its side than balancing precariously on its knife edge of a belly! Even though its evolutionary course was eventually destined to lead it into the complicated and probably costly distortions involved in having two eyes on one side, even though the skate way of being a flat fish might *ultimately* have been the best design for bony fish too, the would-be intermediates that set out along this evolutionary pathway apparently did less well in the short term than their rivals lying on their side. The rivals lying on their side were so much better, in the short term, at hugging the bottom. In genetic hyperspace, there is a smooth trajectory connecting free-swimming ancestral bony fish to flatfish lying on their side with twisted skulls. There is not a smooth trajectory connecting these bony fish ancestors to flatfish lying on their belly. This speculation cannot be the whole truth, because there are some bony fish that have evolved flatness in a symmetrical, skate-like way. Perhaps their free-swimming ancestors were already slightly flattened for some other reason.

My second example of an evolutionary progression that didn't happen because of disadvantageous intermediates, even though it might ultimately have turned out better if it had, concerns the retina of our eyes (and all other vertebrates). Like any nerve, the optic nerve is a trunk cable, a bundle of separate 'insulated' wires, in this case about three million of them. Each of the three million wires leads from one cell in the retina to the brain. You can think of them as the wires leading from a bank of three million photocells (actually three million relay stations gathering information from an even larger number of photocells) to the computer that is to process the information in the brain. They are gathered together from all over the retina into a single bundle, which is the optic nerve for that eye.

Any engineer would naturally assume that the photocells would point towards the light, with their wires leading backwards towards the brain. He would laugh at any suggestion that the photocells might point away from the light, with their wires departing on the side *nearest* the light. Yet this is exactly what happens in all vertebrate retinas. Each photocell is, in effect, wired in backwards, with its wire sticking out on the side nearest the light. The wire has to travel over the surface of the retina, to a point where it dives through a hole in the retina (the so-called 'blind spot') to join the optic nerve. This means that the light, instead of being granted an unrestricted passage to the photocells, has to pass through a forest of connecting wires, presumably suffering at least some attenuation and distortion (actually probably not much but, still, it is the *principle* of the thing that would offend any tidy-minded engineer!).

I don't know the exact explanation for this strange state of affairs. The relevant period of evolution is so long ago. But I am ready to bet that it had something to do with the trajectory, the pathway through the real-life equivalent of Biomorph Land, that would have to be traversed in order to turn the retina the right way round, starting from whatever ancestral organ preceded the eye. There probably is such a trajectory, but that hypothetical trajectory, when realized in actual bodies of intermediate animals, proved disadvantageous — temporarily disadvantageous only, but that is enough. Intermediates could see even less well than their imperfect ancestors, and it is no consolation that they are building better eyesight for their remote descendants! What matters is survival in the here and now.

'Dollo's Law' states that evolution is irreversible. This is often confused with a lot of idealistic nonsense about the inevitability of progress, often coupled with ignorant nonsense about evolution 'violating the Second Law of Thermodynamics' (those that belong to the half of the educated population that, according to the novelist C. P. Snow, know what the Second Law is, will realize that it is no more violated by evolution than it is violated by the growth of a baby). There is no reason why general trends in evolution shouldn't be reversed. If there is a trend towards large antlers for a while in evolution, there can easily be a subsequent trend towards smaller antlers again. Dollo's Law is really just a statement about the statistical improbability of following exactly the same evolutionary trajectory twice (or, indeed, any *particular* trajectory), in either direction. A single mutational step can easily be reversed. But for larger numbers of mutational steps, even in the case of the biomorphs with their nine little genes, the mathematical space of all possible trajectories is so vast that the chance of two trajectories ever arriving at the same point becomes vanishingly small. This is even more true of real animals with their vastly larger numbers of genes. There is nothing mysterious or mystical about Dollo's Law, nor is it something that we go out and 'test' in nature. It follows simply from the elementary laws of probability.

For just the same reason, it is vanishingly improbable that exactly the same evolutionary pathway should ever be travelled twice. And it would seem similarly improbable, for the same statistical reasons, that two lines of evolution should converge on exactly the same endpoint from different starting points.

It is all the more striking a testimony to the power of natural selection, therefore, that numerous examples can be found in real nature, in which independent lines of evolution appear to have converged, from very different starting points, on what looks very like the same endpoint. When we look in detail we find — it would be worrying if we didn't — that the convergence is not total. The different lines of evolution betray their independent origins in numerous points of detail. For instance, octopus eyes are very like ours, but the wires leading from their photocells don't point forwards towards the light, as ours do. Octopus eyes are, in this respect, more 'sensibly' designed. They have arrived at a'similar endpoint, from a very different starting point. And the fact is betrayed in details such as this.

Such superficially convergent resemblances are often extremely striking, and I shall devote the rest of the chapter to some of them. They provide most impressive demonstrations of the power of natural selection to put together good designs. Yet the fact that the superficially similar designs also differ, testifies to their independent evolutionary origins and histories. The basic rationale is that, if a design is good enough to evolve once, the same design *principle* is good enough to evolve twice, from different starting points, in different parts of the animal kingdom. This is nowhere better illustrated than in the case we used for our basic illustration of good design itself — echolocation.

Most of what we know about echolocation comes from bats (and human instruments), but it also occurs in a number of other unrelated groups of animals. At least two separate groups of birds do it, and it has been carried to a very high level of sophistication by dolphins and whales. Moreover, it was almost certainly 'discovered' independently by at least two different groups of bats. The birds that do it are the oil-birds of South America, and the cave swiftlets of the Far East, the ones whose nests are used for birds' nest soup. Both types of bird nest deep in caves where little or no light penetrates, and both navigate through the blackness using echoes from their own vocal clicks. In both cases the sounds are audible to humans, not ultrasonic like the more specialized bat clicks. Indeed, neither bird species seems to have developed echolocation to such a pitch of sophistication as bats have. Their clicks are not FM, nor do they appear suitable for Doppler-shift speed metering. Probably, like the fruit bat *Rousettus*, they just time the silent interval between each click and its echo.

In this case we can be absolutely certain that the two bird species have invented echolocation independently of bats, and independently of each other. The line of reasoning is of a kind that evolutionists frequently use. We look at all the thousands of species of birds, and observe that the vast majority of them don't use echolocation. Just two isolated little genera of birds do it, and those two have nothing else in common with each other except that both live in caves. Although we believe that all birds and bats must have a common ancestor if we trace their lineages back far enough, that common ancestor was also the common ancestor of all mammals (including ourselves) and all birds. The vast majority of mammals and the vast majority of birds don't use echolocation, and it is highly probable that their common ancestor didn't either (nor did it fly – that is another technology that has been independently evolved several times). It follows that the echolocation technology has been independently developed in bats and birds, just as it was independently developed by British, American and German scientists. The same kind of reasoning, on a smaller scale, leads to the conclusion that the common ancestor of the oil-bird and the cave swiftlet also did not use echolocation, and that these two genera have developed the same technology independently of each other.

Within the mammals too, bats are not the only group to have independently developed the echolocation technology. Several different kinds of mammals, for instance shrews, rats and seals, seem to use echoes to a small extent, as blind humans do, but the only animals to rival bats in sophistication are whales. Whales are divided into two main groups, toothed whales and baleen whales. Both, of course, are mammals descended from land-dwelling ancestors, and they may well have 'invented' the whale way of life independently of one another, starting from different land-dwelling ancestors. The toothed whales include sperm whales, killer whales and the various species of dolphins, all of which hunt relatively large prey such as fish and squids, which they catch in their jaws. Several toothed whales, of which only dolphins have been thoroughly studied, have evolved sophisticated echo-sounding equipment in their heads.

Dolphins emit rapid trains of high-pitched clicks, some audible to us, some ultrasonic. It is probable that the 'melon', the bulging dome on the front of a dolphin's head, looking – pleasing coincidence – like the weirdly bulging radar dome of a Nimrod 'advance-warning' surveillance aircraft, has something to do with beaming the sonar signals forwards, but its exact workings are not understood. As in the case of bats, there is a relatively slow 'cruising rate' of clicking, rising to a high-speed (400 clicks per second) buzz when the animal is closing in on prey. Even the 'slow' cruising rate is pretty fast. The river dolphins that live in muddy water are probably the most skilled echolocators, but some open-sea dolphins have been shown in tests to be pretty good too. An Atlantic bottlenose dolphin can discriminate circles, squares and triangles (all of the same standardized area), using only its sonar. It can tell which of two targets is the nearer, when the difference is only 1¼ inches at an overall distance of about 7 yards. It can detect a steel sphere half the size of a golf ball, at a range of 70 yards. This performance is not quite as good as human *vision* in a good light, but probably better than human vision in moonlight.

The intriguing suggestion has been made that dolphins, if they chose to use it, have a potentially effortless means of communicating 'mental pictures' to one another. All that they would have to do is use their highly versatile voices to mimic the pattern of sound that would be produced by echoes from a particular object. In this way they could convey to one another mental pictures of such objects. There is no evidence for this delightful suggestion. Theoretically, bats could do the same thing, but dolphins seem more likely candidates because they are in general more social. They are also probably 'cleverer', but this isn't necessarily a relevant consideration. The instruments that would be needed for communicating echo pictures are no more sophisticated than the instruments that both bats and dolphins already have for echolocating in the first place. And there would seem to be an easy, gradual continuum between using the voice to make echoes and using it to mimic echoes.

At least two groups of bats then, two groups of birds, toothed whales, and probably several other kinds of mammals to a smaller extent, have all independently converged on the technology of sonar, at some time during the last hundred million years. We have no way of knowing whether any other animals now extinct – pterodactyls perhaps? – also evolved the technology independently.

No insects and no fish have so far been found to use sonar, but two quite different groups of fish, one in South America and one in Africa, have developed a somewhat similar navigation system, which appears to be just about as sophisticated and which can be seen as a related, but different, solution to the same problem. These are so-called weakly electric fish. The word 'weakly' is to differentiate them from strongly electric fish, which use electric fields, not to navigate, but to stun their prey. The stunning technique, incidentally, has also been independently invented by several unrelated groups of fish, for example electric 'eels' (which are not true eels but whose shape is convergent on true eels) and electric rays.

The South American and the African weakly electric fish are quite unrelated to each other, but both live in the same kinds of waters in their respective continents, waters that are too muddy for vision to be effective. The physical principle that they exploit – electric fields in water – is even more alien to our consciousness than that of bats and dolphins. We at least have a subjective idea of what an echo is, but we have almost no subjective idea of what it might be like to perceive an electric field. We didn't even know of the existence of electricity until a couple of centuries ago. We cannot as subjective human beings empathize with electric fish, but we can, as physicists, understand them.

It is easy to see on the dinner plate that the muscles down each side of any fish are arranged as a row of segments, a *battery* of muscle units. In most fish they contract successively to throw the body into sinuous waves, which propel it forwards. In electric fish, both strongly and weakly electric ones, they have become a battery in the electric sense. Each segment ('cell') of the battery generates a voltage. These voltages are connected up in series along the length of the fish so that, in a strongly electric fish such as an electric eel, the whole battery generates as much as 1 amp at 650 volts. An electric eel is powerful enough to knock a man out. Weakly electric fish don't need high voltages or currents for their purposes, which are purely information-gathering ones.

The principle of electrolocation, as it has been called, is fairly well understood at the level of physics though not, of course, at the level of what it feels like to be an electric fish. The following account applies equally to African and South American weakly electric fish: the convergence is that thorough. Current flows from the front half of the fish, out into the water in lines that curve back and return to the tail end of the fish. There are not really discrete 'lines' but a continuous 'field', an invisible cocoon of electricity surrounding the fish's body. However, for human visualization it is easiest to think in terms of a family of curved lines leaving the fish through a series of portholes spaced along the front half of the body, all curving round in the water and diving into the fish again at the tip of its tail. The fish has what amounts to a tiny voltmeter monitoring the voltage at each 'porthole'. If the fish is suspended in open water with no obstacles around, the lines are smooth curves. The tiny voltmeters at each porthole all register the voltage as 'normal' for their porthole. But if some obstacle appears in the vicinity, say a rock or an item of food, the lines of current that happen to hit the obstacle will be changed. This will change the voltage at any porthole whose current line is affected, and the appropriate voltmeter will register the fact. So in theory a computer, by comparing the pattern of voltages registered by the voltmeters at all the portholes, could calculate the pattern of obstacles around the fish. This is apparently what the fish brain does. Once again, this doesn't have to mean that the fish are clever mathematicians. They have an apparatus that solves the necessary equations, just as our brains unconsciously solve equations every time we catch a ball.

It is very important that the fish's own body is kept absolutely rigid. The computer in the head couldn't cope with the extra distortions that would be introduced if the fish's body were bending and twisting like an ordinary fish. Electric fish have, at least twice independently, hit upon this ingenious method of navigation, but they have had to pay a price: they have had to give up the normal, highly efficient, fish method of swimming, throwing the whole body into serpentine waves. They have solved the problem by keeping the body stiff as a poker, but they have a single long fin all the way along the length of the body. Then instead of the whole body being thrown into waves, just the long fin is. The fish's progress through the water is rather slow, but it does move, and apparently the sacrifice of fast movement is worth it: the gains in navigation seem to outweigh the losses in speed of swimming. Fascinatingly, the South American electric fish have hit upon almost exactly the same solution as the African ones, but not quite. The difference is revealing. Both groups have developed a single long fin that runs the whole length of the body, but in the African fish it runs along the back whereas in the South American fish it runs along the belly. This kind of difference in detail is very characteristic of convergent evolution, as we have seen. It is characteristic of convergent designs by human engineers too, of course.

Although the majority of weakly electric fish, in both the African and the South American groups, give their electric discharges in discrete pulses and are called 'pulse' species, a minority of species in both groups do it a different way and are called 'wave' species. I shall not discuss the difference further. What is interesting for this chapter is that the pulse/wave split has evolved twice, independently, in the unrelated New World and Old World groups.

One of the most bizarre examples of convergent evolution that I know concerns the so-called periodical cicadas. Before getting to the convergence, I must fill in some background information. Many insects have a rather rigid separation between a juvenile feeding stage, in which they spend most of their lives, and a relatively brief adult reproducing stage. Mayflies, for instance, spend most of their lives as underwater feeding larvae, then emerge into the air for a single day into which they cram the whole of their adult lives. We can think of the adult as analogous to the ephemeral winged seed of a plant like a sycamore, and the larva as analogous to the main plant, the difference being that sycamores make many seeds and shed them over many successive years, while a mayfly larva gives rise to only one adult right at the end of its own life. Anyway, periodical cicadas have carried the mayfly trend to an extreme. The adults live for a few weeks, but the 'juvenile' stage (technically 'nymphs' rather than larvae) lasts for 13 years (in some varieties) or 17 years (in other varieties). The adults emerge at almost exactly the same moment, having spent 13 (or 17) years cloistered underground. Cicada plagues, which occur in any given area exactly 13 (or 17) years apart, are spectacular eruptions that have led to their incorrectly being called 'locusts' in vernacular American speech. The varieties are known, respectively, as 13-year cicadas and 17-year cicadas.

Now here is the really remarkable fact. It turns out that there is not just one 13-year cicada species and one 17-year species. Rather, there are three species, and each one of the three has both a 17-year and a 13-year variety or race. The division into a 13-year race and a 17-year race has been arrived at independently, no fewer than three times. It looks as though the intermediate periods of 14, 15 and 16 years have been shunned convergently, no fewer than three times. Why? We don't know. The only suggestion anyone has come up with is that what is special about 13 and 17, as opposed to 14, 15 and 16, is that they are prime numbers. A prime number is a number that is not exactly divisible by any other number. The idea is that a race of animals that regularly erupts in plagues gains the benefit of alternately 'swamping' and starving its enemies, predators or parasites. And if these plagues are carefully timed to occur a prime number of years apart, it makes it that much more difficult for the enemies to synchronize their own life cycles. If the cicadas erupted every 14 years, for instance, they could be exploited by a parasite species with a 7-year life cycle. This is a bizarre idea, but no more bizarre than the phenomenon itself. We really don't know what is special about 13 and 17 years. What matters for our purposes here is that there must be *something* special about those numbers, because three different species of cicada have independently converged upon them.

Examples of convergence on a large scale occur when two or more continents are isolated from one another for a long time, and a parallel range of 'trades' is adopted by unrelated animals on each of the continents. By 'trades' I mean ways of making a living, such as burrowing for worms, digging for ants, chasing large herbivores, eating leaves up trees. A good example is the convergent evolution of a whole range of mammal trades in the separate continents of South America, Australia, and the Old World.

These continents weren't always separate. Because our lives are measured in decades, and even our civilizations and dynasties are measured only in centuries, we are accustomed to thinking of the map of the world, the outlines of the continents, as fixed. The theory that continents drifted about was proposed long ago by the German geophysicist Alfred Wegener, but most people laughed at him until well after the Second World War. The admitted fact that South America and Africa look a bit like separated pieces of a jigsaw puzzle was assumed to be just an amusing coincidence. In one of the most rapid and complete revolutions science has known, the formerly controversial theory of 'continental drift' has now become universally accepted under the name of plate tectonics. The evidence that the continents have drifted, that South America did indeed break away from Africa for instance, is now literally overwhelming, but this is not a book about geology and I shall not spell it out. For us the important point is that the timescale on which continents have drifted about is the same slow timescale on which animal lineages have evolved, and we cannot ignore continental drift if we are to understand the patterns of animal evolution on those continents.

Up until about 100 million years ago, then, South America was joined to Africa in the east and to Antarctica in the south. Antarctica was joined to Australia, and India was joined to Africa via Madagascar. There was in fact one huge southern continent, which we now call Gondwanaland, consisting of what is now South America, Africa, Madagascar, India, Antarctica and Australia all rolled into one. There was also a single large northern continent called Laurasia consisting of what is now North America, Greenland, Europe and Asia (apart from India). North America was not connected to South America. About 100 million years ago there was a big break-up of the land masses, and the continents have been slowly moving towards their present positions ever since (they will, of course, continue to move in the future). Africa joined up with Asia via Arabia and became part of the huge continent that we now speak of as the Old World. North America drifted away from Europe, Antartica drifted south to its present icy location. India detached itself from Africa and set off across what is now called the Indian Ocean, eventually to crunch into south Asia and raise the Himalayas. Australia drifted away from Antarctica into the open sea to become an island continent miles from anywhere else.

It happens that the break-up of the great southern continent of Gondwanaland began during the age of the dinosaurs. When South America and Australia broke away to begin their long periods of isolation from the rest of the world, they each carried their own cargo of dinosaurs, and also of the less-prominent animals that were to become the ancestors of modern mammals. When, rather later, for reasons that are not understood and are the subject of much profitable speculation, the dinosaurs (with the exception of the group of dinosaurs that we now call birds) went extinct, they went extinct all over the world. This left a vacuum in the 'trades' open to land-dwelling animals. The vacuum was filled, over a period of millions of years of evolution, mostly by mammals. The interesting point for us here is that there were three independent vacuums, and they were independently filled by mammals in Australia, South America and the Old World.

The primitive mammals that happened to be around in the three areas when the dinosaurs more or less simultaneously vacated the great life trades, were all rather small and insignificant, probably nocturnal, previously overshadowed and overpowered by the dinosaurs. They could have evolved in radically different directions in the three areas. To some extent this is what happened. There is nothing in the Old World that resembles the giant ground sloth of South America, alas now extinct. The great range of South American mammals included an extinct giant guinea-pig, the size of a modern rhinoceros but a rodent (I have to say 'modern' rhinoceros because the Old World fauna included a giant rhinoceros the size of a two-storey house). But although the separate continents each produced their unique mammals, the general pattern of evolution in all three areas was the same. In all three areas the mammals that happened to be around at the start fanned out in evolution, and produced a specialist for each trade which, in many cases, came to bear a remarkable resemblance to the corresponding specialist in the other two areas. Each trade, the burrowing trade, the large hunter trade, the plains-grazing trade, and so on, was the subject of independent convergent evolution in two or three separate continents. In addition to these three major sites of independent evolution, smaller islands such as Madagascar have interesting parallel stories of their own, which I shall not go into.

Setting aside the strange egg-laying mammals of Australia – the duck-billed platypus and the spiny anteaters – modern mammals all belong to one of two great groups. These two are the marsupials (whose young are born very small and are then kept in a pouch) and the placentals (all the rest of us). The marsupials came to dominate the Australian story and the placentals the Old World, while the two groups played important roles alongside each other in South America. The South American story is complicated by the fact that it was subject to sporadic waves of invasion by mammals from North America.

Having set the scene, we can now look at some of the trades and convergences themselves. An important trade is concerned with the exploitation of the great grasslands variously known as prairie, pampas, savannah, etc. Practitioners of this trade include horses (of which the main African species are called zebras and the desert models are called donkeys), and cattle, such as the North American bison, now hunted to near-extinction. Herbivores typically have very long guts containing various kinds of fermenting bacteria, since grass is a poor-quality food and needs a lot of digesting. Rather than break their eating up into discrete meals, they typically eat more or less continuously. Huge volumes of plant material flow through them like a river, all the day long. The animals are often very large, and they frequently go about in great herds. Each one of these big herbivores is a mountain of valuable food to any predator that can exploit it. As a consequence of this there is, as we shall see, a whole trade devoted to the difficult task of catching and killing them. These are the predators. Actually, when I say 'a' trade, I really mean a whole lot of 'sub-trades': lions, leopards, cheetahs, wild dogs and hyenas all hunt in their own specialized ways. The same kind of subdivision is found in the herbivores, and in all the other 'trades'.

The herbivores have keen senses with which they are continuously alert for predators, and they are usually capable of running very fast to escape them. To this end they often have long, spindly legs, and they typically run on the tips of their toes, which have become specially elongated and strengthened in evolution. The nails at the ends of these specialized toes have become large and hard, and we call them hooves. Cattle have two enlarged toes at the extremities of each leg: the familiar 'cloven' hooves. Horses do much the same thing except that, probably for reasons of historical accident, they run on only one toe instead of two. It is derived from what was originally the middle one of the five toes. The other toes have almost completely disappeared over evolutionary time, although they occasionally reappear in freakish 'throwbacks'.

Now South America, as we have seen, was isolated during the period in which horses and cattle were evolving in other parts of the world. But South America has its own great grasslands, and it evolved its own separate groups of large herbivores to exploit the resource. There were massive rhino-like Leviathans that had no connection with true rhinos. The skulls of some of the early South American herbivores suggest that they 'invented' the trunk independently of the true elephants. Some resembled camels, some looked like nothing on earth (today), or like weird chimeras of modern animals. The group called the litopterns are almost unbelievably similar to horses in their legs, yet they were utterly unrelated to horses. The superficial resemblance fooled a nineteenth-century Argentinian expert who thought, with pardonable national pride, that they were the ancestors of all horses in the rest of the world. In fact their resemblance to horses was super-ficial, and convergent. Grassland life is much the same the world over, and horses and litopterns independently evolved the same qualities to cope with the problems of grassland life. In particular, the litopterns, like the horses, lost all their toes except the middle one on each leg, which became enlarged as the bottom joint of the leg and developed a hoof. The leg of a litoptern is all but indistinguishable from the leg of a horse, yet the two animals are only distantly related.

In Australia the large grazers and browsers are very different – kangaroos. Kangaroos have the same need to move rapidly, but they have done it in a different way. Instead of developing four-legged galloping to the high pitch of perfection that horses (and presumably litopterns) did, kangaroos have perfected a different gait: two-legged hopping with a large balancing tail. There is little point in arguing over which of these two gaits is 'better'. They are each highly effective if the body evolves in such a way as to exploit them to the full. Horses and litopterns happened to exploit four-legged galloping, and so ended up with almost identical legs. Kangaroos happened to exploit two-legged hopping, and so ended up with their own uniquely (at least since the dinosaurs) massive hind legs and tail. Kangaroos and horses arrived at different endpoints in 'animal space', probably because of some accidental difference in their starting points.

Turning now to the meat-eaters that the great grazers were running away from, we find some more fascinating convergences. In the Old World we are familiar with such large hunters as wolves, dogs, hyenas, and the big cats – lions, tigers, leopards and cheetahs. A big cat that has only recently gone extinct is the sabre-tooth ('tiger'), named after its colossal canine teeth which jutted down from the upper jaw in the front of what must have been a terrifying gape. Until recent times there were no true cats or dogs in Australia or the New World (pumas and jaguars are recently evolved from Old World cats). But in both those continents there were marsupial equivalents. In Australia the thylacine, or marsupial 'wolf' (often called the Tasmanian wolf because it survived in Tasmania for a little longer than in mainland Australia), was tragically driven extinct within living memory, slaughtered in enormous numbers as a 'pest' and for 'sport' by humans (there is a slight hope that it may still survive in remote parts of Tasmania, areas which themselves are now threatened with destruction in the interests of providing 'employment' for humans). It is not to be confused with the dingo, by the way, which is a true dog, introduced to Australia more recently by (aboriginal) man. A ciné film made in the 1930s of the last known thylacine, restlessly pacing its lonely zoo cage, shows an uncannily dog-like animal, its marsupial nature betrayed only by its slightly undog-like way of holding its pelvis and back legs, presumably something to do with accommodating its pouch. To any dog-lover, the contemplation of this alternative approach to the dog design, this evolutionary traveller along a parallel road separated by 100 million years, this part-familiar yet part utterly alien other-worldly dog, is a moving experience. Maybe they were pests to humans, but humans were much bigger pests to them; now there are no thylacines left and a considerable surplus of humans.

In South America, too, there were no true dogs or cats during the long period of isolation that we are discussing but, as in Australia, there were marsupial equivalents. Probably the most spectacular was *Thylacosmilus*, which looked exactly like the recently extinct sabre-tooth 'tiger' of the Old World, only more so if you see what I mean. Its daggered gape was even wider, and I imagine that it was even more terrifying. Its name records its superficial affinity with the sabre-tooth (*Smilodon*) and the Tasmanian wolf (*Thylacinus*), but in terms of ancestry it is very remote from both. It is slightly closer to the thylacine since both are marsupials, but the two have evolved their big carnivore design independently on different continents; independently of each other and of the placental carnivores, the true cats and dogs of the Old World.

Australia, South America and the Old World offer numerous further examples of multiple convergent evolution. Australia has a marsupial 'mole', superficially almost indistinguishable from the familiar moles of other continents, but pouched, making its living in the same way as other moles and with the same enormously strengthened forepaws for digging. There is a pouched mouse in Australia, though in this case the resemblance is not so close and it does not make its living in quite the same way. Anteating (where 'ants' are deemed for convenience to include termites – another convergence as we shall see) is a 'trade' that is filled by a variety of convergent mammals. They may be subdivided into anteaters that burrow, anteaters that climb trees and anteaters that wander over the ground. In Australia, as we might expect, there is a marsupial anteater. Called *Myrmecobius*, it has a long thin snout for poking into ants' nests, and a long sticky tongue with which it mops up its prey. It is a grounddwelling anteater. Australia also has a burrowing anteater, the spiny anteater. This is not a marsupial, but a member of the group of egglaying mammals, the monotremes, so remote from us that marsupials are our close cousins by comparison. The spiny anteater, too, has a long pointed snout, but its spines give it a superficial resemblance to a hedgehog rather than to another typical anteater.

South America could easily have had a marsupial anteater, alongside its marsupial sabre-tooth 'tiger', but as it happens the anteater trade was early filled by placental mammals instead. The largest of today's anteaters is *Myrmecophaga* (which just means anteater in Greek), the large ground-wandering anteater of South America and probably the most extreme anteating specialist in the world. Like the Australian marsupial *Myrmecobius*, it has a long and pointed snout, extremely long and pointed in this case, and an extremely long sticky tongue. South America also has a small tree-climbing anteater, which is a close cousin of *Myrmecophaga* and looks like a miniature and less extreme version of it, and a third, intermediate form. Although placental mammals, these anteaters are very far from any Old World placentals. They belong to a uniquely South American family, which also includes armadillos and sloths. This ancient placental family coexisted with the marsupials from the early days of the continent's isolation.

The Old World anteaters include various species of pangolin in Africa and Asia, ranging from tree-climbing forms to digging forms, all looking a bit like fircones with pointed snouts. Also in Africa is the weird ant-bear or aardvark, which is partially specialized for digging. A feature that characterizes all anteaters, whether marsupial, monotreme or placental, is an extremely low metabolic rate. The metabolic rate is the rate at which their chemical 'fires' burn, most easily measured as the blood temperature. There is a tendency for metabolic rate to depend on body size in mammals generally. Smaller animals tend to have higher metabolic rates, just as the engines of small cars tend to turn over at a higher rate than those of larger cars. But some animals have high metabolic rates for their size, and anteaters, of whatever ancestry and affinities, tend to have very low metabolic rates for their size. It is not obvious why this is, but it is so strikingly convergent among animals that have nothing else in common but their anteating habit, that it almost certainly is somehow related to this habit.

As we have seen, the 'ants' that anteaters eat are often not true ants at all, but termites. Termites are often known as 'white ants', but they are related to cockroaches, rather than to true ants, which are related to bees and wasps. Termites resemble ants superficially because they have convergently adopted the same habits. The same range of habits, I should say, because there are many different branches of the ant/termite trade, and both ants and termites have independently adopted most of them. As so often with convergent evolution, the differences are revealing as well as the similarities.

Both ants and termites live in large colonies consisting mostly of sterile, wingless workers, dedicated to the efficient production of winged reproductive castes which fly off to found new colonies. An interesting difference is that in ants the workers are all sterile females, whereas in termites they are sterile males and sterile females. Both ant and termite colonies have one (or sometimes several) enlarged 'queens', sometimes (in both ants and termites) grotesquely enlarged. In both ants and termites the workers can include specialist castes such as soldiers. Sometimes these are such dedicated fighting machines, especially in their huge jaws (in the case of ants, but 'gun-turrets' for chemical warfare in the case of termites), that they are incapable of feeding themselves and have to be fed by non-soldier workers. Particular species of ants parallel particular species of termites. For example, the habit of fungus-farming has arisen inde-pendently in ants (in the New World) and termites (in Africa). The ants (or termites) forage for plant material that they do not digest them-selves but make into compost on which they grow fungi. It is the fungi that they themselves eat. The fungi, in both cases, grow nowhere else than in the nests of ants or termites, respectively. The fungus-farming habit has also been discovered independently and convergently (more than once) by several species of beetles.

There are also interesting convergences within the ants. Although most ant colonies live a settled existence in a fixed nest, there seems to be a successful living to be made by wandering in enormous pillaging armies. This is called the legionary habit. Obviously all ants walk about and forage, but most kinds return to a fixed nest with their booty, and the queen and the brood are left behind in the nest. The key to the wandering legionary habit, on the other hand, is that the armies take the brood and the queen with them. The eggs and larvae are carried in the jaws of workers. In Africa the legionary habit has been developed by the so-called driver ants. In Central and South America the parallel 'army ants' are very similar to driver ants in habit and appearance. They are not particularly closely related. They have certainly evolved the characteristics of the 'army' trade independently and convergently.

Both driver ants and army ants have exceptionally large colonies, up to a million in army ants, up to about 20 million in driver ants. Both have nomadic phases alternating with 'statary' phases, relatively stable encampments or 'bivouacs'. Army ants and driver ants, or rather their colonies taken together as amoeba-like units, are both ruthless and terrible predators of their respective jungles. Both cut to pieces any-thing animal in their path, and both have acquired a mystique of terror in their own land. Villagers in parts of South America are reputed traditionally to vacate their villages, lock, stock and barrel when a large ant army is approaching, and to return when the legions have marched through, having cleaned out every cockroach, spider and scorpion even from the thatched roofs. I remember as a child in Africa being more frightened of driver ants than of lions or crocodiles. It is worth getting this formidable reputation into perspective by quoting the words of Edward O. Wilson, the world's foremost authority on ants as well as the author of *Sociobiology*:

> In answer to the single question I am asked most frequently about ants, I can give the following answer: No, driver ants are not really the terror of the jungle. Although the driver ant colony is an 'animal' weighing in excess of 20 kg and possessing on the order of 20 million mouths and stings and is surely the most formidable creation of the insect world, it still does not match up to the lurid stories told about it. After all, the swarm can only cover about a metre of ground every three minutes. Any competent bush mouse, not to mention man or elephant, can step aside and contemplate the whole grass-roots frenzy at leisure, an object less of menace than of strangeness and wonder, the culmination of an evolutionary story as different from that of mammals as it is possible to conceive in this world.

As an adult in Panama I have stepped aside and contemplated the New World equivalent of the driver ants that I had feared as a child in Africa, flowing by me like a crackling river, and I can testify to the strangeness and wonder. Hour after hour the legions marched past, walking as much over each others' bodies as over the ground, while I waited for the queen. Finally she came, and hers was an awesome presence. It was impossible to see her body. She appeared only as a moving wave of worker frenzy, a boiling peristaltic ball of ants with linked arms. She was somewhere in the middle of the seething ball of workers, while all around it the massed ranks of soldiers faced threateningly outwards with jaws agape, every one prepared to kill and to die in defence of the queen. Forgive my curiosity to see her: I prodded the ball of workers with a long stick, in a vain attempt to flush out the queen. Instantly 20 soldiers buried their massively muscled pincers in my stick, possibly never to let go, while dozens more swarmed up the stick causing me to let go with alacrity.

I never did glimpse the queen, but somewhere inside that boiling ball she was, the central data bank, the repository of the master DNA of the whole colony. Those gaping soldiers were prepared to die for the queen, not because they loved their mother, not because they had been drilled in the ideals of patriotism, but simply because their brains and their jaws were built by genes stamped from the master die carried in the queen herself. They behaved like brave soldiers because they had inherited the genes of a long line of ancestral queens whose lives, and whose genes, had been saved by soldiers as brave as themselves. My soldiers had inherited the same genes from the present queen as those old soldiers had inherited from the ancestral queens. My soldiers were guarding the master copies of the very instructions that made them do the guarding. They were guarding the wisdom of their ancestors, the Ark of the Covenant. These strange statements will be made plain in the next chapter.

I felt the strangeness then, and the wonder, not unmixed with revivals of half-forgotten fears, but transfigured and enhanced by a mature understanding, which I had lacked as a child in Africa, of what the whole performance was for. Enhanced, too, by the knowledge that this story of the legions had reached the same evolutionary culmination not once but twice. These were not the driver ants of my childhood nightmares, however similar they might be, but remote, New World cousins. They were doing the same thing as the driver ants, and for the same reasons. It was night now and I turned for home, an awestruck child again, but joyful in the new world of understanding that had supplanted the dark, African fears.

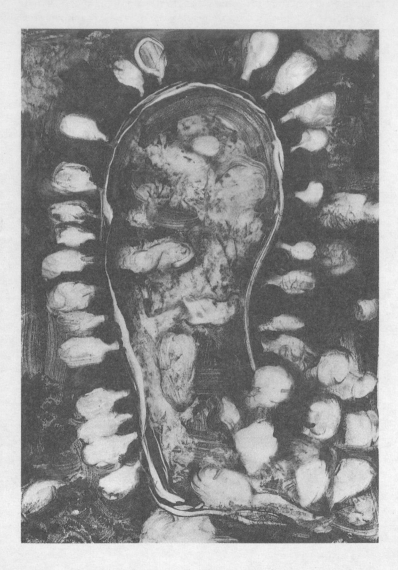

CHAPTER 5

THE POWER
AND THE ARCHIVES

It is raining DNA outside. On the bank of the Oxford canal at the bottom of my garden is a large willow tree, and it is pumping downy seeds into the air. There is no consistent air movement, and the seeds are drifting outwards in all directions from the tree. Up and down the canal, as far as my binoculars can reach, the water is white with floating cottony flecks, and we can be sure that they have carpeted the ground to much the same radius in other directions too. The cotton wool is mostly made of cellulose, and it dwarfs the tiny capsule that contains the DNA, the genetic information. The DNA content must be a small proportion of the total, so why did I say that it was raining DNA rather than cellulose? The answer is that it is the DNA that matters. The cellulose fluff, although more bulky, is just a parachute, to be discarded. The whole performance, cotton wool, catkins, tree and all, is in aid of one thing and one thing only, the spreading of DNA around the countryside. Not just any DNA, but DNA whose coded characters spell out specific instructions for building willow trees that will shed a new generation of downy seeds. Those fluffy specks are, literally, spreading instructions for making themselves. They are there because their ancestors succeeded in doing the same. It is raining instructions out there; it's raining programs; it's raining tree-growing, fluff-spreading, algorithms. That is not a metaphor, it is the plain truth. It couldn't be any plainer if it were raining floppy discs.

It is plain and it is true, but it hasn't long been understood. A few years ago, if you had asked almost any biologist what was special about living things as opposed to nonliving things, he would have told you about a special substance called protoplasm. Protoplasm wasn't like any other substance; it was vital, vibrant, throbbing, pulsating, 'irritable' (a schoolmarmish way of saying responsive). If you took a living body and cut it up into ever smaller pieces, you would eventually come down to specks of pure protoplasm. At one time in the last century, a real-life counterpart of Arthur Conan Doyle's Professor Challenger thought that the 'globigerina ooze' at the bottom of the sea was pure protoplasm. When I was a schoolboy, elderly textbook authors still wrote about protoplasm although, by then, they really should have known better. Nowadays you never hear or see the word. It is as dead as phlogiston and the universal aether. There is nothing special about the substances from which living things are made. Living things are collections of molecules, like everything else.

What is special is that these molecules are put together in much more complicated patterns than the molecules of nonliving things, and this putting together is done by following programs, sets of instructions for how to develop, which the organisms carry around inside themselves. Maybe they do vibrate and throb and pulsate with 'irritability', and glow with 'living' warmth, but these properties all emerge incidentally. What lies at the heart of every living thing is not a fire, not warm breath, not a 'spark of life'. It is information, words, instructions. If you want a metaphor, don't think of fires and sparks and breath. Think, instead, of a billion discrete, digital characters carved in tablets of crystal. If you want to understand life, don't think about vibrant, throbbing gels and oozes, think about information technology. It is this that I was hinting at in the previous chapter, when I referred to the queen ant as the central data bank.

The basic requirement for an advanced information technology is some kind of storage medium with a large number of memory locations. Each location must be capable of being in one of a discrete number of states. This is true, anyway, of the *digital* information technology that now dominates our world of artifice. There is an alternative kind of information technology based upon *analogue* information. The information on an ordinary gramophone record is analogue. It is stored in a wavy groove. The information on a modern laser disc (often called 'compact disc', which is a pity, because the name is uninformative and also usually mispronounced with the stress on the first syllable) is digital, stored in a series of tiny pits, each of which is either definitely there or definitely not there: there are no half measures. That is the diagnostic feature of a digital system: its fundamental elements are either definitely in one state or definitely in another state, with no half measures and no intermediates or compromises.

The information technology of the genes is digital. This fact was discovered by Gregor Mendel in the last century, although he wouldn't have put it like that. Mendel showed that we don't blend our inheritance from our two parents. We receive our inheritance in discrete particles. As far as each particle is concerned, we either inherit it or we don't. Actually, as R. A. Fisher, one of the founding fathers of what is now called neo-Darwinism, has pointed out, this fact of particulate inheritance has always been staring us in the face, every time we think about sex. We inherit attributes from a male and a female parent, but each of us is either male or female, not hermaphrodite. Each new baby born has an approximately equal *probability* of inheriting maleness or femaleness, but any one baby inherits only one of these, and doesn't combine the two. We now know that the same goes for all our particles of inheritance. They don't blend, but remain discrete and separate as they shuffle and reshuffle their way down the generations. Of course there is often a powerful appearance of blending in the effects that the genetic units have on bodies. If a tall person mates with a short person, or a black person with a white person, their offspring are often intermediate. But the appearance of blending applies only to effects on bodies, and is due to the summed small effects of large numbers of particles. The particles themselves remain separate and discrete when it comes to being passed on to the next generation.

The distinction between blending inheritance and particulate inheritance has been of great importance in the history of evolutionary ideas. In Darwin's time everybody (except Mendel who, tucked away in his monastery, was unfortunately ignored until after his death) thought that inheritance was blending. A Scottish engineer called Fleeming Jenkin pointed out that the fact (as it was thought to be) of blending inheritance all but ruled out natural selection as a plausible theory of evolution. Ernst Mayr rather unkindly remarks that Jenkin's article 'is based on all the usual prejudices and misunderstandings of the physical scientists'. Nevertheless, Darwin was deeply worried by Jenkin's argument. It was most colourfully embodied in a parable of a white man shipwrecked on an island inhabited by 'negroes':

> grant him every advantage which we can conceive a white to possess over the native; concede that in the struggle for existence his chance of a long life will be much superior to that of the native chiefs; yet from all these admissions, there does not follow the conclusion that, after a limited or unlimited number of generations, the inhabitants of the island will be white. Our shipwrecked hero would probably become king; he would kill a great many blacks in the struggle for existence; he would have a great many wives and children, while many of his subjects would live and die as bachelors . . . Our white's qualities would certainly tend very much to

preserve him to a good old age, and yet he would not suffice in any number of generations to turn his subjects' descendants white . . . In the first generation there will be some dozens of intelligent young mulattoes, much superior in average intelligence to the negroes. We might expect the throne for some generations to be occupied by a more or less yellow king; but can any one believe that the whole island will gradually acquire a white, or even a yellow population, or that the islanders would acquire the energy, courage, ingenuity, patience, self-control, endurance, in virtue of which qualities our hero killed so many of their ancestors, and begot so many children; these qualities, in fact, which the struggle for existence would select, if it could select anything?

Don't be distracted by the racist assumptions of white superiority. These were as unquestioned in the time of Jenkin and Darwin as our speciesist assumptions of *human rights*, *human* dignity, and the sacredness of *human* life are unquestioned today. We can rephrase Jenkin's argument in a more neutral analogy. If you mix white paint and black paint together, what you get is grey paint. If you mix grey paint and grey paint together, you can't reconstruct either the original white or the original black. Mixing paints is not so far from the pre-Mendelian vision of heredity, and even today popular culture frequently expresses heredity in terms of a mixing of 'bloods'. Jenkin's argument is an argument about swamping. As the generations go by, under the assumption of blending inheritance, variation is bound to become swamped. Greater and greater uniformity will prevail. Eventually there will be no variation left for natural selection to work upon.

Plausible as this argument must have sounded, it is not only an argument against natural selection. It is more an argument against inescapable facts about heredity itself! It manifestly isn't *true* that variation disappears as the generations go by. People are *not* more similar to each other today than they were in their grandparents' time. Variation is maintained. There is a pool of variation for selection to work on. This was pointed out mathematically in 1908 by W. Weinberg, and independently by the eccentric mathematician G. H. Hardy, who incidentally, as the betting book of his (and my) college records, once took a bet from a colleague of 'One half penny to his fortune till death, that the sun will rise tomorrow'. But it took R. A. Fisher and his colleagues, the founders of modern population genetics, to develop the full answer to Fleeming Jenkin in terms of Mendel's theory of *particle* genetics. This was an irony at the time, because, as we shall see in Chapter 11, the leading followers of Mendel in the early twentieth century thought of themselves as anti-Darwinian. Fisher and his colleagues showed that Darwinian selection made sense, and

Jenkin's problem was elegantly solved, if what changed in evolution was the relative *frequency* of discrete hereditary particles, or genes, each of which was either there or not there in any particular individual body. Darwinism post-Fisher is called neo-Darwinism. Its digital nature is not an incidental fact that happens to be true of genetic information technology. Digitalness is probably a necessary precondition for Darwinism itself to work.

In our electronic technology the discrete, digital locations have only two states, conventionally represented as 0 and 1 although you can think of them as high and low, on and off, up and down: all that matters is that they should be distinct from one another, and that the pattern of their states can be 'read out' so that it can have some influence on something. Electronic technology uses various physical media for storing 1s and 0s, including magnetic discs, magnetic tape, punched cards and tape, and integrated 'chips' with lots of little semiconductor units inside them.

The main storage medium inside willow seeds, ants and all other living cells is not electronic but chemical. It exploits the fact that certain kinds of molecule are capable of 'polymerizing', that is joining up in long chains of indefinite length. There are lots of different kinds of polymer. For example, 'polythene' is made of long chains of the small molecule called ethylene – polymerized ethylene. Starch and cellulose are polymerized sugars. Some polymers, instead of being uniform chains of one small molecule like ethylene, are chains of two or more different kinds of small molecule. As soon as such heterogeneity enters into a polymer chain, information technology becomes a theoretical possibility. If there are two kinds of small molecule in the chain, the two can be thought of as 1 and 0 respectively, and immediately any amount of information, of any kind, can be stored, provided only that the chain is long enough. The particular polymers used by living cells are called polynucleotides. There are two main families of polynucleotides in living cells, called DNA and RNA for short. Both are chains of small molecules called nucleotides. Both DNA and RNA are heterogeneous chains, with four different kinds of nucleotides. This, of course, is where the opportunity for information storage lies. Instead of just the two states 1 and 0, the information technology of living cells uses four states, which we may conventionally represent as A, T, C and G. There is very little difference, in principle, between a two-state binary information technology like ours, and a four-state information technology like that of the living cell.

As I mentioned at the end of Chapter 1, there is enough information capacity in a single human cell to store the *Encyclopaedia Britannica*, all 30 volumes of it, three or four times over. I don't know the comparable figure for a willow seed or an ant, but it will be of the same order of staggeringness. There is enough storage capacity in the DNA of a single lily seed or a single salamander sperm to store the *Encyclopaedia Britannica* 60 times over. Some species of the unjustly called 'primitive' amoebas have as much information in their DNA as 1,000 *Encyclopaedia Britannica*s.

Amazingly, only about 1 per cent of the genetic information in, for example, human cells, seems to be actually used: roughly the equivalent of one volume of the *Encyclopaedia Britannica*. Nobody knows why the other 99 per cent is there. In a previous book I suggested that it might be parasitic, freeloading on the efforts of the 1 per cent, a theory that has more recently been taken up by molecular biologists under the name of 'selfish DNA'. A bacterium has a smaller information capacity than a human cell, by a factor of about 1,000, and it probably uses nearly all of it: there is little room for parasites. Its DNA could 'only' hold one copy of the New Testament!

Modern genetic engineers already have the technology to write the New Testament or anything else into a bacterium's DNA. The 'meaning' of the symbols in any information technology is arbitrary, and there is no reason why we should not assign combinations, say triplets, from DNA's 4-letter alphabet, to letters of our own 26-letter alphabet (there would be room for all the upper and lower-case letters with 12 punctuation characters). Unfortunately, it would take about five man-centuries to write the New Testament into a bacterium, so I doubt if anybody will bother. If they did, the rate of reproduction of bacteria is such that 10 million copies of the New Testament could be run off in a single day, a missionary's dream if only people could read the DNA alphabet but, alas, the characters are so small that all 10 million copies of the New Testament could simultaneously dance upon the surface of a pin's head.

Electronic computer memory is conventionally classified into ROM and RAM. ROM stands for 'read only' memory. More strictly it is 'write once, read many times' memory. The pattern of 0s and 1s is 'burned' into it once and for all on manufacture. It then remains unchanged throughout the life of the memory, and the information can be read out any number of times. Other electronic memory, called RAM, can be 'written to' (one soon gets used to this inelegant computer jargon) as well as read. RAM can therefore do everything that ROM can do, and more. What the letters R A M actually stand for is misleading, so I won't mention it. The point about RAM is that you can put any pattern of 1s and 0s into any part of it that you like, on as many occasions as you like. Most of a computer's memory is RAM. As I type these words they are going straight into RAM, and the word-processing program controlling things is also in RAM, although it could theoretically be burned into ROM and then never subsequently altered. ROM is used for a fixed repertoire of standard programs, which are needed again and again, and which you can't change even if you wanted to.

DNA is ROM. It can be read millions of times over, but only written to once – when it is first assembled at the birth of the cell in which it resides. The DNA in the cells of any individual is 'burned in', and is never altered during that individual's lifetime, except by very rare random deterioration. It can be copied, however. It is duplicated every time a cell divides. The pattern of A,T,C and G nucleotides is faithfully copied into the DNA of each of the trillions of new cells that are made as a baby grows. When a new individual is conceived, a new and unique pattern of data is 'burned into' his DNA ROM, and he is then stuck with that pattern for the rest of his life. It is copied into all his cells (except his reproductive cells, into which a random half of his DNA is copied, as we shall see).

All computer memory, whether 'ROM' or 'RAM', is *addressed*. This means that every location in the memory has a label, usually a number but this is an arbitrary convention. It is important to understand the distinction between the *address* and the *contents* of a memory location. Each location is known by its address. For instance the first two letters of this chapter, 'It', are at this moment sitting in RAM locations 6446 and 6447 of my computer, which has 65,536 RAM locations altogether. At another time, the contents of those two locations will be different. The contents of a location is whatever was most recently written in that location. Each ROM location also has an address and a contents. The difference is that each location is stuck with its contents, once and for all.

The DNA is arranged along stringy chromosomes, like long computer tapes. All the DNA in each of our cells is addressed in the same sense as computer ROM, or indeed computer tape, is addressed. The exact numbers or names that we use to label a given address are arbitrary, just as they are for computer memory. What matters is that a particular location in my DNA corresponds precisely to one particular location in your DNA: they have the same address. The *contents* of my DNA location 321762 may or may not be the same as the contents of your location 321762. But my location 321762 is in precisely the same position in my cells as your location 321762 is in your cells. 'Position' here means position along the length of a particular chromosome. The exact physical position of a chromosome in a cell doesn't matter. Indeed, it floats about in fluid so its physical position varies, but every location along the chromosome is precisely addressed in terms of linear order along the length of the chromosome, just as every location along a computer tape is precisely addressed, even if the tape is strewn around the floor rather than being neatly rolled up. All of us, all human beings, have the same set of DNA *addresses*, but not necessarily the same *contents* of those addresses. That is the main reason why we are all different from each other.

Other species don't have the same set of *addresses*. Chimpanzees, for instance, have 48 chromosomes compared to our 46. Strictly speaking it is not possible to compare contents, address by address, because addresses don't correspond to each other across species barriers. Closely related species, however, like chimps and humans, have such large chunks of adjacent contents in common that we can easily identify them as basically the same, even though we can't use quite the same addressing system for the two species. The thing that defines a species is that all members have the same addressing system for their DNA. Give or take a few minor exceptions, all members have the same number of chromosomes, and every location along the length of a chromosome has its exact opposite number in the same position along the length of the corresponding chromosome in all other members of the species. What can differ among the members of a species is the contents of those locations.

The differences in contents in different individuals come about in the following manner, and here I must stress that I am talking about sexually reproducing species such as our own. Our sperms or eggs each contain 23 chromosomes. Each addressed location in one of my sperms corresponds to a particular addressed location in every other one of my sperms, and in every one of your eggs (or sperms). All my other cells contain 46 – a double set. The same addresses are used twice over in each of these cells. Every cell contains two chromosome 9s, and two versions of location 7230 along chromosome 9. The contents of the two may or may not be the same, just as they may or may not be the same in other members of the species. When a sperm, with its 23 chromosomes, is made from a body cell with its 46 chromosomes, it only gets one of the two copies of each addressed location. Which one it gets can be treated as random. The same goes for eggs. The result is that every sperm produced and every egg produced is unique in terms of the *contents* of their locations, although their addressing system is identical in all members of one species (with minor exceptions that need not concern us] When a sperm fertilizes an egg, a full complement of 46 chromosomes is, of course, made up; and all 46 are then duplicated in all the cells of the developing embryo.

I said that ROM cannot be written to except when it is first manufactured, and that is true also of the DNA in cells, except for occasional random errors in copying. But there is a sense in which the collective data bank consisting of the ROMs of an entire species can be constructively written to. The nonrandom survival and reproductive success of individuals within the species effectively 'writes' improved instructions for survival into the collective genetic memory of the species as the generations go by. Evolutionary change in a species largely consists of changes in how many copies there are of each of the various possible *contents* at each addressed DNA location, as the generations pass. Of course, at any particular time, every copy has to be inside an individual body. But what matters in evolution is changes in frequency of alternative possible contents at each address in *populations*. The addressing system remains the same, but the statistical profile of location contents changes as the centuries go by.

Once in a blue moon the addressing system itself changes. Chimpanzees have 24 pairs of chromosomes and we have 23. We share a common ancestor with chimpanzees, so at some point in either our ancestry or chimps' there must have been a change in chromosome number. Either we lost a chromosome (two merged), or chimps gained one (one split). There must have been at least one individual who had a different number of chromosomes from his parents. There are other occasional changes in the entire genetic system. Whole lengths of code, as we shall see, may occasionally be copied to completely different chromosomes. We know this because we find, scattered around the chromosomes, long strings of DNA text that are identical.

When the information in a computer memory has been read from a particular location, one of two things may happen to it. It can either simply be written somewhere else, or it can become involved in some 'action'. Being written somewhere else means being copied. We have already seen that DNA is readily copied from one cell to a new cell, and that chunks of DNA may be copied from one individual to another individual, namely its child. 'Action' is more complicated. In computers, one kind of action is the execution of program instructions. In my computer's ROM, location numbers 64489, 64490 and 64491, taken together, contain a particular pattern of contents – 1s and 0s which – when interpreted as instructions, result in the computer's little loudspeaker uttering a blip sound. This bit pattern is 10101101 00110000 11000000. There is nothing inherently blippy or noisy about that bit pattern. Nothing about it tells you that it will have that effect on the loudspeaker. It has that effect only because of the way the rest of the computer is wired up. In the same way, patterns in the DNA four-letter code have effects, for instance on eye colour or behaviour, but these effects are not inherent in the DNA data patterns themselves. They have their effects only as a result of the way the rest of the embryo develops, which in turn is influenced by the effects of patterns in other parts of the DNA. This interaction between genes will be a main theme of Chapter 7.

Before they can be involved in any kind of action, the code symbols of DNA have to be translated into another medium. They are first transcribed into exactly corresponding RNA symbols. RNA also has a four-letter alphabet. From here, they are translated into a different kind of polymer called a polypeptide or protein. It might be called a polyamino acid, because the basic units are amino acids. There are 20 kinds of amino acids in living cells. All biological proteins are chains made of these 20 basic building-blocks. Although a protein is a chain of amino acids, most of them don't remain long and stringy. Each chain coils up into a complicated knot, the precise shape of which is determined by the order of amino acids. This knot shape therefore never varies for any given sequence of amino acids. The sequence of amino acids in turn is precisely determined by the code symbols in a length of DNA (via RNA as an intermediary). There is a sense, therefore, in which the three-dimensional coiled shape of a protein is determined by the one-dimensional sequence of code symbols in the DNA.

The translation procedure embodies the celebrated three-letter 'genetic code'. This is a dictionary, in which each of the 64 ($4 \times 4 \times 4$) possible *triplets* of DNA (or RNA) symbols is translated into one of the 20 amino acids or a 'stop reading' symbol. There are three of these 'stop reading' punctuation marks. Many of the amino acids are coded by more than one triplet (as you might have guessed from the fact that there are 64 triplets and only 20 amino acids). The whole translation, from strictly sequential DNA ROM to precisely invariant three-dimensional protein shape, is a remarkable feat of digital information technology. Subsequent steps by which genes influence bodies are a little less obviously computer-like.

Every living cell, even a single bacterial cell, can be thought of as a gigantic chemical factory. DNA patterns, or genes, exert their effects by influencing the course of events in the chemical factory, and they do this via their influence on the three-dimensional shape of protein molecules. The word gigantic may seem surprising for a cell, especially when you remember that 10 million bacterial cells could sit on the surface of a pin's head. But you will also remember that each of these cells is capable of holding the whole text of the New Testament and, moreover, it *is* gigantic when measured by the number of sophisticated machines that it contains. Each machine is a large protein molecule, put together under the influence of a particular stretch of DNA. Protein molecules called enzymes are machines in the sense that each one causes a particular chemical reaction to take place. Each kind of protein machine churns out its own particular chemical product. To do this it uses raw materials that are drifting around in the cell, being, very probably, the products of other protein machines. To get an idea of the size of these protein machines, each one is made of about 6,000 atoms, which is very large by molecular standards. There are about a million of these large pieces of apparatus in a cell, and there are more than 2,000 different kinds of them, each kind specialized to do a particular operation in the chemical factory – the cell. It is the characteristic chemical products of such enzymes that give a cell its individual shape and behaviour.

Since all body cells contain the same genes, it might seem surprising that all body cells aren't the same as each other. The reason is that a different subset of genes is *read* in different kinds of cells, the others being ignored. In liver cells, those parts of the DNA ROM specifically relevant to the building of kidney cells are not read, and vice versa. The shape and behaviour of a cell depend upon which genes inside that cell are being read and translated into their protein products. This in turn depends on the chemicals already in the cell, which depends partly on which genes have previously been read in the cell, and partly on neighbouring cells. When one cell divides into two, the two daughter cells aren't necessarily the same as each other. In the original fertilized egg, for instance, certain chemicals congregate at one end of the cell, others at the other end. When such a polarized cell divides, the two daughter cells receive different chemical allocations. This means that different genes will be read in the two daughter cells, and a kind of self-reinforcing divergence gets going. The final shape of the whole body, the size of its limbs, the wiring up of its brain, the timing of its behaviour patterns, are all the indirect consequences of interactions between different kinds of cells, whose differences in their turn arise through different genes being read. These diverging processes are best thought of as locally autonomous in the manner of the 'recursive' procedure of Chapter 3, rather than as coordinated in some grand central design.

'Action', in the sense used in this chapter, is what a geneticist is talking about when he mentions the 'phenotypic effect' of a gene. DNA has effects upon bodies, upon eye colour, hair crinkliness, strength of aggressive behaviour and thousands of other attributes, all of which are called phenotypic effects. DNA exerts these effects initially locally, after being read by RNA and translated into protein chains, which then affect cell shape and behaviour. This is one of the two ways in which the information in the pattern of DNA can be read out. The other way is that it can be duplicated into a new DNA strand. This is the copying that we discussed earlier.

There is a fundamental distinction between these two routes of transmission of the DNA information, vertical and horizontal transmission. The information is transmitted vertically to other DNA in cells (that make other cells) that make sperms or eggs. Hence it is transmitted vertically to the next generation and then, vertically again, to an indefinite number of future generations. I shall call this 'archival DNA'. It is potentially immortal. The succession of cells along which archival DNA travels is called the germ line. The germ line is that set of cells, within a body, which is ancestral to sperms or eggs and hence ancestral to future generations. DNA is also transmitted *sideways* or horizontally: to DNA in non-germ-line cells such as liver cells or skin cells; within such cells to RNA, thence to protein and various effects on embryonic development and therefore on adult form and behaviour. You can think of horizontal transmission and vertical transmission as corresponding to the two sub-programs called DEVELOPMENT and REPRODUCTION in Chapter 3.

Natural selection is all about the differential success of rival DNA in getting itself transmitted vertically in the species archives. 'Rival DNA' means alternative contents of particular addresses in the chromosomes of the species. Some genes are more successful than rival genes at remaining in the archives. Although *vertical* transmission down the archives of the species is ultimately what 'success' means, the criterion for success is normally the *action* that the genes have on bodies, by means of their *sideways* transmission. This, too, is just like the biomorph computer model. For instance, suppose that in tigers there is a particular gene which, by means of its sideways influence in cells of the jaw, causes the teeth to be a little sharper than those that would be grown under the influence of a rival gene. A tiger with extra-sharp teeth can kill prey more efficiently than a normal tiger; hence it has more offspring; hence it passes on, vertically, more copies of the gene that makes sharp teeth. It passes on all its other genes at the same time, of course, but only the specific 'sharp-teeth gene' will find itself, *on average*, in the bodies of sharp-toothed tigers. The gene itself benefits, in terms of its vertical transmission, from the average effects that it has on a whole series of bodies.

DNA's performance as an archival medium is spectacular. In its capacity to preserve a message it far outdoes tablets of stone. Cows and pea plants (and, indeed, all the rest of us) have an almost identical gene called the histone H4 gene. The DNA text is 306 characters long. We can't say that it occupies the same addresses in all species, because we can't meaningfully compare address labels across species. But what we can say is that there is a length of 306 characters in cows, which is virtually identical to a length of 306 characters in peas. Cows and peas differ from each other in only two characters out of these 306. We don't know exactly how long ago the common ancestor of cows and peas lived, but fossil evidence suggests that it was somewhere between 1,000 and 2,000 million years ago. Call it 1.5 billion years ago. Over this unimaginably (for humans) long time, each of the two lineages that branched from that remote ancestor has preserved 305 out of the 306 characters (on average: it could be that one lineage has preserved all 306 of them and the other has preserved 304). Letters carved on gravestones become unreadable in mere hundreds of years.

In a way the conservation of the histone-H4 DNA document is even more impressive because, unlike tablets of stone, it is not the same physical structure that lasts and preserves the text. It is repeatedly being copied and recopied as the generations go by, like the Hebrew scriptures which were ritually copied by scribes every 80 years to forestall their wearing-out. It is hard to estimate exactly how many times the histone H4 document has been recopied in the lineage leading to cows from the common ancestor with peas, but it is probably as many as 20 billion times. It is also hard to find a yardstick with which to compare the preservation of more than 99 per cent of information in 20 billion successive copyings. We can try using a version of the game of grandmothers' whispers. Imagine 20 billion typists sitting in a row. The line of typists would reach right round the Earth 500 times. The first typist writes a page of a document and hands it to his neighbour. He copies it and hands his copy to the next one. He copies it again and hands it on to the next, and so on. Eventually, the message reaches the end of the line, and we read it (or rather our 12,000th great grandchildren do, assuming that all the typists have a speed typical of a good secretary). How faithful a rendering of the original message would it be?

To answer this we have to make some assumption about the accuracy of the typists. Let's twist the question round the other way. How good would each typist have to be, in order to match the DNA's performance? The answer is almost too ludicrous to express. For what it is worth, every typist would have to have an error rate of about one in a trillion; that is, he would have to be accurate enough to make only a single error in typing the Bible 250,000 times at a stretch. A good secretary in real life has an error rate of about one per page. This is about half a billion times the error rate of the histone H4 gene. A line of real-life secretaries would degrade a text to 99 per cent of its original letters by the 20th member of the line of 20 billion. By the 10,000th member of the line, less than 1 per cent of the original text would survive. This point of near total degradation would be reached before 99.9995 per cent of the typists had even seen it.

This whole comparison has been a bit of a cheat, but in an interesting and revealing respect. I gave the impression that what we are measuring is copying errors. But the histone H4 document hasn't just been copied, it has been subjected to natural selection. Histone is vitally important for survival. It is used in the structural engineering of chromosomes. Maybe lots more mistakes in *copying* the histone H4 gene occurred, but the mutant organisms did not survive, or at least did not reproduce. To make the comparison fair, we should have to assume that built into each typist's chair is a gun, wired up so that if he makes a mistake he is summarily shot, his place being taken by a reserve typist (squeamish readers may prefer to imagine a spring-loaded ejector seat gently catapulting miscreant typists out of the line, but the gun gives a more realistic picture of natural selection).

So, this method of measuring the conservatism of DNA, by looking at the number of changes that have actually occurred during geological time, compounds genuine copying fidelity with the filtering effects of natural selection. We see only the descendants of successful DNA changes. The ones that led to death are obviously not with us. Can we measure the actual copying fidelity on the ground, before natural selection gets to work on each new generation of genes? Yes, this is the inverse of what is known as the mutation rate, and it can be measured. The probability of any particular letter being miscopied on any one copying occasion turns out to be a little more than one in a billion. The difference between this, the mutation rate, and the lower rate at which change has actually been incorporated in the histone gene during evolution, is a measure of the effectiveness of natural selection in preserving this ancient document.

The histone gene's conservatism over the aeons is exceptional by genetic standards. Other genes change at a higher rate, presumably because natural selection is more tolerant of variations in them. For instance, genes coding the proteins known as fibrinopeptides change in evolution at a rate that closely approximates the basic mutation rate. This probably means that mistakes in the details of these proteins (they are produced during the clotting of blood) don't matter much for the organism. Haemoglobin genes have a rate of changing that is intermediate betwen histones and fibrinopeptides. Presumably natural selection's tolerance of their errors is intermediate. Haemoglobin is doing an important job in the blood, and its details really matter; but several alternative variants of it seem capable of doing the job equally well.

Here we have something that seems a little paradoxical, until we think about it further. The slowest-evolving molecules, like histones, turn out to be the ones that have been most subject to natural selection. Fibrinopeptides are the fastest-evolving molecules because natural selection almost completely ignores them. They are free to evolve at the mutation rate. The reason this seems paradoxical is that we place so much emphasis on natural selection as the driving force of evolution. If there is no natural selection, therefore, we might expect that there would be no evolution. Conversely, strong 'selection pressure', we could be forgiven for thinking, might be expected to lead to rapid evolution. Instead, what we find is that natural selection exerts a braking effect on evolution. The baseline rate of evolution, in the absence of natural selection, is the maximum possible rate. That is synonymous with the mutation rate.

This isn't really paradoxical. When we think about it carefully, we see that it couldn't be otherwise. Evolution by natural selection could not be faster than the mutation rate, for mutation is, ultimately, the only way in which new variation enters the species. All that natural selection can do is accept certain new variations, and reject others. The mutation rate is bound to place an upper limit on the rate at which evolution can proceed. As a matter of fact, most of natural selection is concerned with preventing evolutionary change rather than with driving it. This doesn't mean, I hasten to insist, that natural selection is a purely destructive process. It can construct too, in ways that Chapter 7 will explain.

Even the mutation rate is pretty slow. This is another way of saying that, even without natural selection, the performance of the DNA code in accurately preserving its archive is very impressive. A conservative estimate is that, in the absence of natural selection, DNA replicates so accurately that it takes five million replication generations to miscopy 1 per cent of the characters. Our hypothetical typists are still hopelessly outclassed by DNA, even if there is no natural selection. To match DNA with no natural selection, the typists would each have to be able to type the whole of the New Testament with only one error. That is, they would each have to be about 450 times more accurate than a typical real-life secretary. This is obviously much less than the comparable figure of half a billion, which is the factor by which the histone H4 gene *after natural selection* is more accurate than a typical secretary; but it is still a very impressive figure.

But I have been unfair to the typists. I assumed, in effect, that they are not capable of noticing their mistakes and correcting them. I have assumed a complete absence of proofreading. In reality, of course, they do proofread. My line of billions of typists wouldn't, therefore, cause the original message to degenerate in quite the simple way that I portrayed. The DNA-copying mechanism does the same kind of error-correction automatically. If it didn't, it wouldn't achieve anything like the stupendous accuracy that I have described. The DNA-copying procedure incorporates various 'proofreading' drills. This is all the more necessary because the letters of the DNA code are by no means static, like hieroglyphs carved in granite. On the contrary, the molecules involved are so small – remember all those New Testaments fitting on a pin's head – that they are under constant assault from the ordinary jostling of molecules that goes on due to heat. There is a constant flux, a turnover of letters in the message. About 5,000 DNA letters degenerate per day in every human cell, and are immediately replaced by repair mechanisms. If the repair mechanisms weren't there and ceaselessly working, the message would steadily dissolve. Proofreading of newly copied text is just a special case of normal repair work. It is mainly proofreading that is responsible for DNA's remarkable accuracy and fidelity of information storage.

We have seen that DNA molecules are the centre of a spectacular information technology. They are capable of packing an immense amount of precise, digital information into a very small space; and they are capable of preserving this information – with astonishingly few errors, but still some errors – for a very long time, measured in millions of years. Where are these facts leading us? They are leading us in the direction of a central truth about life on Earth, the truth that I alluded to in my opening paragraph about willow seeds. This is that living organisms exist for the benefit of DNA rather than the other way around. This won't be obvious yet, but I hope to persuade you of it. The messages that DNA molecules contain are all but eternal when seen against the time scale of individual lifetimes. The lifetimes of DNA messages (give or take a few mutations) are measured in units ranging from millions of years to hundreds of millions of years; or, in other words, ranging from 10,000 individual lifetimes to a trillion individual lifetimes. Each individual organism should be seen as a temporary vehicle, in which DNA messages spend a tiny fraction of their geological lifetimes.

The world is full of things that exist . . . ! No disputing that, but is it going to get us anywhere? Things exist either because they have recently come into existence or because they have qualities that made them unlikely to be destroyed in the past. Rocks don't come into existence at a high rate, but once they exist they are hard and durable. If they were not they wouldn't be rocks, they would be sand. Indeed, some of them are, which is why we have beaches! It is the ones that happen to be durable that exist as rocks. Dewdrops, on the other hand, exist, not because they are durable, but because they have only just come into existence and have not yet had time to evaporate. We seem to have two kinds of 'existenceworthiness': the dewdrop kind, which can be summed up as 'likely to come into existence but not very durable'; and the rock kind, which can be summed up as 'not very likely to come into existence but likely to last for a long time once there'. Rocks have durability and dewdrops have 'generatability'. (I've tried to think of a less ugly word but I can't.)

DNA gets the best of both worlds. DNA molecules themselves, as physical entities, are like dewdrops. Under the right conditions they come into existence at a great rate, but no one of them has existed for long, and all will be destroyed within a few months. They are not durable like rocks. But the *patterns* that they bear in their sequences are as durable as the hardest rocks. They have what it takes to exist for millions of years, and that is why they are still here today. The essential difference from dewdrops is that new dewdrops are not begotten by old dewdrops. Dewdrops doubtless resemble other dewdrops, but they don't specifically resemble their own 'parent' dewdrops. Unlike DNA molecules, they don't form lineages, and therefore can't pass on messages. Dewdrops come into existence by spontaneous generation, DNA messages by replication.

Truisms like 'the world is full of things that have what it takes to be in the world' are trivial, almost silly, until we come to apply them to a special kind of durability, durability in the form of lineages of multiple copies. DNA messages have a different kind of durability from that of rocks, and a different kind of generatability from that of dewdrops. For DNA molecules, 'what it takes to be in the world' comes to have a meaning that is anything but obvious and tautological. 'What it takes to be in the world' turns out to include the ability to build machines like you and me, the most complicated things in the known universe. Let us see how this can be so.

Fundamentally, the reason is that the properties of DNA that we have identified turn out to be the basic ingredients necessary for any process of cumulative selection. In our computer models in Chapter 3, we deliberately built into the computer the basic ingredients of cumulative selection. If cumulative selection is really to happen in the world, some entities have got to arise whose properties constitute those basic ingredients. Let us look, now, at what those ingredients are. As we do so, we shall keep in mind the fact that these very same ingredients, at least in some rudimentary form, must have arisen spontaneously on the early Earth, otherwise cumulative selection, and therefore life, would never have got started in the first place. We are talking here not specifically about DNA, but about the basic ingredients needed for life to arise anywhere in the universe.

When the prophet Ezekiel was in the valley of bones he prophesied to the bones and made them join up together. Then he prophesied to them and made flesh and sinews come around them. But still there was no breath in them. The vital ingredient, the ingredient of life, was missing. A dead planet has atoms, molecules and larger lumps of matter, jostling and nestling against each other at random, according to the laws of physics. Sometimes the laws of physics cause the atoms and molecules to join up together like Ezekiel's dry bones, sometimes they cause them to split apart. Quite large accretions of atoms can form, and they can crumble and break apart again. But still there is no breath in them.

Ezekiel called upon the four winds to put living breath into the dry bones. What is the vital ingredient that a dead planet like the early Earth must have, if it is to have a chance of eventually coming alive, as our planet did? It is not breath, not wind, not any kind of elixir or potion. It is not a substance at all, it is a *property*, the property of self-replication. This is the basic ingredient of cumulative selection. There must somehow, as a consequence of the ordinary laws of physics, come into being *self-copying* entities or, as I shall call them, *replicators*. In modern life this role is filled, almost entirely, by DNA molecules, but anything of which copies are made would do. We may suspect that the first replicators on the primitive Earth were not DNA molecules. It is unlikely that a fully fledged DNA molecule would spring into existence without the aid of other molecules that normally exist only in living cells. The first replicators were probably cruder and simpler than DNA.

There are two other necessary ingredients, which will normally arise automatically from the first ingredient, self-replication itself. There must be occasional errors in the self-copying; even the DNA system very occasionally makes mistakes, and it seems likely that the first replicators on Earth were much more erratic. And at least some of the replicators should exert *power* over their own future. This last ingredient sounds more sinister than it actually is. All it means is that some properties of the replicators should have an influence over their probability of being replicated. At least in a rudimentary form, this is likely to be an inevitable consequence of the basic facts of self-replication itself.

Each replicator, then, has copies of itself made. Each copy is the same as the original, and has the same properties as the original. Among these properties, of course, is the property of making (sometimes with errors) *more* copies of itself. So each replicator is potentially the 'ancestor' of an indefinitely long line of descendant replicators, stretching into the distant future, and branching to produce, potentially, an exceedingly large number of descendant replicators. Each new copy must be made from raw materials, smaller building blocks knocking around. Presumably the replicators act as some kind of mould or template. Smaller components fall together into the mould in such a way that a duplicate of the mould is made. Then the duplicate breaks free and is able to act as a mould in its own right. Hence we have a potentially growing *population* of replicators. The population will not grow indefinitely, because eventually the supply of raw materials, the smaller elements that fall into the moulds, will become limiting.

Now we introduce our second ingredient into the argument. Sometimes the copying will not be perfect. Mistakes will happen. The possibility of errors can never be totally eliminated from any copying process, although their probability can be reduced to low levels. This is what the manufacturers of hi-fi equipment are striving towards all the time, and the DNA-replication process, as we have seen, is spectactulary good at reducing errors. But modern DNA replication is a high-technology affair, with elaborate proofreading techniques that have been perfected over many generations of cumulative selection. As we have seen, the first replicators probably were relatively crude, low-fidelity contraptions in comparison.

Now go back to our population of replicators, and see what the effect of erratic copying will be. Obviously, instead of there being a uniform population of identical replicators, we shall have a mixed population. Probably many of the products of erratic copying will be found to have lost the property of self-replication that their 'parent' had. But a few will retain the property of self-replication, while being different from the parent in some other respect. So we shall have copies of errors being duplicated in the population.

When you read the word 'error', banish from your mind all pejorative associations. It simply means an error from the point of view of high-fidelity copying. It is possible for an error to result in an improvement. I dare say many an exquisite new dish has been created because a cook made a mistake while trying to follow a recipe. Insofar as I can claim to have had any original scientific ideas, these have sometimes been misunderstandings, or misreadings, of other peoples' ideas. To return to our primeval replicators, while most miscopyings probably resulted in diminished copying effectiveness, or total loss of the self-copying property, a few might actually have turned out to be *better* at self-replication than the parent replicator that gave rise to them.

What does 'better' mean? Ultimately it means more efficient at self-replication, but what might this mean in practice? This brings us to our third 'ingredient'. I referred to this as 'power', and you'll see why in a moment. When we discussed replication as a moulding process, we saw that the last step in the process must be the new copy's breaking free of the old mould. The time that this occupies may be influenced by a property which I shall call the 'stickiness' of the old mould. Suppose that in our population of replicators, which vary because of old copying errors back in their 'ancestry', some varieties happen to be more sticky than others. A very sticky variety clings to each new copy for an average time of more than an hour before it finally breaks free and the process can begin again. A less-sticky variety lets go of each new copy within a split second of its formation. Which of these two varieties will come to predominate in the population of replicators? There is no doubt about the answer. If this is the only property by which the two varieties differ, the sticky one is bound to become far less numerous in the population. The non-sticky one is churning out copies of non-sticky ones at thousands of times the rate that the sticky one is making copies of sticky ones. Varieties of intermediate stickiness will have intermediate rates of self-propagation. There will be an 'evolutionary trend' towards reduced stickiness.

Something like this kind of elementary natural selection has been duplicated in the test-tube. There is a virus called Q-beta which lives as a parasite of the gut bacterium *Escherichia coli*. Q-beta has no DNA but it does contain, indeed it largely consists of, a single strand of the related molecule RNA. RNA is capable of being replicated in a similar way to DNA.

In the normal cell, protein molecules are assembled to the specification of RNA plans. These are working copies of plans, run off from the DNA masters held in the cell's precious archives. But it is theoretically possible to build a special machine – a protein molecule like the rest of the cellular machines – that runs off RNA copies from other RNA copies. Such a machine is called an RNA-replicase molecule. The bacterial cell itself normally has no use for these machines, and doesn't build any. But since the replicase is just a protein molecule like any other, the versatile protein-building machines of the bacterial cell can easily turn to building them, just as the machine tools in a car factory can quickly be turned over in time of war to making munitions: all they need is to be fed the right blueprints. This is where the virus comes in.

The business part of the virus is an RNA plan. Superficially it is indistinguishable from any of the other RNA working blueprints that are floating around, after being run off the bacterium's DNA master. But if you read the small print of the viral RNA you will find something devilish written there. The letters spell out a plan for making RNA-replicase, for making machines that make more copies of the very same RNA plans, that make more machines that make more copies of the plans, that make more . . .

So the factory is hijacked by these self-interested blueprints. In a sense it was crying out to be hijacked. If you fill your factory with machines so sophisticated that they can make anything that any blueprint tells them to make, it is hardly surprising if sooner or later a blueprint arises that tells these machines to make copies of itself. The factory fills up with more and more of these rogue machines, each churning out rogue blueprints for making more machines that will make more of themselves. Finally, the unfortunate bacterium bursts and releases millions of viruses that infect new bacteria. So much for the normal life cycle of the virus in nature.

I have called RNA-replicase and RNA respectively a machine and a blueprint. So they are, in a sense (to be disputed on other grounds in a later chapter), but they are also molecules, and it is possible for human chemists to purify them, bottle them and store them on a shelf. This is what Sol Spiegelman and his colleagues did in America in the 1960s. Then they put the two molecules together in solution, and a fascinating thing happened. In the test-tube, the RNA molecules acted as templates for the synthesis of copies of themselves, aided by the presence of the RNA-replicase. The machine tools and the blueprints had been extracted and put into cold storage, separately from one another. Then, as soon as they were given access to each other, and also to the small molecules needed as raw materials, in water, both got back to their old tricks even though they were no longer in a living cell but in a test tube.

It is but a short step from this to natural selection and evolution in the laboratory. It is just a chemical version of the computer biomorphs. The experimental method is basically to lay out a long row of test-tubes each containing a solution of RNA-replicase, and also of raw materials, small molecules that can be used for RNA synthesis. Each test-tube contains the machine tools and the raw material, but so far it is sitting idle, doing nothing because it lacks a blueprint to work from. Now a tiny amount of RNA itself is dropped into the first test-tube. The replicase apparatus immediately gets to work and manufactures lots of copies of the newly introduced RNA molecules, which spread through the test-tube. Now a drop of the solution in the first test-tube is removed and put into the second test-tube. The process repeats itself in the second test-tube and then a drop is removed and used to seed the third test-tube, and so on.

Occasionally, because of random copying errors, a slightly different, mutant RNA molecule spontaneously arises. If, for any reason, the new variety is competitively superior to the old one, superior in the sense that, perhaps because of its low 'stickiness', it gets itself replicated faster or otherwise more effectively, the new variety will obviously spread through the test-tube in which it arose, outnumbering the parental type that gave rise to it. Then, when a drop of solution is removed from that test-tube to seed the next test-tube, it will be the new mutant variety that does the seeding. If we examine the RNAs in a long succession of test-tubes, we see what can only be called evolutionary change. Competitively superior varieties of RNA produced at the end of several test-tube 'generations' can be bottled and named for future use. One variety for example, called V2, replicates much more rapidly than normal Q-beta RNA, probably because it is smaller. Unlike Q-beta RNA, it doesn't have to 'bother' to contain the plans for making replicase. Replicase is provided free by the experimenters. V2 RNA was used as the starting point for an interesting experiment by Leslie Orgel and his colleagues in California, in which they imposed a 'difficult' environment.

They added to their test-tubes a poison called ethidium bromide which inhibits the synthesis of RNA: it gums up the works of the machine tools. Orgel and colleagues began with a weak solution of the poison. At first, the rate of synthesis was slowed down by the poison, but after evolving through about nine test-tube transfer 'generations', a new strain of RNA that was resistant to the poison had been selected. Rate of RNA synthesis was now comparable to that of normal V2 RNA in the absence of poison. Now Orgel and his colleagues doubled the concentration of poison. Again the rate of RNA replication dropped, but after another 10 or so test-tube transfers a strain of RNA had

evolved that was immune even to the higher concentration of poison.
Then the concentration of the poison was doubled again. In this way,
by successive doublings, they managed to evolve a strain of RNA that
could self-replicate in very high concentrations of ethidium bromide,
10 times as concentrated as the poison that had inhibited the original
ancestral V2 RNA. They called the new, resistant RNA V40. The
evolution of V40 from V2 took about 100 test-tube transfer
'generations' (of course, many actual RNA-replication generations go
on between each test-tube transfer).

Orgel has also done experiments in which no enzyme was provided.
He found that RNA molecules can replicate themselves spontaneously
under these conditions, albeit very slowly. They seem to need some
other catalyzing substance, such as zinc. This is important because, in
the early days of life when replicators first arose, we cannot suppose
that there were enzymes around to help them to replicate. There
probably was zinc, though.

The complementary experiment was carried out a decade ago in the
laboratory of the influential German school working on the origin of
life under Manfred Eigen. These workers provided replicase and RNA
building blocks in the test-tube, but they did *not* seed the solution
with RNA. Nevertheless, a particular large RNA molecule evolved
spontaneously in the test-tube, and the same molecule re-evolved
itself again and again in subsequent independent experiments! Careful
checking showed that there was no possibility of chance infection by
RNA molecules. This is a remarkable result when you consider the
statistical improbability of the same large molecule spontaneously
arising twice. It is very much more improbable than the spontaneous
typing of METHINKS IT IS LIKE A WEASEL. Like that phrase in our
computer model, the particular favoured RNA molecule was built up
by gradual, *cumulative* evolution.

The variety of RNA produced, repeatedly, in these experiments was
of the same size and structure as the molecules that Spiegelman had
produced. But whereas Spiegelman's had evolved by 'degeneration'
from naturally occurring, larger, Q-beta viral RNA, those of the Eigen
group had built themselves up from almost nothing. This particular
formula is well adapted to an environment consisting of test-tubes
provided with ready-made replicase. It therefore is converged upon by
cumulative selection from two very different starting points. The
larger, Q-beta RNA molecules are less well adapted to a test-tube
environment but better adapted to the environment provided by *E.coli*
cells.

Experiments such as these help us to appreciate the entirely automatic and non-deliberate nature of natural selection. The replicase 'machines' don't 'know' why they make RNA molecules: it is just a byproduct of their shape that they do. And the RNA molecules themselves don't work out a strategy for getting themselves duplicated. Even if they could think, there is no obvious reason why any thinking entity should be motivated to make copies of itself. If I knew how to make copies of myself, I'm not sure that I would give the project high priority in competition with all the other things I want to do: why should I? But motivation is irrelevant for molecules. It is just that the structure of the viral RNA *happens* to be such that it makes cellular machinery churn out copies of itself. And if any entity, anywhere in the universe, happens to have the property of being good at making more copies of itself, then automatically more and more copies of that entity *will* obviously come into existence. Not only that but, since they automatically form lineages and are occasionally miscopied, later versions tend to be 'better' at making copies of themselves than earlier versions, because of the powerful processes of cumulative selection. It is all utterly simple and automatic. It is so predictable as to be almost inevitable.

A 'successful' RNA molecule in a test-tube is successful because of some direct, intrinsic property of itself, something analogous to the 'stickiness' of my hypothetical example. But properties like 'stickiness' are rather boring. They are elementary properties of the replicator itself, properties that have a direct effect on its probability of being replicated. What if the replicator has some effect upon something else, which affects something else, which affects something else, which . . . eventually, indirectly affects the replicator's chance of being replicated? You can see that, if long chains of causes like this existed, the fundamental truism would still hold. Replicators that happen to have what it takes to get replicated would come to predominate in the world, *no matter how long and indirect* the chain of causal links by which they influence their probability of being replicated. And, by the same token, the world will come to be filled with the links in this causal chain. We shall see those links, and marvel at them.

In modern organisms we see them all the time. They are eyes and skins and bones and toes and brains and instincts. These things are the tools of DNA replication. They are caused by DNA, in the sense that differences in eyes, skins, bones, instincts, etc. are caused by differences in DNA. They exert an influence over the replication of the DNA that caused them, in that they affect the survival and reproduction of their bodies – which contain that same DNA, and whose fate is therefore shared by the DNA. Therefore, the DNA itself exerts an influence over its own replication, via the attributes of bodies. DNA can be said to exert power over its own future, and bodies and their organs and behaviour patterns are the instruments of that power.

When we talk about power, we are talking about consequences of replicators that affect their own future, however indirect those consequences might be. It doesn't matter how many links there are in the chain from cause to effect. If the cause is a self-replicating entity, the effect, be it ever so distant and indirect, can be subject to natural selection. I shall summarize the general idea by telling a particular story about beavers. In detail it is hypothetical, but it certainly cannot be far from the truth. Although nobody has done research upon the development of brain connections in the beaver, they have done this kind of research on other animals, like worms. I am borrowing the conclusions and applying them to beavers, because beavers are more interesting and congenial to many people than worms.

A mutant gene in a beaver is just a change in one letter of the billion-letter text; a change in a particular gene G. As the young beaver grows, the change is copied, together with all the other letters in the text, into all the beaver's cells. In most of the cells the gene G is not read; other genes, relevant to the workings of the other cell types, are. G is read, however, in some cells in the developing brain. It is read and transcribed into RNA copies. The RNA working copies drift around the interior of the cells, and eventually some of them bump into protein-making machines called ribosomes. The protein-making machines read the RNA working plans, and turn out new protein molecules to their specification. These protein molecules curl up into a particular shape determined by their own amino-acid sequence, which in turn is governed by the DNA code sequence of the gene G. When G mutates, the change makes a crucial difference to the amino-acid sequence normally specified by the gene G, and hence to the coiled-up shape of the protein molecule.

These slightly altered protein molecules are mass-produced by the protein-making machines inside the developing brain cells. They in turn act as enzymes, machines that manufacture other compounds in the cells, the gene products. The products of the gene G find their way into the membrane surrounding the cell, and are involved in the processes whereby the cell makes connections with other cells. Because of the slight alteration in the original DNA plans, the production-rate of certain of these membrane compounds is changed. This in turn changes the way in which certain developing brain cells connect up with one another. A subtle alteration in the wiring diagram of a particular part of the beaver's brain has occurred, the indirect, indeed far-removed, consequence of a change in the DNA text.

Now it happens that this particular part of the beaver's brain, because of its position in the total wiring diagram, is involved in the beaver's dam-building behaviour. Of course, large parts of the brain are involved whenever the beaver builds a dam but, when the G mutation affects this particular part of the brain's wiring diagram, the change has a specific effect on the behaviour. It causes the beaver to hold its head higher in the water while swimming with a log in its jaws. Higher, that is, than a beaver without the mutation. This makes it a little less likely that mud, attached to the log, will wash off during the journey. This increases the stickiness of the log, which in turn means that, when the beaver thrusts it into the dam, the log is more likely to stay there. This will tend to apply to all the logs placed by any beaver bearing this particular mutation. The increased stickiness of the logs is a consequence, again a very indirect consequence, of an alteration in the DNA text.

The increased stickiness of the logs makes the dam a sounder structure, less likely to break up. This in turn increases the size of the lake created by the dam, which makes the lodge in the centre of the lake more secure against predators. This tends to increase the number of offspring successfully reared by the beaver. If we look at the whole population of beavers, those that possess the mutated gene will, on average, tend therefore to rear more offspring than those not possessing the mutated gene. Those offspring will tend to inherit archive copies of the self-same altered gene from their parents. Therefore, in the population, this form of the gene will become more numerous as the generations go by. Eventually it will become the norm, and will no longer deserve the title 'mutant'. Beaver dams in general will have improved another notch.

The fact that this particular story is hypothetical, and that the details may be wrong, is irrelevant. The beaver dam evolved by natural selection, and therefore what happened cannot be very different, except in practical details, from the story I have told. The general implications of this view of life are explained and elaborated in my book *The Extended Phenotype*, and I shan't repeat the arguments here. You will notice that in this hypothetical story there were no fewer than 11 links in the causal chain linking altered gene to improved survival. In real life there might be even more. Every one of those links, whether it is an effect on the chemistry inside a cell, a later effect on how brain cells wire themselves together, an even later effect on behaviour, or a final effect on lake size, is correctly regarded as *caused* by a change in the DNA. It wouldn't matter if there were 111 links. Any effect that a change in a gene has on its own replication probability is fair game for natural selection. It is all perfectly simple, and delightfully automatic and unpremeditated. Something like it is well-nigh inevitable, once the fundamental ingredients of cumulative selection – replication, error and power – have come into existence in the first place. But how did this happen? How did they come into existence on Earth, before life was there? We shall see how this difficult question might be answered, in the next chapter.

CHAPTER 6

ORIGINS AND MIRACLES

Chance, luck, coincidence, miracle. One of the main topics of this chapter is miracles and what we mean by them. My thesis will be that events that we commonly call miracles are not supernatural, but are part of a spectrum of more-or-less improbable natural events. A miracle, in other words, if it occurs at all, is a tremendous stroke of luck. Events don't fall neatly into natural events *versus* miracles.

There are some would-be events that are too improbable to be contemplated, but we can't know this until we have done a calculation. And to do the calculation, we must know how much *time* was available, more generally how many *opportunities* were available, for the event to occur. Given infinite time, or infinite opportunities, anything is possible. The large numbers proverbially furnished by astronomy, and the large timespans characteristic of geology, combine to turn topsy-turvy our everyday estimates of what is expected and what is miraculous. I shall build up to this point using a specific example which is the other main theme of this chapter. This example is the problem of how life originated on Earth. To make the point clearly, I shall arbitrarily concentrate on one particular theory of the origin of life, although any one of the modern theories would have served the purpose.

We can accept a certain amount of luck in our explanations, but not too much. The question is, *how* much? The immensity of geological time entitles us to postulate more improbable coincidences than a court of law would allow but, even so, there are limits. Cumulative selection is the key to all our modern explanations of life. It strings a series of acceptably lucky events (random mutations) together in a nonrandom sequence so that, at the end of the sequence, the finished product carries the illusion of being very very lucky indeed, far too improbable to have come about by chance alone, even given a timespan millions of times longer than the age of the universe so far. Cumulative selection is the key but it had to get started, and we cannot escape the need to postulate a *single-step* chance event in the origin of cumulative selection itself.

And that vital first step was a difficult one because, at its heart, there lies what seems to be a paradox. The replication processes that we know seem to need complicated machinery to work. In the presence of a replicase 'machine tool', fragments of RNA will evolve, repeatedly and convergently, towards the same endpoint, an endpoint whose 'probability' seems vanishingly small until you reflect on the power of cumulative selection. But we have to assist this cumulative selection to get started. It won't go unless we provide a catalyst, such as the replicase 'machine tool' of the previous chapter. And that catalyst, it seems, is unlikely to come into existence spontaneously, except under the direction of other RNA molecules. DNA molecules replicate in the complicated machinery of the cell, and written words replicate in Xerox machines, but neither seem capable of spontaneous replication in the absence of their supporting machinery. A Xerox machine is capable of copying its own blueprints, but it is not capable of springing spontaneously into existence. Biomorphs readily replicate in the environment provided by a suitably written computer program, but they can't write their own program or build a computer to run it. The theory of the blind watchmaker is extremely powerful given that we are allowed to assume replication and hence cumulative selection. But if replication needs complex machinery, since the only way we know for complex machinery ultimately to come into existence is cumulative selection, we have a problem.

Certainly the modern cellular machinery, the apparatus of DNA replication and protein synthesis, has all the hallmarks of a highly evolved, specially fashioned machine. We have seen how staggeringly impressive it is as an accurate data storage device. At its own level of ultra-miniaturization, it is of the same order of elaborateness and complexity of design as the human eye is at a grosser level. All who have given thought to the matter agree that an apparatus as complex as the human eye could not possibly come into existence through single-step selection. Unfortunately, the same seems to be true of at least parts of the apparatus of cellular machinery whereby DNA replicates itself, and this applies not just to the cells of advanced creatures like ourselves and amoebas, but also to relatively more primitive creatures like bacteria and blue-green algae.

So, cumulative selection can manufacture complexity while single-step selection cannot. But cumulative selection cannot work unless there is some minimal machinery of replication and replicator power, and the only machinery of replication that we know seems too complicated to have come into existence by means of anything less than many generations of cumulative selection! Some people see this as a fundamental flaw in the whole theory of the blind watchmaker. They see it as the ultimate proof that there must originally have been a designer, not a *blind* watchmaker but a far-sighted supernatural watchmaker. Maybe, it is argued, the Creator does not control the day-to-day succession of evolutionary events; maybe he did not frame the tiger and the lamb, maybe he did not make a tree, but he *did* set up the original machinery of replication and replicator power, the original machinery of DNA and protein that made cumulative selection, and hence all of evolution, possible.

This is a transparently feeble argument, indeed it is obviously self-defeating. Organized complexity is the thing that we are having difficulty in explaining. Once we are allowed simply to *postulate* organized complexity, if only the organized complexity of the DNA/protein replicating engine, it is relatively easy to invoke it as a generator of yet more organized complexity. That, indeed, is what most of this book is about. But of course any God capable of intelligently designing something as complex as the DNA/protein replicating machine must have been at least as complex and organized as that machine itself. Far more so if we suppose him *additionally* capable of such advanced functions as listening to prayers and forgiving sins. To explain the origin of the DNA/protein machine by invoking a supernatural Designer is to explain precisely nothing, for it leaves unexplained the origin of the Designer. You have to say something like 'God was always there', and if you allow yourself that kind of lazy way out, you might as well just say 'DNA was always there', or 'Life was always there', and be done with it.

The more we can get away from miracles, major improbabilities, fantastic coincidences, large chance events, and the more thoroughly we can break large chance events up into a cumulative series of small chance events, the more satisfying to rational minds our explanations will be. But in this chapter we are asking *how* improbable, how *miraculous*, a single event we are allowed to postulate. What is the largest single event of sheer naked coincidence, sheer unadulterated miraculous luck, that we are allowed to get away with in our theories, and still say that we have a satisfactory explanation of life? In order for a monkey to write 'Methinks it is like a weasel' by chance, it needs a very large amount of luck, but it is still measurable. We calculated the odds against it as about 10 thousand million million million million million million (10^{40}) to 1 against. Nobody can really comprehend or imagine such a large number, and we just think of this degree of improbability as synonymous with impossible. But although we can't comprehend these levels of improbability in our minds, we shouldn't just run away from them in terror. The number 10^{40} may be very large but we can still write it down, and we can still use it in calculations. There are, after all, even larger numbers: 10^{46}, for instance, is not just larger; you must add 10^{40} to itself a million times in order to obtain 10^{46}. What if we could somehow muster a gang of 10^{46} monkeys each with its own typewriter? Why, lo and behold, one of them would solemnly type 'Methinks it is like a weasel', and another would almost certainly type 'I think therefore I am'. The problem is, of course, that we couldn't assemble that many monkeys. If all the matter in the universe were turned into monkey flesh, we still couldn't get enough monkeys. The miracle of a monkey typing 'Methinks it is like a weasel' is quantitatively too great, *measurably* too great, for us to admit it to our theories about what actually happens. But we couldn't know this until we sat down and did the calculation.

So, there are some levels of sheer luck, not only too great for puny human imaginations, but too great to be allowed in our hard-headed calculations about the origin of life. But, to repeat the question, how great a level of luck, how much of a miracle, *are* we allowed to postulate? Don't let's run away from this question just because large numbers are involved. It is a perfectly valid question, and we can at least write down what we would need to know in order to calculate the answer.

Now here is a fascinating thought. The answer to our question – of how much luck we are allowed to postulate – depends upon whether our planet is the only one that has life, or whether life abounds all around the universe. The one thing we know for certain is that life has arisen once, here on this very planet. But we have no idea at all whether there is life anywhere else in the universe. It is entirely possible that there isn't. Some people have calculated that there must be life elsewhere, on the following grounds (I won't point out the fallacy until afterwards). There are probably at least 10^{20} (i.e. 100 billion billion) roughly suitable planets in the universe. We know that life has arisen here, so it can't be *all* that improbable. Therefore it is almost inescapable that at least some among all those billions of billions of other planets have life.

The flaw in the argument lies in the inference that, *because life has arisen here*, it can't be too terribly improbable. You will notice that this inference contains the built-in assumption that whatever went on on Earth is likely to have gone on elsewhere in the universe, and this begs the whole question. In other words, that kind of statistical argument, that there must be life elsewhere in the universe because there is life here, builds in, as an assumption, what it is setting out to prove. This doesn't mean that the conclusion that life exists all around the universe is necessarily wrong. My guess is that it is probably right. It simply means that that particular argument that led up to it is no argument at all. It is just an assumption.

Let us, for the sake of discussion, entertain the alternative assumption that life has arisen only once, ever, and that was here on Earth. It is tempting to object to this assumption on the following emotional grounds. Isn't there something terribly medieval about it? Doesn't it recall the time when the church taught that our Earth was the centre of the universe, and the stars just little pinpricks of light set in the sky for our delight (or, even more absurdly presumptuous, that the stars go out of their way to exert astrological influences on our little lives)? How very conceited to assume that, out of all the billions of billions of planets in the universe, our own little backwater of a world, in our own local backwater of a solar system, in our own local backwater of a galaxy, should have been singled out for life? Why, for goodness sake, should it have been *our* planet?

I am genuinely sorry, for I am heartily thankful that we have escaped from the small-mindedness of the medieval church and I despise modern astrologers, but I am afraid that the rhetoric about backwaters in the previous paragraph is just empty rhetoric. It is *entirely* possible that our backwater of a planet is literally the only one that has ever borne life. The point is that if there *were* only one planet that had ever borne life, then it would *have* to be our planet, for the very good reason that 'we' are here discussing the question! If the origin of life *is* such an improbable event that it happened on only one planet in the universe, then our planet has to be that planet. So, we can't use the fact that Earth has life to conclude that life must be probable enough to have arisen on another planet. Such an argument would be circular. We have to have some independent arguments about how easy or difficult it is for life to originate on a planet, before we can even begin to answer the question of how many other planets in the universe have life.

But that isn't the question we set out with. Our question was, how much luck are we allowed to assume in a theory of the origin of life on Earth? I said that the answer depends upon whether life has arisen only once, or many times. Begin by giving a name to the probability, however low it is, that life will originate on any randomly designated planet of some particular type. Call this number the spontaneous generation probability or SGP. It is the SGP that we shall arrive at if we sit down with our chemistry textbooks, or strike sparks through plausible mixtures of atmospheric gases in our laboratory, and calculate the odds of replicating molecules springing spontaneously into existence in a typical planetary atmosphere. Suppose that our best guess of the SGP is some very very small number, say one in a billion. This is obviously such a small probability that we haven't the faintest hope of duplicating such a fantastically lucky, miraculous event as the origin of life in our laboratory experiments. Yet if we assume, as we are perfectly entitled to do for the sake of argument, that life has originated only once in the universe, it follows that we are *allowed* to postulate a very large amount of luck in a theory, because there are so many planets in the universe where life *could* have originated. If, as one estimate has it, there are 100 billion billion planets, this is 100 billion times greater than even the very low SGP that we postulated. To conclude this argument, the maximum amount of luck that we are allowed to assume, before we reject a particular theory of the origin of life, has odds of one in N, where N is the number of suitable planets in the universe. There is a lot hidden in that word 'suitable', but let us put an upper limit of 1 in 100 billion billion for the maximum amount of luck that this argument entitles us to assume.

Think about what this means. We go to a chemist and say: get out your textbooks and your calculating machine; sharpen your pencil and your wits; fill your head with formulae, and your flasks with methane and ammonia and hydrogen and carbon dioxide and all the other gases that a primeval nonliving planet can be expected to have; cook them all up together; pass strokes of lightning through your simulated atmospheres, and strokes of inspiration through your brain; bring all your clever chemist's methods to bear, and give us your best chemist's estimate of the probability that a typical planet will spontaneously generate a self-replicating molecule. Or, to put it another way, how long would we have to wait before random chemical events on the planet, random thermal jostling of atoms and molecules, resulted in a self-replicating molecule?

Chemists don't know the answer to this question. Most modern chemists would probably say that we'd have to wait a long time by the standards of a human lifetime, but perhaps not all that long by the standards of cosmological time. The fossil history of earth suggests that we have about a billion years – one 'aeon', to use a convenient modern definition – to play with, for this is roughly the time that elapsed between the origin of the Earth about 4.5 billion years ago and the era of the first fossil organisms. But the point of our 'numbers of planets' argument is that, even if the chemist said that we'd have to wait for a 'miracle', have to wait a billion billion years – far longer than the universe has existed, we can still accept this verdict with equanimity. There are probably more than a billion billion available planets in the universe. If each of them lasts as long as Earth, that gives us about a billion billion billion planet years to play with. That will do nicely! A miracle is translated into practical politics by a multiplication sum.

There is a concealed assumption in this argument. Well, actually there are lots, but there's one in particular that I want to talk about. This is that, once life (i.e. replicators and cumulative selection) originates at all, it always advances to the point where its creatures evolve enough intelligence to speculate about their origins. If this is not so, our estimate of the amount of luck that we are allowed to postulate must be reduced accordingly. To be more precise, the maximum odds against the origin of life on any one planet that our theories are allowed to postulate, is the number of available planets in the universe divided by the odds that life, once started, will evolve sufficient intelligence to speculate about its own origins.

It may seem a little strange that 'sufficient intelligence to speculate about its own origins' is a relevant variable. To understand why it is, consider an alternative assumption. Suppose that the origin of life was quite a probable event, but the subsequent evolution of intelligence was exceedingly improbable, demanding a huge stroke of luck. Suppose the origin of intelligence is so improbable that it has happened on only one planet in the universe, even though life has started on many planets. Then, since we know we are intelligent enough to discuss the question, we know that Earth must be that one planet. Now suppose that the origin of life, *and* the origin of intelligence given that life is there, are *both* highly improbable events. Then the probability of any one planet, such as Earth, enjoying both strokes of luck is the *product* of the two low probabilities, and this is a far smaller probability.

It is as though, in our theory of how we came to exist, we are allowed to postulate a certain ration of luck. This ration has, as its upper limit, the number of eligible planets in the universe. Given our ration of luck, we can then 'spend' it as a limited commodity over the course of our explanation of our own existence. If we use up almost all our ration of luck in our theory of how life gets started on a planet in the first place, then we are allowed to postulate very little more luck in subsequent parts of our theory, in, say, the cumulative evolution of brains and intelligence. If we don't use up all our ration of luck in our theory of the origin of life, we have some left over to spend on our theories of subsequent evolution, after cumulative selection has got going. If we want to use up most of our ration of luck in our theory of the origin of intelligence, then we haven't much left over to spend on our theory of the origin of life: we must come up with a theory that makes the origin of life almost inevitable. Alternatively, if we don't need our whole luck ration for these two stages of our theory, we can, in effect, use the surplus to postulate life elsewhere in the universe.

My personal feeling is that, once cumulative selection has got itself properly started, we need to postulate only a relatively small amount of luck in the subsequent evolution of life and intelligence. Cumulative selection, once it has begun, seems to me powerful enough to make the evolution of intelligence probable, if not inevitable. This means that we can, if we want to, spend virtually our entire ration of postulatable luck in one big throw, in our theory of the origin of life on a planet. Therefore we have at our disposal, if we want to use it, odds of 1 in 100 billion billion as an upper limit (or 1 in however many available planets we think there are) to spend in our theory of the origin of life. This is the maximum amount of luck we are allowed to postulate in our theory. Suppose we want to suggest, for instance, that life began when both DNA and its protein-based replication machinery spontaneously chanced to come into existence. We can allow ourselves the luxury of such an extravagant theory, provided that the odds against this coincidence occurring on a planet do not exceed 100 billion billion to one.

This allowance may seem large. It is probably ample to accommodate the spontaneous arising of DNA or RNA. But it is nowhere near enough to enable us to do without cumulative selection altogether. The odds against assembling a well-designed body that flies as well as a swift, or swims as well as a dolphin, or sees as well as a falcon, in a single blow of luck – single-step selection – are stupendously greater than the number of atoms in the universe, let alone the number of planets! No, it is certain that we are going to need a hefty measure of cumulative selection in our explanations of life.

But although we are entitled, in our theory of the origin of life, to spend a maximum ration of luck amounting, perhaps, to odds of 100 billion billion to one against, my hunch is that we aren't going to need more than a small fraction of that ration. The origin of life on a planet can be a very improbable event indeed by our everyday standards, or indeed by the standards of the chemistry laboratory, and still be sufficiently probable to have occurred, not just once but many times, all over the universe. We can regard the statistical argument about numbers of planets as an argument of last resort. At the end of the chapter I shall make the paradoxical point that the theory we are looking for may actually *need* to seem improbable, even miraculous, to our subjective judgement (because of the way our subjective judgement has been made). Nevertheless, it is still sensible for us to begin by seeking that theory of the origin of life with the least degree of improbability. If the theory that DNA and its copying machinery arose spontaneously is so improbable that it obliges us to assume that life is very rare in the universe, and may even be unique to Earth, our first resort is to try to find a more probable theory. So, can we come up with any speculations about relatively *probable* ways in which cumulative selection might have got its start?

The word 'speculate' has pejorative overtones, but these are quite uncalled for here. We can hope for nothing more than speculation when the events we are talking about took place four billion years ago and took place, moreover, in a world that must have been radically different from that which we know today. For instance, there almost certainly was no free oxygen in the atmosphere. Though the chemistry of the world may have changed, the *laws* of chemistry have not changed (that's why they are called laws), and modern chemists know enough about those laws to make some well-informed speculations, speculations that have to pass rigorous tests of plausibility imposed by the laws. You can't just speculate wildly and irresponsibly, allowing your imagination to run riot in the manner of such unsatisfying space fiction panaceas as 'hyperdrives', 'time warps' and 'infinite improbability drives'. Of all possible speculations about the origin of life, most run foul of the laws of chemistry and can be ruled out, even if we make full use of our statistical fall-back argument about numbers of planets. Careful selective speculation is therefore a constructive exercise. But you do have to be a chemist to do it.

I am a biologist not a chemist, and I must rely on chemists to get their sums right. Different chemists prefer different pet theories, and there is no shortage of theories. I could attempt to lay all these theories before you impartially. That would be the proper thing to do in a student textbook. This isn't a student textbook. The basic idea of *The Blind Watchmaker* is that we don't need to postulate a designer in order to understand life, or anything else in the universe. We are here concerned with the *kind* of solution that must be found, because of the kind of problem we are faced with. I think that this is best explained, not by looking at lots of particular theories, but by looking at *one* as an example of how the basic problem – how cumulative selection got its start – *might* be solved.

Now, which theory to choose as my representative sample? Most textbooks give greatest weight to the family of theories based on an organic 'primeval soup'. It seems probable that the atmosphere of Earth before the coming of life was like that of other planets which are still lifeless. There was no oxygen, plenty of hydrogen and water, carbon dioxide, very likely some ammonia, methane and other simple organic gases. Chemists know that oxygen-free climates like this tend to foster the spontaneous synthesis of organic compounds. They have set up in flasks miniature reconstructions of conditions on the early Earth. They have passed through the flasks electric sparks simulating lightning, and ultraviolet light, which would have been much stronger before the Earth had an ozone layer shielding it from the sun's rays. The results of these experiments have been exciting. Organic molecules, some of them of the same general types as are normally only found in living things, have spontaneously assembled themselves in these flasks. Neither DNA nor RNA has appeared, but the building blocks of these large molecules, called purines and pyrimidines, have. So have the building blocks of proteins, amino acids. The missing link for this class of theories is still the origin of replication. The building blocks haven't come together to form a self-replicating chain like RNA. Maybe one day they will.

But, in any case, the organic primeval-soup theory is not the one I have chosen for my illustration of the kind of solution that we must look for. I did choose it in my first book, *The Selfish Gene*, so I thought that here I would fly a kite for a somewhat less-fashionable theory (although it recently has started gaining ground), which seems to me to have at least a sporting chance of being right. Its audacity is appealing, and it does illustrate well the properties that any satisfying theory of the origin of life must have. This is the 'inorganic mineral' theory of the Glasgow chemist Graham Cairns-Smith, first proposed 20 years ago and since developed and elaborated in three books, the latest of which, *Seven Clues to the Origin of Life*, treats the origin of life as a mystery needing a Sherlock Holmes solution.

Cairns-Smith's view of the DNA/protein machinery is that it probably came into existence relatively recently, perhaps as recently as three billion years ago. Before that there were many generations of cumulative selection, based upon some quite different replicating entities. Once DNA was there, it proved to be so much more efficient as a replicator, and so much more powerful in its effects on its own replication, that the original replication system that spawned it was cast off and forgotten. The modern DNA machinery, according to this view, is a late-comer, a recent usurper of the role of fundamental replicator, having taken over that role from an earlier and cruder replicator. There may even have been a whole series of such usurpations, but the original replication process must have been sufficiently simple to have come about through what I have dubbed 'single-step selection'.

Chemists divide their subject into two main branches, organic and inorganic. Organic chemistry is the chemistry of one particular element, carbon. Inorganic chemistry is all the rest. Carbon is important and deserves to have its own private branch of chemistry, partly because life chemistry is all carbon-chemistry, and partly because those same properties that make carbon-chemistry suitable for life also make it suitable for industrial processes, such as those of the plastics industry. The essential property of carbon atoms that makes them so suitable for life and for industrial synthetics, is that they join together to form a limitless repertoire of different kinds of very large molecules. Another element that has some of these same properties is silicon. Although the chemistry of modern Earth-bound life is all carbon-chemistry, this may not be true all over the universe, and it may not always have been true on this Earth. Cairns-Smith believes that the original life on this planet was based on self-replicating inorganic crystals such as silicates. If this is true, organic replicators, and eventually DNA, must later have taken over or usurped the role.

He gives some arguments for the general plausibility of this idea of 'takeover'. An arch of stones, for instance, is a stable structure capable of standing for many years even if there is no cement to bind it. Building a complex structure by evolution is like trying to build a mortarless arch if you are allowed to touch only one stone at a time. Think about the task naïvely, and it can't be done. The arch will stand once the last stone is in place, but the intermediate stages are unstable. It's quite easy to build the arch, however, if you are allowed to subtract stones as well as add them. Start by building a solid heap of stones, then build the arch resting on top of this solid foundation. Then, when the arch is all in position, including the vital keystone at the top, carefully remove the supporting stones and, with a modicum of luck, the arch will remain standing. Stonehenge is incomprehensible until we realize that the builders used some kind of scaffolding, or perhaps ramps of earth, *which are no longer there*. We can see only the end-product, and have to infer the vanished scaffolding. Similarly, DNA and protein are two pillars of a stable and elegant arch, which persists once all its parts simultaneously exist. It is hard to imagine it arising by any step-by-step process unless some earlier scaffolding has completely disappeared. That scaffolding must itself have been built by an earlier form of cumulative selection, at whose nature we can only guess. But it must have been based upon replicating entities with power over their own future.

Cairns-Smith's guess is that the original replicators were crystals of inorganic materials, such as those found in clays and muds. A crystal is just a large orderly array of atoms or molecules in the solid state. Because of properties that we can think of as their 'shape', atoms and small molecules tend naturally to pack themselves together in a fixed and orderly manner. It is almost as though they 'want' to slot together in a particular way, but this illusion is just an inadvertent consequence of their properties. Their 'preferred' way of slotting together shapes the whole crystal. It also means that, even in a large crystal such as a diamond, any part of the crystal is *exactly* the same as any other part, except where there are flaws. If we could shrink ourselves to the atomic scale, we would see almost endless rows of atoms, stretching to the horizon in straight lines – galleries of geometric repetition.

Since it is replication we are interested in, the first thing we must know is, can crystals replicate their structure? Crystals are made of myriads of layers of atoms (or equivalent), and each layer builds upon the layer below. Atoms (or ions; the difference needn't concern us) float around free in solution, but if they happen to encounter a crystal they have a natural tendency to slot into position on the surface of the crystal. A solution of common salt contains sodium ions and chloride ions jostling about in a more or less chaotic fashion. A crystal of common salt is a packed, orderly array of sodium ions alternating with chloride ions at right angles to one another. When ions floating in the water happen to bump into the hard surface of the crystal, they tend to stick. And they stick in just the right places to cause a new layer to be added to the crystal just like the layer below. So once a crystal gets started it grows, each layer being the same as the layer below.

Sometimes crystals spontaneously start to form in solution. At other times they have to be 'seeded', either by particles of dust or by small crystals dropped in from elsewhere. Cairns-Smith invites us to perform the following experiment. Dissolve a large quantity of photographer's 'hypo' fixer in very hot water. Then let the solution cool down, being careful not to let any dust drop in. The solution is now 'supersaturated', ready and waiting to make crystals, but with no seed crystals to start the process going. I quote from Cairns-Smith's *Seven Clues to the Origin of Life*:

> Carefully take the lid off the beaker, drop one tiny piece of 'hypo' crystal onto the surface of the solution, and watch amazed at what happens. Your crystal grows visibly: it breaks up from time to time and the pieces also grow . . . Soon your beaker is crowded with crystals, some several centimetres long. Then after a few minutes it all stops. The magic solution has lost its power – although if you want another performance just re-heat and re-cool the beaker . . . to be supersaturated means to have more dissolved than there ought to be . . . the cold supersaturated solution almost literally did not know what to do. It had to be 'told' by adding a piece of crystal that already had its units (billions and billions of them) packed together in the way that is characteristic for 'hypo' crystals. The solution had to be seeded.

Some chemical substances have the potential to crystallize in two alternative ways. Graphite and diamonds, for instance, are both crystals of pure carbon. Their atoms are identical. The two substances differ from each other only in the geometric pattern with which the carbon atoms are packed. In diamonds, the carbon atoms are packed in a tetrahedral pattern which is extremely stable. This is why diamonds are so hard. In graphite the carbon atoms are arranged in flat hexagons layered on top of each other. The bonding between layers is weak, and they therefore slide over each other, which is why graphite feels slippery and is used as a lubricant. Unfortunately you can't crystallize diamonds out of a solution by seeding them, as you can with hypo. If you could, you'd be rich; no on second thoughts you wouldn't, because any fool could do the same.

Now suppose we have a supersaturated solution of some substance, like hypo in that it was eager to crystallize out of solution, and like carbon in that it was capable of crystallizing in either of two ways. One way might be somewhat like graphite, with the atoms arranged in layers, leading to little flat crystals; while the other way gives chunky, diamond-shaped crystals. Now we simultaneously drop into our supersaturated solution a tiny flat crystal and a tiny chunky crystal. We can describe what would happen in an elaboration of Cairns-Smith's description of his hypo experiment. You watch amazed at what happens. Your two crystals grow visibly: they break up from time to time and the pieces also grow. Flat crystals give rise to a population of flat crystals. Chunky crystals give rise to a population of chunky crystals. If there is any tendency for one type of crystal to grow and split more quickly than the other, we shall have a simple kind of natural selection. But the process still lacks a vital ingredient in order to give rise to evolutionary change. That ingredient is hereditary variation, or something equivalent to it. Instead of just two types of crystal, there must be a whole range of minor variants that form lineages of like shape, and that sometimes 'mutate' to produce new shapes. Do real crystals have something corresponding to hereditary mutation?

Clays and muds and rocks are made of tiny crystals. They are abundant on Earth and probably always have been. When you look at the surface of some types of clay and other minerals with a scanning electron microscope you see an amazing and beautiful sight. Crystals grow like rows of flowers or cactuses, gardens of inorganic rose petals, tiny spirals like cross-sections of succulent plants, bristling organ pipes, complicated angular shapes folded as if in miniature crystalline origami, writhing growths like worm casts or squeezed toothpaste. The ordered patterns become even more striking at greater levels of magnification. At levels that betray the actual position of atoms, the surface of a crystal is seen to have all the regularity of a machine-woven piece of herringbone tweed. But – and here is the vital point – there are flaws. Right in the middle of an expanse of orderly herringbone there can be a patch, identical to the rest except that it is twisted round at a different angle so that the 'weave' goes off in another direction. Or the weave may lie in the same direction, but each row has 'slipped' half a row to one side. Nearly all naturally occurring crystals have flaws. And once a flaw has appeared, it tends to be copied as subsequent layers of crystal encrust themselves on top of it.

Flaws can occur anywhere over the surface of a crystal. If you like thinking about capacity for information storage (I do), you can imagine the enormous number of different patterns of flaws that could be created over the surface of a crystal. All those calculations about packing the New Testament into the DNA of a single bacterium could be done just as impressively for almost any crystal. What DNA has over normal crystals is a means by which its information can be read. Leaving aside the problem of read-out, you could easily devise an arbitrary code whereby flaws in the atomic structure of the crystal denote binary numbers. You could then pack several New Testaments into a mineral crystal the size of a pin's head. On a larger scale, this is essentially how music information is stored on the surface of a laser ('compact') disc. The musical notes are converted, by computer, into binary numbers. A laser is used to etch a pattern of tiny flaws in the otherwise glassy smooth surface of the disc. Each little hole etched corresponds to a binary 1 (or a 0, the labels are arbitrary). When you play the disc, another laser beam 'reads' the pattern of flaws, and a special-purpose computer built into the player turns the binary numbers back into sound vibrations, which are amplified so that you can hear them.

Although laser discs are used today mainly for music, you could pack the whole *Encyclopaedia Britannica* onto one of them, and read it out using the same laser technique. Flaws in crystals at the atomic level are far smaller than the pits etched in a laser disc's surface, so crystals can potentially pack more information into a given area. Indeed DNA molecules, whose capacity for storing information has already impressed us, are something close to crystals themselves. Although clay crystals theoretically could store the same prodigious quantities of information as DNA or laser discs can, nobody is suggesting that they ever did. The role of clay and other mineral crystals in the theory is to act as the original 'low-tech' replicators, the ones that were eventually replaced by high-tech DNA. They form spontaneously in the waters of our planet without the elaborate 'machinery' that DNA needs; and they develop flaws spontaneously, some of which can be replicated in subsequent layers of crystal. If fragments of suitably flawed crystal later broke away, we could imagine them acting as 'seeds' for new crystals, each one 'inheriting' its 'parent's' pattern of flaws.

So, we have a speculative picture of mineral crystals on the primeval Earth showing some of the properties of replication, multiplication, heredity and mutation that would have been necessary in order for a form of cumulative selection to get started. There is still the missing ingredient of 'power': the nature of the replicators must somehow have influenced their own likelihood of being replicated. When we were talking about replicators in the abstract, we saw that 'power' might simply be direct properties of the replicator itself, intrinsic properties like 'stickiness'. At this elementary level, the name 'power' seems scarcely justified. I use it only because of what it can become in later stages of evolution: the power of a snake's fang, for instance, to propagate (by its indirect consequences on snake survival) DNA coding for fangs. Whether the original low-tech replicators were mineral crystals or organic direct forerunners of DNA itself, we may guess that the 'power' they exercised was direct and elementary, like stickiness. Advanced levers of power, like a snake's fang or an orchid's flower, came far later.

What might 'power' mean to a clay? What incidental properties of the clay could influence the likelihood that it, the same variety of clay, would be propagated around the countryside? Clays are made from chemical building blocks such as silicic acid and metal ions, which are in solution in rivers and streams having been dissolved – 'weathered' –

out of rocks further upstream. If conditions are right they crystallize out of solution again downstream, forming clays. (Actually the 'stream', in this case, is more likely to mean the seeping and trickling of the groundwater than a rushing open river. But, for simplicity, I shall continue to use the general word stream.) Whether or not a particular type of clay crystal is allowed to build up depends, among other things, upon the rate and pattern of flow of the stream. But deposits of clay can also *influence* the flow of the stream. They do this inadvertently by changing the level, shape and texture of the ground through which the water is flowing. Consider a variant of clay that just happens to have the property of reshaping the structure of the soil so that the flow speeds up. The consequence is that the clay concerned gets washed away again. This kind of clay, by definition, is not very 'successful'. Another unsuccessful clay would be one that changed the flow in such a way that a rival variant of clay was favoured.

We aren't, of course, suggesting that clays 'want' to go on existing. Always we are talking only about incidental consequences, events which follow from properties that the replicator just happens to have. Consider yet another variant of clay. This one happens to slow down the flow in such a way that future deposition of its *own* kind of clay is enhanced. Obviously this second variant will tend to become common, because it happens to manipulate streams to its own 'advantage'. This will be a 'successful' variant of clay. But so far we are dealing only with single-step selection. Could a form of cumulative selection get going?

To speculate a little further, suppose that a variant of a clay improves its own chances of being deposited, by damming up streams. This is an inadvertent consequence of the peculiar defect structure of the clay. In any stream in which this kind of clay exists, large, stagnant shallow pools form above dams, and the main flow of water is diverted into a new course. In these still pools, more of the same kind of clay is laid down. A succession of such shallow pools proliferates along the length of any stream that happens to be 'infected' by seeding crystals of this kind of clay. Now, because the main flow of the stream is diverted, during the dry season the shallow pools tend to dry up. The clay dries and cracks in the sun, and the top layers are blown off as dust. Each dust particle inherits the characteristic defect structure of the parent clay that did the damming, the structure that gave it its damming properties. By analogy with the genetic information raining down on the canal from my willow tree, we could say that the dust carries 'instructions' for how to dam streams and eventually make more dust. The dust spreads far and wide in the wind, and there is a good chance that some particles of it will happen to land in another stream, hitherto not 'infected' with the seeds of this kind of dam-making clay. Once infected by the right sort of dust, a new stream starts to grow crystals of dam-making clay, and the whole depositing, damming, drying, eroding cycle begins again.

To call this a 'life' cycle would be to beg an important question, but it is a cycle of a sort, and it shares with true life cycles the ability to initiate cumulative selection. Because streams are infected by dust 'seeds' blown from other streams, we can arrange the streams in an order of 'ancestry' and 'descent'. The clay that is damming up pools in stream B arrived there in the form of dust crystals blown from stream A. Eventually, the pools of stream B will dry up and make dust, which will infect streams F and P. With respect to the source of their dam-making clay, we can arrange streams into 'family trees'. Every infected stream has a 'parent' stream, and it may have more than one 'daughter' stream. Each stream is analogous to a body, whose 'development' is influenced by dust seed 'genes', a body that eventually spawns new dust seeds. Each 'generation' in the cycle starts when seed crystals break away from the parent stream in the form of dust. The crystalline structure of each particle of dust is copied from the clay in the parent stream. It passes on that crystalline structure to the daughter stream, where it grows and multiplies and finally sends 'seeds' out again.

The ancestral crystal structure is preserved down the generations unless there is an occasional mistake in crystal growth, an occasional alteration in the pattern of laying down of atoms. Subsequent layers of the same crystal will copy the same flaw, and if the crystal breaks in two it will give rise to a sub-population of altered crystals. Now if the alteration makes the crystal either less or more efficient in the damming/drying/erosion cycle, this will affect how many copies it has in subsequent 'generations'. Altered crystals might, for instance, be more likely to split ('reproduce'). Clay formed from altered crystals might have greater damming power in any of a variety of detailed ways. It might crack more readily in a given amount of sun. It might crumble into dust more readily. The dust particles might be better at catching the wind, like fluff on a willow seed. Some crystal types might induce a shortening of the 'life cycle', consequently a speeding up of their 'evolution'. There are many opportunities for successive 'generations' to become progressively 'better' at getting passed to subsequent generations. In other words, there are many opportunities for rudimentary cumulative selection to get going.

These little flights of fancy, embellishments of Cairns-Smith's own, concern only one of several kinds of mineral 'life cycle' that could have started cumulative selection along its momentous road. There are others. Different varieties of crystals might earn their passage to new streams, not by crumbling into dust 'seeds', but by dissecting their streams into lots of little streamlets that spread around, eventually joining and infecting new river systems. Some varieties might engineer waterfalls that wear down the rocks faster, and hence speed into solution the raw materials needed to make new clays further downstream. Some varieties of crystal might better themselves by making conditions hard for 'rival' varieties that compete for raw materials. Some varieties might become 'predatory', breaking up rival varieties and using their elements as raw materials. Keep holding in mind that there is no suggestion of 'deliberate' engineering, either here or in modern, DNA-based life. It is just that the world automatically tends to become full of those varieties of clay (or DNA) that *happen* to have properties that make them persist and spread themselves about.

Now to move on to the next stage of the argument. Some lineages of crystals might happen to catalyse the synthesis of new substances that assist in their passage down the 'generations'. These secondary substances would not (not at first, anyway) have had their own lineages of ancestry and descent, but would have been manufactured anew by each generation of primary replicators. They could be seen as tools of the replicating crystal lineages, the beginnings of primitive 'phenotypes'. Cairns-Smith believes that *organic* molecules were prominent among non-replicating 'tools' of his inorganic crystalline replicators. Organic molecules frequently are used in the commercial inorganic chemical industry because of their effects on the flow of fluids, and on the break-up or growth of inorganic particles: just the sorts of effects, in short, that could have influenced the 'success' of lineages of replicating crystals. For instance, a clay mineral with the lovely name montmorillonite tends to break up in the presence of small amounts of an organic molecule with the less-lovely name carboxymethyl cellulose. Smaller quantities of carboxymethyl cellulose, on the other hand, have just the opposite effect, helping to stick montmorillonite particles together. Tannins, another kind of organic molecule, are used in the oil industry to make muds easier to drill. If oil-drillers can exploit organic molecules to manipulate the flow and drillability of mud, there is no reason why cumulative selection should not have led to the same kind of exploitation by self-replicating minerals.

At this point Cairns-Smith's theory gets a sort of free bonus of added plausibility. It so happens that other chemists, supporting more conventional organic 'primeval soup' theories, have long accepted that clay minerals would have been a help. To quote one of them (D. M. Anderson), 'It is widely accepted that some, perhaps many, of the abiotic chemical reactions and processes leading to the origin on Earth of replicating micro-organisms occurred very early in the history of Earth in close proximity to the surfaces of clay minerals and other inorganic substrates.' This writer goes on to list five 'functions' of clay minerals in assisting the origin of organic life, for instance 'Concentration of chemical reactants by adsorption'. We needn't spell the five out here, or even understand them. From our point of view, what matters is that each of these five 'functions' of clay minerals can be twisted round the other way. It shows the close association that can exist between organic chemical synthesis and clay surfaces. It is therefore a bonus for the theory that clay replicators synthesized organic molecules and used them for their own purposes.

Cairns-Smith discusses, in more detail than I can accommodate here, early uses that his clay-crystal replicators might have had for proteins, sugars and, most important of all, nucleic acids like RNA. He suggests that RNA was first used for purely structural purposes, as oil drillers use tannins or we use soap and detergents. RNA-like molecules, because of their negatively charged backbones, would tend to coat the outsides of clay particles. This is getting us into realms of chemistry that are beyond our scope. For our purposes what matters is that RNA, or something like it, was around for a long time before it became self-replicating. When it finally did become self-replicating, this was a device evolved by the mineral crystal 'genes' to improve the efficiency of manufacture of the RNA (or similar molecule). But, once a new self-replicating molecule had come into existence, a new kind of cumulative selection could get going. Originally a side-show, the new replicators turned out to be so much more efficient than the original crystals that they took over. They evolved further, and eventually perfected the DNA code that we know today. The original mineral replicators were cast aside like worn-out scaffolding, and all modern life evolved from a relatively recent common ancestor, with a single, uniform genetic system and a largely uniform biochemistry.

In *The Selfish Gene* I speculated that we may now be on the threshold of a new kind of genetic takeover. DNA replicators built 'survival machines' for themselves – the bodies of living organisms including ourselves. As part of their equipment, bodies evolved onboard computers – brains. Brains evolved the capacity to communicate with other brains by means of language and cultural traditions. But the new milieu of cultural tradition opens up new possibilities for self-replicating entities. The new replicators are not DNA and they are not clay crystals. They are patterns of information that can thrive only in brains or the artificially manufactured products of brains – books, computers, and so on. But, given that brains, books and computers exist, these new replicators, which I called memes to distinguish them from genes, can propagate themselves from brain to brain, from brain to book, from book to brain, from brain to computer, from computer to computer. As they propagate they can change – mutate. And perhaps 'mutant' memes can exert the kinds of influence that I am here calling 'replicator power'. Remember that this means any kind of influence affecting their own likelihood of being propagated. Evolution under the influence of the new replicators – memic evolution – is in its infancy. It is manifested in the phenomena that we call cultural evolution. Cultural evolution is many orders of magnitude faster than DNA-based evolution, which sets one even more to thinking of the idea of 'takeover'. And if a new kind of replicator takeover is beginning, it is conceivable that it will take off so far as to leave its parent DNA (and its grandparent clay if Cairns-Smith is right) far behind. If so, we may be sure that computers will be in the van.

Could it be that one far-off day intelligent computers will speculate about their own lost origins? Will one of them tumble to the heretical truth, that they have sprung from a remote, earlier form of life, rooted in organic, carbon chemistry, rather than the silicon-based electronic principles of their own bodies? Will a robotic Cairns-Smith write a book called *Electronic Takeover?* Will he rediscover some electronic equivalent of the metaphor of the arch, and realize that computers could not have sprung spontaneously into existence but must have originated from some earlier process of cumulative selection? Will he go into detail and reconstruct DNA as a plausible early replicator, victim of electronic usurpation? And will he be far-sighted enough to guess that even DNA may itself have been a usurper of yet more remote and primitive replicators, crystals of inorganic silicates? If he is of a poetic turn of mind, will he even see a kind of justice in the eventual return to silicon-based life, with DNA no more than an interlude, albeit one that lasted longer than three aeons?

That is science fiction, and it probably sounds far-fetched. That doesn't matter. Of more immediate moment is that Cairns-Smith's own theory, and indeed all other theories of the origin of life, may sound far-fetched to you and hard to believe. Do you find both Cairns-Smith's clay theory, and the more orthodox organic primeval-soup theory, wildly improbable? Does it sound to you as though it would need a miracle to make randomly jostling atoms join together into a self-replicating molecule? Well, at times it does to me too. But let's look more deeply into this matter of miracles and improbability. By doing so, I shall demonstrate a point which is paradoxical but all the more interesting for that. This is that we should, as scientists, be even a little worried if the origin of life did *not* seem miraculous to our own human consciousness. An apparently (to ordinary human consciousness) miraculous theory is *exactly* the kind of theory we should be looking for in this particular matter of the origin of life. This argument, which amounts to a discussion of what we mean by a miracle, will occupy the rest of this chapter. In a way it is an extension of the argument we made earlier about billions of planets.

So, what do we mean by a miracle? A miracle is something that happens, but which is exceedingly surprising. If a marble statue of the Virgin Mary suddenly waved its hand at us we should treat it as a miracle, because all our experience and knowledge tells us that marble doesn't behave like that. I have just uttered the words 'May I be struck by lightning this minute'. If lightning did strike me in the same minute, it would be treated as a miracle. But actually neither of these two occurrences would be classified by science as utterly impossible. They would simply be judged very improbable, the waving statue much more improbable than the lightning. Lightning does strike people. Any one of us might be struck by lightning, but the probability is pretty low in any one minute (although the *Guinness Book of Records* has a charming picture of a Virginian man, nicknamed the human lightning conductor, recovering in hospital from his seventh lightning strike, with an expression of apprehensive bewilderment on his face). The only thing miraculous about my hypothetical story is the *coincidence* between my being struck by lightning and my verbal invocation of the disaster.

Coincidence means multiplied improbability. The probability of my being struck by lightning in any one minute of my life is perhaps 1 in 10 million as a conservative estimate. The probability of my inviting a lightning strike in any particular minute is also very low. I have just done it for the only time in the 23,400,000 minutes of my life so far, and I doubt if I'll do it again, so call these odds one in 25 million. To calculate the joint probability of the coincidence occurring in any one minute we multiply the two separate probabilities. For my rough calculation this comes to about one in 250 trillion. If a coincidence of this magnitude happened to me, I should call it a miracle and would watch my language in future. But although the odds against the coincidence are extremely high, we can still calculate them. They are not literally zero.

In the case of the marble statue, molecules in solid marble are continuously jostling against one another in random directions. The jostlings of the different molecules cancel one another out, so the whole hand of the statue stays still. But if, by sheer coincidence, all the molecules just happened to move in the same direction at the same moment, the hand would move. If they then all reversed direction at the same moment the hand would move back. In this way it is *possible* for a marble statue to wave at us. It could happen. The odds against such a coincidence are unimaginably great but they are not incalculably great. A physicist colleague has kindly calculated them for me. The number is so large that the entire age of the universe so far is too short a time to write out all the noughts! It is theoretically possible for a cow to jump over the moon with something like the same improbability. The conclusion to this part of the argument is that we can *calculate* our way into regions of miraculous improbability far greater than we can *imagine* as plausible.

Let's look at this matter of what we think is plausible. What we can imagine as plausible is a narrow band in the middle of a much broader spectrum of what is actually possible. Sometimes it is narrower than what is actually there. There is a good analogy with light. Our eyes are built to cope with a narrow band of electromagnetic frequencies (the ones we call light), somewhere in the middle of the spectrum from long radio waves at one end to short X-rays at the other. We can't see the rays outside the narrow light band, but we can do calculations about them, and we can build instruments to detect them. In the same way, we know that the scales of size and time extend in both directions far outside the realm of what we can visualize. Our minds can't cope with the large distances that astronomy deals in or with the small distances that atomic physics deals in, but we can represent those distances in mathematical symbols. Our minds can't imagine a time span as short as a picosecond, but we can do calculations about picoseconds, and we can build computers that can complete calculations within picoseconds. Our minds can't imagine a timespan as long as a million years, let alone the thousands of millions of years that geologists routinely compute.

Just as our eyes can see only that narrow band of electromagnetic frequencies that natural selection equipped our ancestors to see, so our brains are built to cope with narrow bands of sizes and times. Presumably there was no need for our ancestors to cope with sizes and times outside the narrow range of everyday practicality, so our brains never evolved the capacity to imagine them. It is probably significant that our own body size of a few feet is roughly in the middle of the range of sizes we can imagine. And our own lifetime of a few decades is roughly in the middle of the range of times we can imagine.

We can say the same kind of thing about improbabilities and miracles.

Picture a graduated scale of improbabilities, analogous to the scale of sizes from atoms to galaxies, or to the scale of times from picoseconds to aeons. On the scale we mark off various landmark points. At the far left-hand end of the scale are events which are all but certain, such as the probability that the sun will rise tomorrow – the subject of G. H. Hardy's halfpenny bet. Near this left-hand end of the scale are things that are only slightly improbable, such as shaking a double six in a single throw of a pair of dice. The odds of this happening are 1 in 36. I expect we've all done it quite often. Moving towards the right-hand end of the spectrum, another landmark point is the probability of a perfect deal in bridge, where each of the four players receives a complete suit of cards. The odds against this happening are 2,235,197,406,895,366,368,301,559,999 to 1. Let us call this one dealon, the unit of improbability. If something with an improbability of one dealon was predicted and then happened, we should diagnose a miracle unless, which is more probable, we suspected fraud. But it *could* happen with a fair deal, and it is far far far more probable than the marble statue's waving at us. Nevertheless, even this latter event, as we have seen, has its rightful place along the spectrum of events that could happen. It is measurable, albeit in units far larger than gigadealons. Between the double-six dice throw, and the perfect deal at bridge, is a range of more or less improbable events that do sometimes happen, including any one individual's being struck by lightning, winning a big prize on the football pools, scoring a hole-in-one at golf, and so on. Somewhere in this range, too, are those coincidences that give us an eerie spine-tingling feeling, like dreaming of a particular person for the first time in decades, then waking up to find that they died in the night. These eerie coincidences are very impressive when they happen to us or to one of our friends, but their improbability is measured in only picodealons.

Having constructed our mathematical scale of improbabilities, with its benchmark or landmark points marked on it, let us now turn a spotlight on that subrange of the scale with which we, in our ordinary thought and conversation, can cope. The width of the spotlight's beam is analogous to the narrow range of electromagnetic frequencies that our eyes can see, or to the narrow range of sizes or times, close to our own size and longevity, that we can imagine. On the spectrum of improbabilities, the spotlight turns out to illuminate only the narrow range from the left-hand end (certainty) up to minor miracles, like a hole-in-one or a dream that comes true. There is a vast range of mathematically calculable improbabilities way outside the range of the spotlight.

Our brains have been built by natural selection to assess probability and risk, just as our eyes have been built to assess electromagnetic wavelength. We are equipped to make mental calculations of risk and odds, within the range of improbabilities that would be useful in human life. This means risks of the order of, say, being gored by a buffalo if we shoot an arrow at it, being struck by lightning if we shelter under a lone tree in a thunderstorm, or drowning if we try to swim across a river. These acceptable risks are commensurate with our lifetimes of a few decades. If we were biologically capable of living for a million years, and wanted to do so, we should assess risks quite differently. We should make a habit of not crossing roads, for instance, for if you crossed a road every day for half a million years you would undoubtedly be run over.

Evolution has equipped our brains with a subjective consciousness of risk and improbability suitable for creatures with a lifetime of less than one century. Our ancestors have always needed to take decisions involving risks and probabilities, and natural selection has therefore equipped our brains to assess probabilities against a background of the short lifetime that we can, in any case, expect. If on some planet there are beings with a lifetime of a million centuries, their spotlight of comprehensible risk will extend that much farther towards the right-hand end of the continuum. They will expect to be dealt a perfect bridge hand from time to time, and will scarcely trouble to write home about it when it happens. But even they will blench if a marble statue waves at them, for you would have to live dealions of years longer than even they do to see a miracle of this magnitude.

What has all this to do with theories of the origin of life? Well, we began this argument by agreeing that Cairns-Smith's theory, and the primeval-soup theory, sound a bit far-fetched and improbable to us. We naturally feel inclined to reject these theories for that reason. But 'we', remember, are beings whose brains are equipped with a spotlight of comprehensible risk that is a pencil-thin beam illuminating the far left-hand end of the mathematical continuum of calculable risks. Our subjective judgement of what seems like a good bet is irrelevant to what is actually a good bet. The subjective judgement of an alien with a lifetime of a million centuries will be quite different. He will judge as quite plausible an event, such as the origin of the first replicating molecule as postulated by some chemist's theory, which we, kitted up by evolution to move in a world of a few decades' duration, would judge to be an astounding miracle. How can we decide whose point of view is the right one, ours or the long-lived alien's?

There is a simple answer to this question. The long-lived alien's point of view is the right one for looking at the plausibility of a theory like Cairns-Smith's or the primeval-soup theory. This is because those two theories postulate a particular event – the spontaneous arising of a self-replicating entity – as occurring only once in about a billion years, once per aeon. One and a half aeons is about the time that elapsed between the origin of the Earth and the first bacteria-like fossils. For our decade-conscious brains, an event that happens only once per aeon is so rare as to seem a major miracle. For the long-lived alien, it will seem less of a miracle than a golf hole-in-one seems to us – and most of us probably know somebody who knows somebody who has scored a hole-in-one. In judging theories of the origin of life, the long-lived alien's subjective timescale is the relevant one, because it is approximately the timescale involved in the origin of life. Our own subjective judgement about the plausibility of a theory of the origin of life is likely to be wrong by a factor of a hundred million.

In fact our subjective judgement is probably wrong by an even greater margin. Not only are our brains equipped by nature to assess risks of things in a short time; they are also equipped to assess risks of things happening to us personally, or to a narrow circle of people that we know. This is because our brains didn't evolve under conditions dominated by mass media. Mass reporting means that, if an improbable thing happens to anybody, anywhere in the world, we shall read about it in our newspapers or in the *Guinness Book of Records*. If an orator, anywhere in the world, publicly challenged the lightning to strike him if he lied, and it promptly did so, we should read about it and be duly impressed. But there are several billion people in the world to whom such a coincidence *could* happen, so the apparent coincidence is actually not as great as it seems. Our brains are probably equipped by nature to assess the risks of things happening to ourselves, or to a few hundred people in the small circle of villages within drum-range that our tribal ancestors could expect to hear news about. When we read in a newspaper about an amazing coincidence happening to somebody in Valparaiso or Virginia, we are more impressed by it than we should be. More impressed by a factor of perhaps a hundred million, if that is the ratio between the world population surveyed by our newspapers, and the tribal population about whom our evolved brains 'expect' to hear news.

This 'population calculation' is also relevant to our judgement of the plausibility of theories of the origin of life. Not because of the population of people on Earth, but because of the population of planets in the universe, the population of planets where life *could* have originated. This is just the argument we met earlier in this chapter, so there is no need to dwell on it here. Go back to our mental picture of a graduated scale of improbable events with its benchmark coincidences of bridge hands and dice throws. On this graduated scale of dealions and microdealions, mark the following three new points. Probability of life arising on a planet (in, say, a billion years), if we assume that life arises at a rate of about once per solar system. Probability of life arising on a planet if life arises at a rate of about once per galaxy. Probability of life on a randomly selected planet if life arose only once in the universe. Label these three points respectively the Solar System Number, the Galaxy Number and the Universe Number. Remember that there are about 10,000 million galaxies. We don't know how many solar systems there are in each galaxy because we can only see stars, not planets, but we earlier used an estimate that there may be 100 billion billion planets in the universe.

When we assess the improbability of an event postulated by, for instance the Cairns-Smith theory, we should assess it, not against what we subjectively think of as probable or improbable, but against numbers like these three numbers, the Solar System Number, the Galaxy Number and the Universe Number. Which of these three numbers is the most appropriate depends upon which of the following three statements we think is nearest the truth:

1. Life has arisen in only one planet in the entire universe (and that planet, as we saw earlier, then has to be Earth).

2. Life has arisen on about one planet per galaxy (in our galaxy, Earth is the lucky planet).

3. The origin of life is a sufficiently probable event that it tends to arise about once per solar system (in our solar system Earth is the lucky planet).

These three statements represent three benchmark views about the uniqueness of life. The actual uniqueness of life probably lies somewhere between the extremes represented by Statement 1 and Statement 3. Why do I say that? Why, in particular, should we rule out a fourth possibility, that the origin of life is a far *more* probable event than is suggested by Statement 3? It isn't a strong argument, but, for what it is worth, it goes like this. If the origin of life were a much more probable event than is suggested by the Solar System Number we should expect, by now, to have encountered extraterrestrial life, if not in (whatever passes for) the flesh, at least by radio.

It is often pointed out that chemists have failed in their attempts to duplicate the spontaneous origin of life in the laboratory. This fact is used as if it constituted evidence against the theories that those chemists are trying to test. But actually one can argue that we should be worried if it turned out to be very easy for chemists to obtain life spontaneously in the test-tube. This is because chemists' experiments last for years not thousands of millions of years, and because only a handful of chemists, not thousands of millions of chemists, are engaged in doing these experiments. If the spontaneous origin of life turned out to be a probable enough event to have occurred during the few man-decades in which chemists have done their experiments, then life should have arisen many times on Earth, and many times on planets within radio range of Earth. Of course all this begs important questions about whether chemists have succeeded in duplicating the conditions of the early Earth but, even so, given that we can't answer these questions, the argument is worth pursuing.

If the origin of life were a probable event by ordinary human standards, then a substantial number of planets within radio range should have developed a radio technology long enough ago (bearing in mind that radio waves travel at 186,000 miles per second) for us to have picked up at least one transmission during the decades that we have been equipped to do so. There are probably about 50 stars within radio range if we assume that they have had radio technology for only as long as we have. But 50 years is just a fleeting instant, and it would be a major coincidence if another civilization were so closely in step with us. If we embrace in our calculation those civilizations that had radio technology 1,000 years ago, there will be something like a million stars within radio range (together with however many planets circle round each one of them). If we include those whose radio technology goes back 100,000 years, the whole trillion-star galaxy would be within radio range. Of course, broadcast signals would become pretty attenuated over such huge distances.

So we have arrived at the following paradox. If a theory of the origin of life is sufficiently 'plausible' to satisfy our subjective judgement of plausibility, it is then *too* 'plausible' to account for the paucity of life in the universe as we observe it. According to this argument, the theory we are looking for has *got* to be the kind of theory that seems implausible to our limited, Earth-bound, decade-bound imaginations. Seen in this light, both Cairns-Smith's theory and the primeval-soup theory seem if anything in danger of erring on the side of being too plausible! Having said all this I must confess that, because there is so much uncertainty in the calculations, if a chemist *did* succeed in creating spontaneous life I would not actually be disconcerted!

We still don't know exactly how natural selection began on Earth.

This chapter has had the modest aim of explaining only the *kind* of way in which it must have happened. The present lack of a definitely accepted account of the origin of life should certainly not be taken as a stumbling block for the whole Darwinian world view, as it occasionally – probably with wishful thinking – is. The earlier chapters have disposed of other alleged stumbling blocks, and the next chapter takes up yet another one, the idea that natural selection can only destroy, never construct.

CHAPTER 7

CONSTRUCTIVE EVOLUTION

People sometimes think that natural selection is a purely negative force, capable of weeding out freaks and failures, but not capable of building up complexity, beauty and efficiency of design. Does it not merely subtract from what is already there, and shouldn't a truly creative process add something too? One can partially answer this by pointing to a statue. Nothing is added to the block of marble. The sculptor only subtracts, but a beautiful statue emerges nevertheless. But this metaphor can mislead, for some people leap straight to the wrong part of the metaphor – the fact that the sculptor is a conscious designer – and miss the important part: the fact that the sculptor works by subtraction rather than addition. Even this part of the metaphor should not be taken too far. Natural selection may only subtract, but mutation can add. There are ways in which mutation and natural selection together can lead, over the long span of geological time, to a building up of complexity that has more in common with addition than with subtraction. There are two main ways in which this build-up can happen. The first of these goes under the name of 'coadapted genotypes'; the second under the name of 'arms races'. The two are superficially rather different from one another, but they are united under the headings of 'coevolution' and 'genes as each others' environments'.

First, the idea of 'coadapted genotypes'. A gene has the particular effect that it does *only* because there is an existing structure upon which to work. A gene can't affect the wiring up of a brain unless there is a brain being wired up in the first place. There won't be a brain being wired up in the first place, unless there is a complete developing embryo. And there won't be a complete developing embryo unless there is a whole program of chemical and cellular events, under the influence of lots and lots of other genes, and lots and lots of other, non-genetic, causal influences. The particular effects that genes have are not intrinsic properties of those genes. They are properties of embryological processes, *existing* processes whose details may be *changed* by genes, acting in particular places and at particular times during embryonic development. We saw this message demonstrated, in elementary form, by the development of the computer biomorphs.

In a sense, the whole process of embryonic development can be looked upon as a cooperative venture, jointly run by thousands of genes together. Embryos are put together by all the working genes in the developing organism, in collaboration with one another. Now comes the key to understanding how such collaborations come about. In natural selection, genes are always selected for their capacity to flourish in the environment in which they find themselves. We often think of this environment as the outside world, the world of predators and climate. But from each gene's point of view, perhaps the most important part of its environment *is all the other genes that it encounters.* And where does a gene 'encounter' other genes? Mostly in the cells of the successive individual bodies in which it finds itself. Each gene is selected for its capacity to cooperate successfully with the population of other genes that it is likely to meet in bodies.

The true population of genes, which constitutes the working environment of any given gene, is not just the temporary collection that happens to have come together in the cells of any particular individual body. At least in sexually reproducing species, it is the set of all genes in the population of interbreeding individuals – the gene 'pool'. At any given moment, any particular copy of a gene, in the sense of a particular collection of atoms, must be sitting in one cell of one individual. But the set of atoms that is any one copy of a gene is not of permanent interest. It has a life-expectancy measured only in months. As we have seen, the long-lived gene as an evolutionary unit is not any particular physical structure but the textual archival *information* that is copied on down the generations. This textual replicator has a distributed existence. It is widely distributed in space among different individuals, and widely distributed in time over many generations. When looked at in this distributed way, any one gene can be said to 'meet' another when they find themselves sharing a body. It can 'expect' to meet a variety of other genes in different bodies at different times in its distributed existence, and in its march through geological time. A successful gene will be one that does well in the environments provided by these other genes that it is likely to meet in lots of different bodies. 'Doing well' in such environments will turn out to be equivalent to 'collaborating' with these other genes. It is most directly seen in the case of biochemical pathways.

Biochemical pathways are sequences of chemicals that constitute successive stages in some useful process, like the release of energy or the synthesis of an important substance. Each step in the pathway needs an enzyme – one of those large molecules that is shaped to act like a machine in a chemical factory. Different enzymes are needed for different steps in the chemical pathway. Sometimes there are two, or more, alternative chemical pathways to the same useful end. Although both pathways culminate in the identical useful result, they have different intermediate stages leading up to that end, and they normally have different starting points. Either of the two alternative pathways will do the job, and it doesn't matter which one is used. The important thing for any particular animal is to avoid trying to do both at once, for chemical confusion and inefficiency would result.

Now suppose that Pathway 1 needs the succession of enzymes A1, B1 and C1, in order to synthesize a desired chemical D, while Pathway 2 needs enzymes A2, B2 and C2 in order to arrive at the same desirable end-product. Each enzyme is made by a particular gene. So, in order to evolve the assembly line for Pathway 1, a species needs the genes coding for A1, B1 and C1 all to coevolve together. In order to evolve the alternative assembly line for Pathway 2, a species would need the genes coding for A2, B2 and C2 to coevolve with one another. The choice between these two coevolutions doesn't come about through advance planning. It comes about simply through each gene being selected by virtue of its compatibility with the other genes *that already happen to dominate the population*. If the population happens to be already rich in genes for B1 and C1, this will set up a climate favouring the A1 gene rather than the A2 gene. Conversely, if the population is already rich in genes for B2 and C2 this will set up a climate in which the A2 gene is favoured by selection rather than the A1 gene.

It will not be as simple as that, but you will have got the idea: one of the most important aspects of the 'climate' in which a gene is favoured or disfavoured is the other genes that are already numerous in the population; the other genes, therefore, with which it is likely to have to share bodies. Since the same will obviously be true of these 'other' genes themselves, we have a picture of teams of genes all evolving towards cooperative solutions to problems. The genes themselves don't evolve, they merely survive or fail to survive in the gene pool. It is the 'team' that evolves. Other teams might have done the job just as well, or even better. But once one team has started to dominate the gene pool of a species it thereby has an automatic advantage. It is difficult for a minority team to break in, even a minority team which would, in the end, have done the job more efficiently. The majority team has an automatic resistance to being displaced, simply by virtue of being in the majority. This doesn't mean that the majority team can never be displaced. If it couldn't, evolution would grind to a halt. But it does mean that there is a kind of built-in inertia.

Obviously this kind of argument is not limited to biochemistry. We could make the same kind of case for clusters of compatible genes building the different parts of eyes, ears, noses, walking limbs, all the cooperating parts of an animal's body. Genes for making teeth suitable for chewing meat tend to be favoured in a 'climate' dominated by genes making guts suitable for digesting meat. Conversely, genes for making plant-grinding teeth tend to be favoured in a climate dominated by genes that make guts suitable for digesting plants. And vice versa in both cases. Teams of 'meat-eating genes' tend to evolve together, and teams of 'plant-eating genes' tend to evolve together. Indeed, there is a sense in which most of the working genes in a body can be said to cooperate with each other as a team, because over evolutionary time they (i.e. ancestral copies of themselves) have each been part of the environment in which natural selection has worked on the others. If we ask why the ancestors of lions took to meat-eating, while the ancestors of antelopes took to grass-eating, the answer could be that originally it was an accident. An accident, in the sense that it could have been the ancestors of lions that took up grass-eating, and the ancestors of antelopes that took up meat-eating. But once one lineage had *begun* to build up a team of genes for dealing with meat rather than grass, the process was self-reinforcing. And once the other lineage had begun to build up a team of genes for dealing with grass rather than meat, *that* process was self-reinforcing in the other direction.

One of the main things that must have happened in the early evolution of living organisms was an increase in the numbers of genes participating in such cooperatives. Bacteria have far fewer genes than animals and plants. The increase may have come about through various kinds of gene duplication. Remember that a gene is just a length of coded symbols, like a file on a computer disc; and genes can be copied to different parts of the chromosomes, just as files can be copied to different parts of the disc. On my disc that holds this chapter there are officially just three files. By 'officially' I mean that the computer's operating system tells me that there are just three files. I can ask it to read one of these three files, and it presents me with a one-dimensional array of alphabetical characters, including the characters that you are now reading. All very neat and orderly, it seems. But in fact, on the disc itself, the arrangement of the text is anything but neat and orderly. You can see this if you break away from the discipline of the computer's own official operating system, and write your own private programs to decipher what is actually written on every sector of the disc. It turns out that fragments of each of my three files are dotted around, interleaved with each other and with fragments of old, dead files that I erased long ago and had forgotten. Any given fragment may turn up, word for word the same, or with minor differences, in half a dozen different places all around the disc.

The reason for this is interesting, and worth a digression because it provides a good genetic analogy. When you tell a computer to delete a file, it appears to obey you. But it doesn't actually wipe out the text of that file. It simply wipes out all *pointers* to that file. It is as though a librarian, ordered to destroy *Lady Chatterley's Lover*, simply tore up the card from the card index, leaving the book itself on the shelf. For the computer, this is a perfectly economical way to do things, because the space formerly occupied by the 'deleted' file is automatically available for new files, as soon as the pointers to the old file have been removed. It would be a waste of time actually to go to the trouble of filling the space itself with blanks. The old file won't itself be finally lost until all its space happens to be used for storing new files.

But this re-using of space occurs piecemeal. New files aren't exactly the same size as old ones. When the computer is trying to save a new file to a disc, it looks for the first available fragment of space, writes as much of the new file as will fit, then looks for another available fragment of space, writes a bit more, and so on until all the file is written *somewhere* on the disc. The human has the illusion that the file is a single, orderly array, only because the computer is careful to keep records 'pointing' to the addresses of all the fragments dotted around. These 'pointers' are like the 'continued on page 94' pointers used by the *New York Times*. The reason many copies of any one fragment of text are found on a disc is that if, like all my chapters, the text has been edited and re-edited many dozens of times, each edit will result in a new saving to the disc of (almost) the same text. The saving may ostensibly be a saving of the same file. But as we have seen, the text will in fact be repeatedly scattered around the available 'gaps' on the disc. Hence multiple copies of a given fragment of text can be found all around the surface of the disc, the more so if the disc is old and much used.

Now the DNA operating system of a species is very very old indeed, and there is evidence that it, seen in the long term, does something a bit like the computer with its disc files. Part of the evidence comes from the fascinating phenomenon of 'introns' and 'exons'. Within the last decade, it has been discovered that any 'single' gene, in the sense of a single continuously read passage of DNA text, is not all stored in one place. If you actually read the code letters as they occur along the chromosome (i.e. if you do the equivalent of breaking out of the discipline of the 'operating system') you find fragments of 'sense', called exons, separated by portions of 'nonsense' called introns. Any one 'gene' in the functional sense, is in fact split up into a sequence of fragments (exons) separated by meaningless introns. It is as if each exon ended with a pointer saying 'continued on page 94'. A complete gene is then made up of a whole series of exons, which are actually strung together only when they are eventually read by the 'official' operating system that translates them into proteins.

Further evidence comes from the fact that the chromosomes are littered with old genetic text that is no longer used, but which still makes recognizable sense. To a computer programmer, the pattern of distribution of these 'genetic fossil' fragments is uncannily reminiscent of the pattern of text on the surface of an old disc that has been much used for editing text. In some animals, a high proportion of the total number of genes is in fact never read. These genes are either complete nonsense, or they are outdated 'fossil genes'.

Just occasionally, textual fossils come into their own again, as I experienced when writing this book. A computer error (or, to be fair, it may have been human error) caused me accidentally to 'erase' the disc containing Chapter 3. Of course the text itself hadn't literally all been erased. All that had been definitely erased were the *pointers* to where each 'exon' began and ended. The 'official' operating system could read nothing, but 'unofficially' I could play genetic engineer and examine all the text on the disc. What I saw was a bewildering jigsaw puzzle of textual fragments, some of them recent, others ancient 'fossils'. By piecing together the jigsaw fragments, I was able to recreate the chapter. But I mostly didn't know which fragments were recent and which were fossil. It didn't matter for, apart from minor details that necessitated some new editing, they were the same. At least some of the 'fossils', or outdated 'introns', had come into their own again. They rescued me from my predicament, and saved me the trouble of re-writing the entire chapter.

There is evidence that, in living species too, 'fossil genes' occasionally come into their own again, and are re-used after lying dormant for a million years or so. To go into detail would carry us too far from the main pathway of this chapter, for you will remember that we are already out on a digression. The main point was that the total genetic capacity of a species may increase due to gene duplication. Re-using of old 'fossil' copies of existing genes is one way in which this can happen. There are other, more immediate, ways in which genes may be copied to widely distributed parts of the chromosomes, like files being duplicated to different parts of a disc, or different discs.

Humans have eight separate genes called globin genes (used for making haemoglobin, among other things), on various different chromosomes. It seems certain that all eight have been copied, ultimately from a single ancestral globin gene. About 1,100 million years ago, the ancestral globin gene duplicated, forming two genes. We can date this event because of independent evidence about how fast globins habitually evolve (see Chapters 5 and 11). Of the two genes produced by this original duplication, one became the ancestor of all the genes that make haemoglobin in vertebrates. The other became the ancestor of all the genes that make myoglobins, a related family of proteins that work in muscles. Various subsequent duplications have given rise to the so-called alpha, beta, gamma, delta, epsilon and zeta globins. The fascinating thing is that we can construct a complete family tree of all the globin genes, and even put dates on all the divergence points (delta and beta globin parted company, for example, about 40 million years ago; epsilon and gamma globins 100 million years ago). Yet the eight globins, descendants as they are of these remote branchings in distant ancestors, are still all present inside every one of us. They diverged to different parts of an ancestor's chromosomes, and we have each inherited them on our different chromosomes. Molecules are sharing the same body with their remote molecular cousins. It is certain that a great deal of such duplication has gone on, all over the chromosomes, and throughout geological time. This is an important respect in which real life is more complicated than the biomorphs of Chapter 3. They all had only nine genes. They evolved by changes in those nine genes, never by increasing the number of genes to ten. Even in real animals, such duplications are rare enough not to invalidate my general statement that all members of a species share the same DNA 'addressing' system.

Duplication within the species isn't the only means by which the number of cooperating genes has increased in evolution. An even rarer, but still possibly very important occurrence, is the occasional incorporation of a gene from another species, even an extremely remote species. There are, for example, haemoglobins in the roots of plants of the pea family. They don't occur in any other plant families, and it seems almost certain that they somehow got into the pea family by cross-infection from animals, viruses perhaps acting as intermediaries.

An especially important event along these lines, according to the increasingly favoured theory of the American biologist Lynn Margulis, took place at the origin of the so-called eukaryotic cell. Eukaryotic cells include all cells except those of bacteria. The living world is divided, fundamentally, into bacteria versus the rest. We are part of the rest, and are collectively called the eukaryotes. We differ from bacteria mainly in that our cells have discrete little mini-cells inside them. These include the nucleus, which houses the chromosomes; the tiny bomb-shaped objects called mitochondria (which we briefly met in Figure 1), filled with intricately folded membranes; and, in the (eukaryotic) cells of plants, chloroplasts. Mitochondria and chloroplasts have their own DNA, which replicates and propagates itself entirely independently of the main DNA in the chromosomes of the nucleus. All the mitochondria in you are descended from the small population of mitochondria that travelled from your mother in her egg. Sperms are too small to contain mitochondria, so mitochondria travel exclusively down the female line, and male bodies are dead ends as far as mitochondrial reproduction is concerned. Incidentally, this means that we can use mitochondria to trace our ancestry, strictly down the female line.

Margulis's theory is that mitochondria and chloroplasts, and a few other structures inside cells, are each descended from bacteria. The eukaryotic cell was formed, perhaps 2 billion years ago, when several kinds of bacteria joined forces because of the benefits that each could obtain from the others. Over the aeons they have become so thoroughly integrated into the cooperative unit that became the eukaryotic cell, that it has become almost impossible to detect the fact, if indeed it is a fact, that they were once separate bacteria.

It seems that, once the eukaryotic cell had been invented, a whole new range of designs became possible. Most interestingly from our point of view, cells could manufacture large bodies comprising many billions of cells. All cells reproduce by splitting into two, both halves getting a full set of genes. As we saw in the case of the bacteria on a pin's head, successive splittings into two can generate a very large number of cells in rather a short time. You start with one and it splits into two. Then each of the two splits, making four. Each of the four splits, making eight. The numbers go up by successive doublings, from 8 to 16, 32, 64, 128, 256, 512, 1,024, 2,048, 4,096, 8,192. After only 20 doublings, which doesn't take very long, we are up in the millions. After only 40 doublings the number of cells is more than a trillion. In the case of bacteria, the enormous numbers of cells produced by successive doublings go their separate ways. The same is true of many eukaryotic cells, for instance protozoa such as amoebas. A major step in evolution was taken when cells that had been produced by successive splittings stuck together instead of going off independently. Higher-order structure could now emerge, just as it did, on an incomparably smaller scale, in the two-way branching computer biomorphs.

Now, for the first time, large body size became a possibility. A human body is a truly colossal population of cells, all descended from one ancestor, the fertilized egg; and all therefore cousins, children, grandchildren, uncles, etc. of other cells in the body. The 10 trillion cells that make up each one of us are the product of a few dozens of generations of cell doublings. These cells are classified into about 210 (according to taste) different kinds, all built by the same set of genes but with different members of the set of genes turned on in different kinds of cells. This, as we have seen, is why liver cells are different from brain cells, and bone cells are different from muscle cells.

Genes working through the organs and behaviour patterns of many-celled bodies can achieve methods of ensuring their own propagation that are not available to single cells working on their own. Many-celled bodies make it possible for genes to manipulate the world, using tools built on a scale that is orders of magnitude larger than the scale of single cells. They achieve these large-scale indirect manipulations via their more direct effects on the miniature scale of cells. For instance, they change the shape of the cell membrane. The cells then interact with one another in huge populations to produce large-scale group effects such as an arm or a leg or (more indirectly) a beaver's dam. Most of the properties of an organism that we are equipped to see with our naked eyes are so-called 'emergent properties'. Even the computer biomorphs, with their nine genes, had emergent properties. In real animals they are produced at the whole-body level by interactions between cells. An organism works as an entire unit, and its genes can be said to have effects on the whole organism, even though each copy of any one gene exerts its immediate effects only within its own cell.

We have seen that a very important part of a gene's environment is the other genes that it is likely to meet in successive bodies as the generations go by. These are the genes that are permuted and combined within the species. Indeed, a sexually reproducing species can be thought of as a device that permutes a discrete set of mutually accustomed genes in different combinations. Species, according to this view, are continually shuffling collections of genes that meet each other within the species, but never meet genes of other species. But there is a sense in which the genes of different species, even if they don't meet at close quarters inside cells, nevertheless constitute an important part of each others' environment. The relationship is often hostile rather than cooperative, but this can be treated as just a reversal of sign. This is where we come to the second major theme of this chapter, 'arms races'. There are arms races between predators and prey, parasites and hosts, even – though the point is a more subtle one and I shan't discuss it further – between males and females within one species.

Arms races are run in evolutionary time, rather than on the timescale of individual lifetimes. They consist of the improvement in one lineage's (say prey animals') equipment to survive, as a direct consequence of improvement in another (say predators') lineage's evolving equipment. There are arms races wherever individuals have enemies with their own capacity for evolutionary improvement. I regard arms races as of the utmost importance because it is largely arms races that have injected such 'progressiveness' as there is in evolution. For, contrary to earlier prejudices, there is nothing inherently progressive about evolution. We can see this if we consider what would have happened if the only problems animals had had to face had been those posed by the weather and other aspects of the nonliving environment.

After many generations of cumulative selection in a particular place, the local animals and plants become well fitted to the conditions, for instance the weather conditions, in that place. If it is cold the animals come to have thick coats of hair, or feathers. If it is dry they evolve leathery or waxy waterproof skins to conserve what little water there is. The adaptations to local conditions affect every part of the body, its shape and colour, its internal organs, its behaviour, and the chemistry in its cells.

If the conditions in which a lineage of animals lives remain constant; say it is dry and hot and has been so without a break for 100 generations, evolution in that lineage is likely to come to a halt, at least as far as adaptations to temperature and humidity are concerned. The animals will become as well fitted as they can be to the local conditions. This doesn't mean that they couldn't be completely redesigned to be even better. It does mean that they can't improve themselves by any *small* (and therefore likely) evolutionary step: none of their *immediate* neighbours in the local equivalent of 'biomorph space' would do any better.

Evolution will come to a standstill until something in the

conditions changes: the onset of an ice age, a change in the average rainfall of the area, a shift in the prevailing wind. Such changes do happen when we are dealing with a timescale as long as the evolutionary one. As a consequence, evolution normally does not come to a halt, but constantly 'tracks' the changing environment. If there is a steady downward drift in the average temperature in the area, a drift that persists over centuries, successive generations of animals will be propelled by a steady selection 'pressure' in the direction, say, of growing longer coats of hair. If, after a few thousand years of reduced temperature the trend reverses and average temperatures creep up again, the animals will come under the influence of a new selection pressure, and will be pushed towards growing shorter coats again.

But so far we have considered only a limited part of the environment, namely the weather. The weather is very important to animals and plants. Its patterns change as the centuries go by, so this keeps evolution constantly in motion as it 'tracks' the changes. But weather patterns change in a haphazard, inconsistent way. There are other parts of an animal's environment that change in more consistently malevolent directions, and that also need to be 'tracked'. These parts of the environment are living things themselves. For a predator such as a hyena, a part of its environment that is at least as important as the weather is its prey, the changing populations of gnus, zebras and antelopes. For the antelopes and other grazers that wander the plains in search of grass, the weather may be important, but the lions, hyenas and other carnivores are important too. Cumulative selection will see to it that animals are well fitted to outrun their predators or outwit their prey, no less than it sees to it that they are well fitted to the prevailing weather conditions. And, just as long-term fluctuations in the weather are 'tracked' by evolution, so long-term changes in the habits or weaponry of predators will be tracked by evolutionary changes in their prey. And vice versa, of course.

We can use the general term 'enemies' of a species, to mean other living things that work to make life difficult. Lions are enemies of zebras. It may seem a little callous to reverse the statement to 'Zebras are enemies of lions'. The role of the zebra in the relationship seems too innocent and wronged to warrant the pejorative 'enemy'. But individual zebras do everything in their power to resist being eaten by lions, and from the lions' point of view this is making life harder for them. If zebras and other grazers all succeeded in their aim, the lions would die of starvation. So by our definition zebras are enemies of lions. Parasites such as tapeworms are enemies of their hosts, and hosts are enemies of parasites since they tend to evolve measures to resist them. Herbivores are enemies of plants, and plants are enemies of herbivores, to the extent that they manufacture thorns, and poisonous or nasty-tasting chemicals.

Lineages of animals and plants will, in evolutionary time, 'track' changes in their enemies no less assiduously than they track changes in average weather conditions. Evolutionary improvements in cheetah weaponry and tactics are, from the gazelles' point of view, like a steady worsening of the climate, and they are tracked in the same kind of way. But there is one enormously important difference between the two. The weather changes over the centuries, but it does *not* change in a specifically malevolent way. It is not out to 'get' gazelles. The average cheetah will change over the centuries, just like the mean annual rainfall changes. But whereas mean annual rainfall will drift up and down, with no particular rhyme or reason, the average cheetah, as the centuries go by, will tend to become *better* equipped to catch gazelles than his ancestors were. This is because the succession of cheetahs, unlike the succession of annual weather conditions, is itself subject to cumulative selection. Cheetahs will tend to become fleeter of foot, keener of eye, sharper of tooth. However 'hostile' the weather and other inanimate conditions may seem to be, they have no necessary tendency to get steadily more hostile. Living enemies, seen over the evolutionary timescale, have exactly that tendency.

The tendency for carnivores to get progressively 'better' would soon run out of steam, as do human arms races (for reasons of economic cost which we shall come to), were it not for the parallel tendency in the prey. And vice versa. Gazelles, no less than cheetahs, are subject to cumulative selection, and they too will tend, as the generations go by, to improve their ability to run fast, to react swiftly, to become invisible by blending into the long grass. They too are capable of evolving in the direction of becoming better enemies, in this case enemies of cheetahs. From the cheetahs' point of view the mean annual temperature does not get systematically better or worse as the years go by, except in so far as any change for a well-adapted animal is a change for the worse. But the mean annual gazelle does tend to get systematically worse – more difficult to catch because better adapted to evade cheetahs. Again, the tendency towards progressive improvement in gazelles would slow to a halt, were it not for the parallel tendency to improvement shown by their predators. One side gets a little better because the other side has. And vice versa. The process goes into a vicious spiral, on a timescale of hundreds of thousands of years.

In the world of nations on their shorter timescale, when two enemies each progressively improve their weaponry in response to the other side's improvements, we speak of an 'arms race'. The evolutionary analogy is close enough to justify borrowing the term, and I make no apology to my pompous colleagues who would purge our language of such illuminating images. I have introduced the idea here in terms of a simple example, gazelles and cheetahs. This was to get across the important difference between a living enemy, which itself is subject to evolutionary change, and an inanimate non-malevolent condition such as the weather, which is subject to change, but not systematic, evolutionary change. But the time has come to admit that in my efforts to explain this one valid point I may have misled the reader in other ways. It is obvious, when you come to think about it, that my picture of an ever-advancing arms race was too simple in at least one respect. Take running speed. As it stands so far, the arms-race idea seems to suggest that cheetahs and gazelles should have gone on, generation after generation, getting ever faster until both travelled faster than sound. This has not happened and it never will. Before resuming the discussion of arms races, it is my duty to forestall misunderstandings.

The first qualification is this. I gave an impression of a steady upward climb in the prey-catching abilities of cheetahs, and the predator-avoiding abilities of gazelles. The reader might have come away with a Victorian idea of the inexorability of progress, each generation better, finer and braver than its parents. The reality in nature is nothing like that. The timescale over which significant improvement might be detected is, in any case, likely to be far longer than could be detected by comparing one typical generation with its predecessor. The 'improvement', moreover, is far from continuous. It is a fitful affair, stagnating or even sometimes going 'backwards', rather than moving solidly 'forwards' in the direction suggested by the arms-race idea. Changes in conditions, changes in the inanimate forces I have lumped under the general heading of 'the weather', are likely to swamp the slow and erratic trends of the arms race, as far as any observer on the ground could be aware. There may well be long stretches of time in which no 'progress' in the arms race, and perhaps no evolutionary change at all, takes place. Arms races sometimes culminate in extinction, and then a new arms race may begin back at square one. Nevertheless, when all this is said, the arms-race idea remains by far the most satisfactory explanation for the existence of the advanced and complex machinery that animals and plants possess. Progressive 'improvement' of the kind suggested by the arms-race image does go on, even if it goes on spasmodically and interruptedly; even if its net rate of progress is too slow to be detected within the lifetime of a man, or even within the timespan of recorded history.

The second qualification is that the relationship that I am calling 'enemy' is more complicated than the simple bilateral relationship suggested by the stories of cheetahs and gazelles. One complication is that a given species may have two (or more) enemies which are even more severe enemies of each other. This is the principle behind the commonly expressed half-truth that grass benefits by being grazed (or mown). Cattle eat grass, and might therefore be thought of as enemies of grass. But grasses also have other enemies in the plant world, competitive weeds, which, if allowed to grow unchecked, might turn out to be even more severe enemies of grasses than cattle. Grasses suffer somewhat from being eaten by cattle, but the competitive weeds suffer even more. Therefore the net effect of cattle on a meadow is that the grasses benefit. The cattle turn out to be, in this sense, friends of grasses rather than enemies.

Nevertheless, cattle are enemies of grass in that it is *still* true that an individual grass plant would be better off not being eaten by a cow than being eaten, and any mutant plant that possessed, say, a chemical weapon that protected it against cows, would set more seed (containing genetic instructions for making the chemical weapon) than rival members of its own species that were more palatable to cows. Even if there is a special sense in which cows are 'friends' of grasses, natural selection does *not* favour individual grass plants that go out of their way to be eaten by cows! The general conclusion to this paragraph is as follows. It may be convenient to think of an arms race between two lineages such as cattle and grass, or gazelles and cheetahs, but we should never lose sight of the fact that both participants have other enemies against whom they are simultaneously running other arms races. I shall not pursue the point here, but it can be developed into one of the explanations for why particular arms races stabilize and do not go on for ever — do not lead to predators pursuing their prey at Mach 2 and so on.

The third 'qualification' to the simple arms-race is not so much a qualification as an interesting point in its own right. In my hypothetical discussion of cheetahs and gazelles I said that cheetahs, unlike the weather, had a tendency as the generations go by to become 'better hunters', to become more severe enemies, better equipped to kill gazelles. But this does not imply that they become more *successful* at killing gazelles. The kernel of the arms-race idea is that both sides in the arms race are improving from their own point of view, while simultaneously making life more difficult for the other side in the arms race. There is no particular reason (or at least none in anything that we have discussed so far) to expect either side in the arms race to become steadily more successful or less successful than the other. In fact the arms-race idea, in its purest form, suggests that there should be absolutely zero progress in the *success rate* on both sides of the arms race, while there is very definite progress in the *equipment* for success on both sides. Predators become better equipped for killing, but at the same time prey become better equipped to avoid being killed, so the net result is no change in the rate of successful killings.

The implication is that if, by the medium of a time machine, predators from one era could meet prey from another era, the later, more 'modern' animals, whether predators or prey, would run rings round the earlier ones. This is not an experiment that can ever be done, although some people assume that certain remote and isolated faunas, such as those of Australia and Madagascar, can be treated as if they were ancient, as if a trip to Australia were like a trip backwards in a time machine. Such people think that native Australian species are usually driven extinct by superior competitors or enemies introduced from the outside world, because the native species are 'older', 'out of date' models, in the same position *vis-à-vis* invading species as a Jutland battleship contending with a nuclear submarine. But the assumption that Australia has a 'living fossil' fauna is hard to justify. Perhaps a good case for it might be made, but it seldom is. I'm afraid it may be no more than the zoological equivalent of chauvinistic snobbery, analogous to the attitude that sees every Australian as an uncouth swagman with not much under his hat and corks dangling round the brim.

The principle of zero change in success *rate*, no matter how great the evolutionary progress in *equipment*, has been given the memorable name of the 'Red Queen effect' by the American biologist Leigh van Valen. In *Through the Looking Glass*, you will remember, the Red Queen seized Alice by the hand and dragged her, faster and faster, on a frenzied run through the countryside, but no matter how fast they ran they always stayed in the same place. Alice was understandably puzzled, saying, 'Well in *our* country you'd generally get to somewhere else – if you ran very fast for a long time as we've been doing.' 'A slow sort of country!' said the Queen. 'Now, *here*, you see, it takes all the running *you* can do, to keep in the same place. If you want to get somewhere else, you must run at least twice as fast as that!'

The Red Queen label is amusing, but it can be misleading if taken (as it sometimes is) to mean something mathematically precise, literally zero relative progress. Another misleading feature is that in the Alice story the Red Queen's statement is genuinely paradoxical, irreconcilable with common sense in the real physical world. But van Valen's evolutionary Red Queen effect is not paradoxical at all. It is entirely in accordance with common sense, so long as common sense is intelligently applied. If not paradoxical, however, arms races can give rise to situations that strike the economically minded human as wasteful.

Why, for instance, are trees in forests so tall? The short answer is that all the other trees are tall, so no one tree can afford not to be. It would be overshadowed if it did. This is essentially the truth, but it offends the economically minded human. It seems so pointless, so wasteful. When all the trees are the full height of the canopy, all are approximately equally exposed to the sun, and none could afford to be any shorter. But if only they were *all* shorter; if only there could be some sort of trade-union agreement to lower the recognized height of the canopy in forests, *all* the trees would benefit. They would be competing with each other in the canopy for exactly the same sunlight, but they would all have 'paid' much smaller growing costs to get into the canopy. The total economy of the forest would benefit, and so would every individual tree. Unfortunately, natural selection doesn't care about total economies, and it has no room for cartels and agreements. There has been an arms race in which forest trees became larger as the generations went by. At every stage of the arms race there was no intrinsic benefit in being tall for its own sake. At every stage of the arms race the only point in being tall was to be relatively taller than neighbouring trees.

As the arms race wore on, the average height of trees in the forest canopy went up. But the benefit that the trees obtained from being tall did not go up. It actually deteriorated because of the enhanced costs of growing. Successive generations of trees got taller and taller, but at the end they might better, in one sense, have stayed where they started. Here, then, is the connection with Alice and the Red Queen, but you can see that in the case of the trees it is not really paradoxical. It is generally characteristic of arms races, including human ones, that although all would be better off if *none* of them escalated, so long as one of them escalates none can afford *not* to. Once again, by the way, I should stress that I have told the story too simply. I do not mean to suggest that in every literal generation trees are taller than their counterparts in the previous generation, nor that the arms race is necessarily still going on.

Another point illustrated by the trees is that arms races do not necessarily have to be between members of different species. Individual trees are just as likely to be harmfully overshadowed by members of their own species as by members of other species. Probably more so in fact, for all organisms are more seriously threatened by competition from their own species than from others. Members of one's own species are competitors for the same resources, to a much more detailed extent, than members of other species. There are also arms races within species between male roles and female roles, and between parent roles and offspring roles. I have discussed these in *The Selfish Gene*, and will not pursue them further here.

The tree story allows me to introduce an important general distinction between two kinds of arms race, called symmetric and asymmetric arms races. A symmetric arms race is one between competitors trying to do roughly the same thing as each other. The arms race between forest trees struggling to reach the light is an example. The different species of trees are not all making their livings in exactly the same way, but as far as the particular race we are talking about is concerned – the race for the sunlight above the canopy – they are competitors for the same resource. They are taking part in an arms race in which success on one side is felt by the other side as failure. And it is a symmetric arms race because the nature of the success and failure on the two sides is the same: attainment of sunlight and being overshadowed, respectively.

The arms race between cheetahs and gazelles, however, is asymmetric. It is a true arms race in which success on either side is felt as failure by the other side, but the nature of the success and failure on the two sides is very different. The two sides are 'trying' to do very different things. Cheetahs are trying to eat gazelles. Gazelles are not trying to eat cheetahs, they are trying to avoid being eaten by cheetahs. From an evolutionary point of view asymmetric arms races are more interesting, since they are more likely to generate highly complex weapons systems. We can see why this is by taking examples from human weapons technology.

I could use the USA and the USSR as examples, but there is really no need to mention specific nations. Weapons manufactured by companies in any of the advanced industrial countries may end up being bought by any of a wide variety of nations. The existence of a successful offensive weapon, such as the Exocet type of surface skimming missile, tends to 'invite' the invention of an effective counter, for instance a radio jamming device to 'confuse' the control system of the missile. The counter is more likely than not to be manufactured by an enemy country, but it could be manufactured by the same country, even by the same company! No company, after all, is better equipped to design a jamming device for a particular missile than the company that made the missile in the first place. There is nothing inherently improbable about the same company producing both and selling them to opposite sides in a war. I am cynical enough to suspect that it probably happens, and it vividly illustrates the point about *equipment* improving while its net *effectiveness* stands still (and its costs increase).

From my present point of view the question of whether the manufacturers on opposite sides of a human arms race are enemies of each other or identical with each other is irrelevant, and interestingly so. What matters is that, regardless of their manufacturers, the devices themselves are enemies of each other in the special sense I have defined in this chapter. The missile, and its specific jamming device, are enemies of each other in that success in one is synonymous with failure in the other. Whether their designers are also enemies of each other is irrelevant, although it will probably be easier to assume that they are.

So far I have discussed the example of the missile and its specific antidote without stressing the evolutionary, progressive aspect, which is, after all, the main reason for bringing it into this chapter. The point here is that not only does the present design of a missile invite, or call forth, a suitable antidote, say a radio jamming device. The antimissile device, in its turn, invites an improvement in the design of the missile, an improvement that specifically counters the antidote, an anti-antimissile device. It is almost as though each improvement in the missile stimulates the next improvement *in itself*, via its effect on the antidote. Improvement in equipment feeds on itself. This is a recipe for explosive, runaway evolution.

At the end of some years of this ding-dong invention and counter-invention, the current version of both the missile and its antidote will have attained a very high degree of sophistication. Yet at the same time – here is the Red Queen effect again – there is no general reason for expecting either side in the arms race to be any more successful at doing its job than it was at the beginning of the arms race. Indeed if both the missile and its antidote have been improving at the same rate, we can expect that the latest, most advanced and sophisticated versions, and the earliest, most primitive and simplest versions will be exactly as successful as each other, against their contemporary counter-devices. There has been progress in design, but no progress in accomplishment, specifically because there has been equal progress in design on both sides of the arms race. Indeed, it is precisely *because* there has been approximately equal progress on both sides that there has been so much progress in the level of sophistication of design. If one side, say the antimissile jamming device, pulled too far ahead in the design race, the other side, the missile in this case, would simply cease to be used and manufactured: it would go 'extinct'. Far from being paradoxical like Alice's original example, the Red Queen effect in its arms-race context turns out to be fundamental to the very idea of progressive advancement.

I said that asymmetric arms races were more likely to lead to interesting progressive improvements than symmetric ones, and we can now see why this is, using human weapons to illustrate the point. If one nation has a 2-megaton bomb, the enemy nation will develop a 5-megaton bomb. This provokes the first nation into developing a 10-megaton bomb, which in turn provokes the second into making a 20-megaton bomb, and so on. This is a true progressive arms race: each advance on one side provokes the counter-advance on the other, and the result is a steady increase in some attribute as time goes by – in this case, explosive power of bombs. But there is no detailed, one-to-one correspondence between the designs in such a symmetric arms race, no 'meshing' or 'interlocking' of design details as there is in an asymmetric arms race such as that between missile and missile-jamming device. The missile-jamming device is designed specificially to overcome particular detailed features of the missile; the designer of the antidote takes into account minute details of the design of the missile. Then in designing a counter to the antidote, the designer of the next generation of missiles makes use of his knowledge of the detailed design of the antidote to the previous generation. This is not true of the bombs of ever-increasing megatonnage. To be sure, designers on one side may pirate good ideas, may imitate design features, from the other side. But if so, this is incidental. It is not a *necessary* part of the design of a Russian bomb that it should have detailed, one-to-one correspondences with specific details of an American bomb. In the case of an asymmetric arms race, between a lineage of weapons and the specific antidotes to those weapons, it is the one-to-one correspondences that, over the successive 'generations', lead to ever greater sophistication and complexity.

In the living world too, we shall expect to find complex and sophisticated design wherever we are dealing with the end-products of a long, asymmetric arms race in which advances on one side have always been matched, on a one-to-one, point-for-point basis, by equally successful *antidotes* (as opposed to competitors) on the other. This is conspicuously true of the arms races between predators and their prey, and, perhaps even more, of arms races between parasites and hosts. The electronic and acoustic weapons systems of bats, which we discussed in Chapter 2, have all the finely tuned sophistication that we expect from the end-products of a long arms race. Not surprisingly, we can trace this same arms race on the other side. The insects that bats prey upon have a comparable battery of sophisticated electronic and acoustic gear. Some moths even emit bat-like (ultra-) sounds that seem to put the bats off. Almost all animals are either in danger of being eaten by other animals or in danger of failing to eat other animals, and an enormous number of detailed facts about animals makes sense only when we remember that they are the end-products of long and bitter arms races. H. B. Cott, author of the classic book *Animal Coloration*, put the point well in 1940, in what may be the first use in print of the arms-race analogy in biology:

> Before asserting that the deceptive appearance of a grasshopper or butterfly is unnecessarily detailed, we must first ascertain what are the powers of perception and discrimination of the insects' natural enemies. Not to do so is like asserting that the armour of a battle-cruiser is too heavy, or the range of her guns too great, without inquiring into the nature and effectiveness of the enemy's armament. The fact is that in the primeval struggle of the jungle, as in the refinements of civilized warfare, we see in progress a great evolutionary armament race – whose results, for defence, are manifested in such devices as speed, alertness, armour, spinescence, burrowing habits, nocturnal habits, poisonous secretions, nauseous taste, and [camouflage and other kinds of protective coloration]; and for offence, in such counter-attributes as speed, surprise, ambush, allurement, visual acuity, claws, teeth, stings, poison fangs, and [lures]. Just as greater speed in the pursued has developed in relation to increased speed in the pursuer; or defensive armour in relation to aggressive weapons; so the perfection of concealing devices has evolved in response to increased powers of perception.

Arms races in human technology are easier to study than their biological equivalents because they are so much faster. We can actually see them going on, from year to year. In the case of a biological arms race, on the other hand, we can usually see only the end-products. Very rarely a dead animal or plant fossilizes, and it is then sometimes possible to see progressive stages in an animal arms race a little more directly. One of the most interesting examples of this concerns the electronic arms race, as shown in the brain sizes of fossil animals.

Brains themselves do not fossilize but skulls do, and the cavity in which the brain was housed – the braincase – if interpreted with care, can give a good indication of brain size. I said 'if interpreted with care', and the qualification is an important one. Among the many problems is the following. Big animals tend to have big brains partly just because they are big, but this doesn't necessarily mean that they are, in any interesting sense, 'cleverer'. Elephants have bigger brains than humans but, probably with some justice, we like to think that we are cleverer than elephants and that our brains are 'really' bigger if you make allowance for the fact that we are much smaller animals. Certainly our brains occupy a much larger *proportion* of our body than elephants' brains do, as is evident from the bulging shape of our skulls. This is not *just* species vanity. Presumably a substantial fraction of any brain is needed to perform routine caretaking operations around the body, and a big body automatically needs a big brain for this. We must find some way of 'taking out' of our calculations that fraction of brain that can be attributed simply to body size, so that we can compare what is left over as the true 'braininess' of animals. This is another way of saying that we need some good way of defining exactly what we mean by true braininess. Different people are at liberty to come up with different methods of doing the calculations, but probably the most authoritative index is the 'encephalization quotient' or EQ used by Harry Jerison, a leading American authority on brain history.

The EQ is actually calculated in a somewhat complicated way, taking logarithms of brain weight and body weight, and standardizing against the average figures for a major group such as the mammals as a whole. Just as the 'intelligence quotient' or IQ used (or it may be misused) by human psychologists is standardized against the average for a whole population, the EQ is standardized against, say, the whole of the mammals. Just as an IQ of 100 means, by definition, an IQ identical to the average for a whole population, so an EQ of 1 means, by definition, an EQ identical to the average for, say, mammals of that size. The details of the mathematical technique don't matter. In words, the EQ of a given species such as a rhino or a cat, is a measure of how much bigger (or smaller) the animal's brain is than we should *expect* it to be, given the animal's body size. How that expectation is calculated is certainly open to debate and criticism. The fact that humans have an EQ of 7 and hippos an EQ of 0.3 may not literally mean that humans are 23 times as clever as hippos! But the EQ as measured is probably telling us *something* about how much 'computing power' an animal has in its head, over and above the irreducible minimum of computing power needed for the routine running of its large or small body.

Measured EQs among modern mammals are very varied. Rats have an EQ of about 0.8, slightly below the average for all mammals. Squirrels are somewhat higher, about 1.5. Perhaps the three-dimensional world of trees demands extra computing power for controlling precision leaps, and even more for thinking about efficient paths through a maze of branches that may or may not connect farther on. Monkeys are well above average, and apes (especially ourselves) even higher. Within the monkeys it turns out that some types have higher EQs than others and that, interestingly, there is some connection with how they make their living: insect-eating and fruit-eating monkeys have bigger brains, for their size, than leaf-eating monkeys. It makes some sense to argue that an animal needs less computing power to find leaves, which are abundant all around, than to find fruit, which may have to be searched for, or to catch insects, which take active steps to get away. Unfortunately, it is now looking as though the true story is more complicated, and that other variables, such as metabolic rate, may be more important. In the mammals as a whole, carnivores typically have a slightly higher EQ than the herbivores upon which they prey. The reader will probably have some ideas about why this might be, but it is hard to test such ideas. Anyway, whatever the reason, it seems to be a fact.

So much for modern animals. What Jerison has done is to reconstruct the probable EQs of extinct animals that now exist only as fossils. He has to estimate brain size by making plaster casts of the insides of braincases. Quite a lot of guesswork and estimation has to go into this, but the margins of error are not so great as to nullify the whole enterprise. The methods of taking plaster casts can, after all, be checked for their accuracy, using modern animals. We make-believe that the dried skull is all that we have from a modern animal, use a plaster cast to estimate how big its brain was from the skull alone, and then check with the real brain to see how accurate our estimate was. These checks on modern skulls encourage confidence in Jerison's estimates of long-dead brains. His conclusion is, firstly, that there is a tendency for brains to become bigger as the millions of years go by. At any given time, the current herbivores tended to have smaller brains than the contemporary carnivores that preyed on them. But later herbivores tended to have larger brains than earlier herbivores, and later carnivores larger brains than earlier carnivores. We seem to be seeing, in the fossils, an arms race, or rather a series of restarting arms races, between carnivores and herbivores. This is a particularly pleasing parallel with human armament races, since the brain is the on-board computer used by both carnivores and herbivores, and electronics is probably the most rapidly advancing element in human weapons technology today.

How do arms races end? Sometimes they may end with one side going extinct, in which case the other side presumably stops evolving in that particular progressive direction, and indeed it will probably even 'regress' for economic reasons soon to be discussed. In other cases, economic pressures may impose a stable halt to an arms race, stable even though one side in the race is, in a sense, permanently ahead. Take running speed, for instance. There must be an ultimate limit to the speed at which a cheetah or a gazelle can run, a limit imposed by the laws of physics. But neither cheetahs nor gazelles have reached that limit. Both have pushed up against a lower limit which is, I believe, economic in character. High-speed technology is not cheap. It demands long leg bones, powerful muscles, capacious lungs. These things can be had by any animal that really needs to run fast, but they must be *bought*. They are bought at a steeply increasing price. The price is measured as what economists call 'opportunity cost'. The opportunity cost of something is measured as the sum of all the other things that you have to forgo in order to have that something. The cost of sending a child to a private, fee-paying school is all the things that you can't afford to buy as a result: the new car that you can't afford, the holidays in the sun that you can't afford (if you're so rich that you can afford all these things easily, the opportunity cost, to you, of sending your child to a private school may be next to nothing). The price, to a cheetah, of growing larger leg muscles is all the other things that the cheetah *could have done* with the materials and energy used to make the leg muscles, for instance make more milk for cubs.

There is no suggestion, of course, that cheetahs do cost-accounting sums in their heads! It is all done automatically by ordinary natural selection. A rival cheetah that doesn't have such big leg muscles may not run quite so fast, but it has resources to spare for making an extra lot of milk and therefore perhaps rearing another cub. More cubs will be reared by cheetahs whose genes equip them with the optimum compromise between running speed, milk production and all the other calls on their budget. It isn't obvious what the optimum trade-off is between, say, milk production and running speed. It will certainly be different for different species, and it may fluctuate within each species. All that is certain is that trade-offs of this kind will be inevitable. When both cheetahs and gazelles reach the maximum running speed that they can 'afford', in their own internal economies, the arms race between them will come to an end.

Their respective economic stopping points may not leave them exactly equally matched. Prey animals may end up spending relatively more of their budget on defensive weaponry than predators do on offensive weaponry. One reason for this is summarized in the Aesopian moral: The rabbit runs faster than the fox, because the rabbit is running for his life, while the fox is only running for his dinner. In economic terms, this means that individual foxes that shift resources into other projects can do better than individual foxes that spend virtually all their resources on hunting technology. In the rabbit population, on the other hand, the balance of economic advantage is shifted towards those individual rabbits that are big spenders on equipment for running fast. The upshot of these economically balanced budgets *within* species is that arms races *between* species tend to come to a mutually stable end, with one side ahead.

We are unlikely to witness arms races in dynamic progress, because they are unlikely to be running at any particular 'moment' of geological time, such as our time. But the animals that are to be seen in our time can be interpreted as the end-products of an arms race that was run in the past.

To summarize the message of this chapter, genes are selected, not for their intrinsic qualities, but by virtue of their interactions with their environments. An especially important component of a gene's environment is other genes. The general reason why this is such an important component is that other genes also change, as generations go by in evolution. This has two main kinds of consequences.

First, it has meant that those genes are favoured that have the property of 'cooperating' with those other genes that they are likely to meet in circumstances that favour cooperation. This is especially, though not exclusively, true of genes within the same species, because genes within one species frequently share cells with one another. It has led to the evolution of large gangs of cooperating genes, and ultimately to the evolution of bodies themselves, as the products of their cooperative enterprise. An individual body is a large vehicle or 'survival machine' built by a gene cooperative, for the preservation of copies of each member of that cooperative. They cooperate because they all stand to gain from the same outcome – the survival and reproduction of the communal body – and because they constitute an important part of the environment in which natural selection works on each other.

Second, circumstances don't always favour cooperation. In their march down geological time, genes also encounter one another in circumstances that favour antagonism. This is especially, though not exclusively, true of genes in different species. The point about different species is that their genes don't mix – because members of different species can't mate with one another. When selected genes in one species provide the environment in which genes in another species are selected, the result is often an evolutionary arms race. Each new genetic improvement selected on one side of the arms race – say predators – changes the environment for selection of genes on the other side of the arms race – prey. It is arms races of this kind that have been mainly responsible for the apparently *progressive* quality of evolution, for the evolution of ever-improved running speed, flying skill, acuity of eyesight, keenness of hearing, and so on. These arms races don't go on forever, but stabilize when, for instance, further improvements become too economically costly to the individual animals concerned.

This has been a difficult chapter, but it had to go into the book. Without it, we would have been left with the feeling that natural selection is only a destructive process, or at best a process of weeding-out. We have seen two ways in which natural selection can be a *constructive* force. One way concerns cooperative relationships between genes within species. Our fundamental assumption must be that genes are 'selfish' entities, working for their own propagation in the gene pool of the species. But because the environment of a gene consists, to such a salient degree, of *other* genes also being selected in the same gene pool, genes will be favoured if they are good at cooperating with other genes in the same gene pool. This is why large bodies of cells, working coherently towards the same cooperative ends, have evolved. This is why bodies exist, rather than separate replicators still battling it out in the primordial soup.

Bodies evolve integrated and coherent purposefulness because genes are selected in the environment provided by other genes *within the same species.* But because genes are also selected in the environment provided by other genes in different species, arms races develop. And arms races constitute the other great force propelling evolution in directions that we recognize as 'progressive', complex 'design'. Arms races have an inherently unstable 'runaway' feel to them. They career off into the future in a way that is, in one sense, pointless and futile, in another sense progressive and endlessly fascinating to us, the observers. The next chapter takes up a particular, rather special case of explosive, runaway evolution, the case that Darwin called sexual selection.

CHAPTER 8

EXPLOSIONS AND SPIRALS

The human mind is an inveterate analogizer. We are compulsively drawn to see meaning in slight similarities between very different processes. I spent much of a day in Panama watching two teeming colonies of leaf-cutter ants fighting, and my mind irresistibly compared the limb-strewn battlefield to pictures I had seen of Passchendaele. I could almost hear the guns and smell the smoke. Shortly after my first book, *The Selfish Gene*, was published, I was independently approached by two clergymen, who both had arrived at the same analogy between ideas in the book and the doctrine of original sin. Darwin applied the idea of evolution in a discriminating way to living organisms changing in body form over countless generations. His successors have been tempted to see evolution in everything; in the changing form of the universe, in developmental 'stages' of human civilizations, in fashions in skirt lengths. Sometimes such analogies can be immensely fruitful, but it is easy to push analogies too far, and get overexcited by analogies that are so tenuous as to be unhelpful or even downright harmful. I have become accustomed to receiving my share of crank mail, and have learned that one of the hallmarks of futile crankiness is overenthusiastic analogizing.

On the other hand, some of the greatest advances in science have come about because some clever person spotted an analogy between a subject that was already understood, and another still mysterious subject. The trick is to strike a balance between too much indiscriminate analogizing on the one hand, and a sterile blindness to fruitful analogies on the other. The successful scientist and the raving crank are separated by the quality of their inspirations. But I suspect that this amounts, in practice, to a difference, not so much in ability to *notice* analogies as in ability to *reject* foolish analogies and pursue helpful ones. Passing over the fact that we have here yet another analogy, which may be foolish or may be fruitful (and is certainly not original), between scientific progress and Darwinian evolutionary selection, let me now come to the point that is relevant to this chapter. This is that I am about to embark on two interwoven analogies which I find inspiring but which can be taken too far if we are not careful. The first is an analogy between various processes that are united by their resemblance to explosions. The second is an analogy between true Darwinian evolution and what has been called cultural evolution. I think that these analogies may be fruitful – obviously, or I would not devote a chapter to them. But the reader is warned.

The property of explosions that is relevant is the one known to engineers as 'positive feedback'. Positive feedback is best understood by comparison with its opposite, negative feedback. Negative feedback is the basis of most automatic control and regulation, and one of its neatest and best-known examples is the Watt steam governor. A useful engine should deliver rotational power at a constant rate, the right rate for the job in hand, milling, weaving, pumping or whatever it happens to be. Before Watt, the problem was that the rate of turning depended upon the steam pressure. Stoke the boiler and you speed up the engine, not a satisfactory state of affairs for a mill or loom that requires uniform drive for its machines. Watt's governor was an automatic valve regulating the flow of steam to the piston.

The clever trick was to link the valve to the rotary motion produced by the engine, in such a way that the faster the engine ran the more the valve shut down the steam. Conversely, when the engine was running slowly, the valve opened up. Therefore an engine going too slowly soon speeded up, and an engine going too fast soon slowed down. The precise means by which the governor measured the speed was simple but effective, and the principle is still used today. A pair of balls on hinged arms spin round, driven by the engine. When they are spinning fast, the balls rise up on their hinges, by centrifugal force. When they are spinning slowly, they hang down. The hinged arms are directly linked to the steam throttle. With suitable fine-tuning, the Watt governor can keep a steam engine turning at an almost constant rate, in the face of considerable fluctuations in the firebox.

The underlying principle of the Watt governor is negative feedback. The output of the engine (rotary motion in this case) is fed back into the engine (via the steam valve). The feedback is *negative* because high output (fast rotation of the balls) has a negative effect upon the input (steam supply). Conversely, low output (slow rotation of the balls) boosts the input (of steam), again reversing the sign. But I introduced the idea of negative feedback only in order to contrast it with positive feedback. Let us take a Watt-governed steam engine, and make one crucial change in it. We reverse the sign of the relationship between the centrifugal ball apparatus and the steam valve. Now when the balls spin fast, the valve, instead of closing as Watt had it, *opens.* Conversely, when the balls spin slowly, the valve, instead of increasing the flow of steam, reduces it. In a normal, Watt-governed engine, an engine that started to slow down would soon correct this tendency and speed up again to the desired speed. But our doctored engine does just the opposite. If it starts to slow down, this makes it slow down even more. It soon throttles itself down to a halt. If, on the other hand, such a doctored engine happens to speed up a little, instead of the tendency being corrected as it would in a proper Watt engine, the tendency is increased. The slight speeding up is reinforced by the inverted governor, and the engine accelerates. The acceleration feeds back positively, and the engine accelerates even more. This continues until either the engine breaks up under the strain and the runaway flywheel careers through the factory wall, or no more steam pressure is available, and a maximum speed is imposed.

Where the original Watt governor makes use of negative feedback, our hypothetical doctored governor exemplifies the opposite process of positive feedback. Positive-feedback processes have an unstable, runaway quality. Slight initial perturbations are increased, and they run away in an ever-increasing spiral, which culminates either in disaster or in an eventual throttling down at some higher level due to other processes. Engineers have found it fruitful to unite a wide variety of processes under the single heading of negative feedback, and another wide variety under the heading of positive feedback. The analogies are fruitful not just in some vague qualitative sense, but because all the processes share the same underlying mathematics. Biologists studying such phenomena as temperature control in the body, and the satiation mechanisms that prevent overeating, have found it helpful to borrow the mathematics of negative feedback from engineers. Positive-feedback systems are used less than negative feedback, both by engineers and by living bodies, but nevertheless it is positive feedbacks that are the subject of this chapter.

The reason engineers and living bodies make more use of negative than positive-feedback systems is, of course, that controlled regulation near an optimum is useful. Unstable runaway processes, far from being useful, can be downright dangerous. In chemistry, the typical positive-feedback process is an explosion, and we commonly use the word explosive as a description of any runaway process. For instance, we may refer to somebody as having an explosive temper. One of my schoolmasters was a cultured, courteous and usually gentle man, but he had occasional explosions of temper, as he himself was aware. When extremely provoked in class he would say nothing at first, but his face showed that something unusual was going on inside. Then, beginning in a quiet and reasonable tone he would say: 'Oh dear. I can't hold it. I'm going to lose my temper. Get down below your desks. I'm warning you. It's coming.' All the time his voice was rising, and at the crescendo he would seize everything within reach, books, wooden-backed blackboard erasers, paperweights, inkpots, and hurl them in quick succession, with the utmost force and ferocity but with wild aim, in the general direction of the boy who had provoked him. His temper then gradually subsided, and next day he would offer the most gracious apology to the same boy. He was aware that he had gone out of control, he had witnessed himself becoming the victim of a positive-feedback loop.

But positive feedbacks don't only lead to runaway increases; they can lead to runaway decreases. I recently attended a debate in Congregation, Oxford University's 'parliament', on whether to offer an honorary degree to somebody. Unusually, the decision was a controversial one. After the vote, during the 15 minutes that it took to count the ballot papers, there was a general hubbub of conversation from those waiting to hear the result. At one point the conversation strangely died away, and there was total silence. The reason was a particular kind of positive feedback. It worked as follows. In any general buzz of conversation there are bound to be chance fluctuations in noise level, both up and down, which we normally don't notice. One of these chance fluctuations, in the direction of quietness, happened to be slightly more marked than usual, with the result that some people noticed it. Since everybody was anxiously waiting for the result of the vote to be announced, those that heard the random decrease in noise level looked up and ceased their conversation. This caused the general noise level to go down a little more, with the result that more people noticed it and stopped their conversation. A positive feedback had been initiated and it continued rather rapidly until there was total silence in the hall. Then, when we realized that it was a false alarm, there was a laugh followed by a slow escalation in noise back up to its former level.

The most noticeable and spectacular positive feedbacks are those that result, not in a decrease, but in a runaway increase in something: a nuclear explosion, a schoolmaster losing his temper, a brawl in a pub, escalating invective at the United Nations (the reader may heed the warning with which I began this chapter). The importance of positive feedbacks in international affairs is implicitly recognized in the jargon word 'escalation': when we say that the Middle East is a 'powder keg', and when we identify 'flashpoints'. One of the best-known expressions of the idea of positive feedback is in St Matthew's Gospel: 'Unto everyone that hath shall be given, and he shall have abundance: but from him that hath not shall be taken away even that which he hath.' This chapter is about positive feedbacks in evolution. There are some features of living organisms that look as though they are the end-products of something like an explosive, positive-feedback-driven, runaway process of evolution. In a mild way the arms races of the previous chapter are examples of this, but the really spectacular examples are to be found in organs of sexual advertisement.

Try to persuade yourself, as they tried to persuade me when I was an undergraduate, that the peacock's fan is a mundanely functional organ like a tooth or a kidney, fashioned by natural selection to do no more than the utilitarian job of labelling the bird, unambiguously as a member of this species and not that. They never persuaded me, and I doubt if you can be persuaded either. For me the peacock's fan has the unmistakable stamp of positive feedback. It is clearly the product of some kind of uncontrolled, unstable explosion that took place in evolutionary time. So thought Darwin in his theory of sexual selection and so, explicitly and in so many words, thought the greatest of his successors, R. A. Fisher. After a short piece of reasoning he concluded (in his book *The Genetical Theory of Natural Selection*):

> plumage development in the male, and sexual preference for such developments in the female, must thus advance together, and so long as the process is unchecked by severe counterselection, will advance with ever-increasing speed. In the total absence of such checks, it is easy to see that the speed of development will be proportional to the development already attained, which will therefore increase with time exponentially, or in geometric progression.

It is typical of Fisher that what he found 'easy to see' was not fully understood by others until half a century later. He did not bother to spell out his assertion that the evolution of sexually attractive plumage might advance with ever-increasing speed, exponentially, explosively. It took the rest of the biological world some 50 years to catch up and finally reconstruct in full the kind of mathematical argument that Fisher must have used, either on paper or in his head, to prove the point to himself. I am going to try to explain, purely in nonmathematical prose, these mathematical ideas which, in their modern form, have largely been worked out by the young American mathematical biologist Russell Lande. While I would not be so pessimistic as Fisher himself who, in the Preface to his 1930 book, said 'No efforts of mine could avail to make the book easy reading', nevertheless, in the words of a kind reviewer of my own first book, 'The reader is warned that he must put on his mental running shoes'. My own understanding of these difficult ideas has been a hard struggle. Here, despite his protests, I must acknowledge my colleague and former student Alan Grafen, whose own mental winged sandals are notoriously in a class of their own, but who has the even rarer ability to take them off and think of the right way to explain things to others. Without his teaching, I simply couldn't have written the middle part of this chapter, which is why I refuse to relegate my acknowledgment to the Preface.

Before we come to these difficult matters, I must back-track and say a little about the origin of the idea of sexual selection. It began, like so much else in this field, with Charles Darwin. Darwin, although he laid his main stress on survival and the struggle for existence, recognized that existence and survival were only means to an end. That end was reproduction. A pheasant may live to a ripe old age, but if it does not reproduce it will not pass its attributes on. Selection will favour qualities that make an animal successful at reproducing, and survival is only part of the battle to reproduce. In other parts of the battle, success goes to those that are most attractive to the opposite sex. Darwin saw that, if a male pheasant or peacock or bird of paradise buys sexual attractiveness, even at the cost of its own life, it may still pass on its sexually attractive qualities through highly successful procreation before its death. He realized that the fan of a peacock must be a handicap to its possessor as far as survival is concerned, and he suggested that this was more than outweighed by the increased sexual attractiveness that it gave the male. With his fondness for the analogy with domestication, Darwin compared the hen to a human breeder directing the course of evolution of domestic animals along the lines of aesthetic whims. We might compare her to the person selecting computer biomorphs in directions of aesthetic appeal.

Darwin simply accepted female whims as given. Their existence was an axiom of his theory of sexual selection, a prior assumption rather than something to be explained in its own right. Partly for this reason his theory of sexual selection fell into disrepute, until it was rescued by Fisher in 1930. Unfortunately, many biologists either ignored or misunderstood Fisher. The objection raised by Julian Huxley and others was that female whims were not legitimate foundations for a truly scientific theory. But Fisher rescued the theory of sexual selection, by treating female preference as a legitimate object of natural selection in its own right, no less than male tails. Female preference is a manifestation of the female nervous system. The female nervous system develops under the influence of her genes, and its attributes are therefore likely to have been influenced by selection over past generations. Where others had thought of male ornaments evolving under the influence of static female preference, Fisher thought in terms of female preference evolving dynamically in step with male ornament. Perhaps you can already begin to see how this is going to link up with the idea of explosive positive feedback.

When we are discussing difficult theoretical ideas, it is often a good idea to keep in mind a particular example from the real world. I shall use the tail of the African long-tailed widow bird as an example. Any sexually selected ornament would have done, and I had a whim to ring the changes and avoid the ubiquitous (in discussions of sexual selection) peacock. The male long-tailed widow bird is a slender black bird with orange shoulder flashes, about the size of an English sparrow except that the main tail feathers, in the breeding season, can be 18 inches long. It is often to be seen performing its spectacular display flight over the grasslands of Africa, wheeling and looping the loop, like an aeroplane with a long advertising streamer. Not surprisingly it can be grounded in wet weather. Even a dry tail that long must be a burdensome load to carry around. We are interested in explaining the evolution of the long tail, which we conjecture has been an explosive evolutionary process. Our starting point, therefore, is an ancestral bird without a long tail. Think of the ancestral tail as about 3 inches long, about a sixth the length of the modern breeding male's tail. The evolutionary change that we are trying to explain is a sixfold increase in tail length.

It is an obvious fact that, when we measure almost anything about animals, although most members of a species are fairly close to the average, some individuals are a little above average, while others are below average. We can be sure that there was a range of tail lengths in the ancestral widow bird, some being longer and some being shorter than the average of 3 inches. It is safe to assume that tail length would have been governed by a large number of genes, each one of small effect, their effects adding up, together with the effects of diet and other environmental variables, to make the actual tail length of an individual. Large numbers of genes whose effects add up are called polygenes. Most measures of ourselves, for instance our height and weight, are affected by large numbers of polygenes. The mathematical model of sexual selection that I am following most closely, that of Russell Lande, is a model of polygenes.

Now we must turn our attention to females, and how they choose their mates. It may seem rather sexist to assume that it is the females that would choose their mates, rather than the other way round. Actually, there are good theoretical reasons for expecting it to be this way round (see *The Selfish Gene*), and as a matter of fact it normally is in practice. Certainly modern long-tailed widow bird males attract harems of half a dozen or so females. This means that there is a surplus of males in the population who do not reproduce. This, in turn, means that females have no difficulty in finding mates, and are in a position to be choosy. A male has a great deal to gain by being attractive to females. A female has little to gain by being attractive to males, since she is bound to be in demand anyway.

So, having accepted the assumption that females do the choosing, we next take the crucial step that Fisher took in confounding Darwin's critics. Instead of simply agreeing that females have whims, we regard female preference as a genetically influenced variable just like any other. Female preference is a quantitative variable, and we can assume that it is under the control of polygenes in just the same kind of way as male tail length itself. These polygenes may act on any of a wide variety of parts of the female's brain, or even on her eyes; on anything that has the effect of altering the female's preference. Female preference doubtless takes account of many parts of a male, the colour of his shoulder patch, the shape of his beak, and so on; but we happen to be interested, here, in the evolution of male tail length, and hence we are interested in female preferences for male tails of different length. We can therefore measure female preference in exactly the same units as we measure male tail length – inches. Polygenes will see to it that there are some females with a liking for longer than average male tails, others with a liking for shorter than average male tails, and others with a liking for tails of about average length.

Now comes one of the key insights in the whole theory. Although genes for female preference only *express* themselves in female behaviour, nevertheless they are present in the bodies of males too. And by the same token, genes for male tail length are present in the bodies of females, whether or not they express themselves in females. The idea of genes failing to express themselves is not a difficult one. If a man has genes for a long penis, he is just as likely to pass those genes on to his daughter as to his son. His son may express those genes whereas his daughter, of course, will not, because she doesn't have a penis at all. But if the man eventually gets grandsons, the sons of his daughter may be just as likely to inherit his long penis as the sons of his son. Genes may be carried in a body but not expressed. In the same way, Fisher and Lande assume that genes for female preference are *carried* in male bodies, even though they are only *expressed* in female bodies. And genes for male tails are carried in female bodies, even if they are not expressed in females.

Suppose we had a special microscope, which enabled us to look inside any bird's cells and inspect its genes. Take a male who happens to have a longer than average tail, and look inside his cells at his genes. Looking first at the genes for tail length itself, it comes as no surprise to discover that he has genes that make a long tail: this is obvious, since he *has* a long tail. But now look at his genes for tail *preference*. Here we have no clue from the outside, since such genes only express themselves in females. We have to look with our microscope. What would we see? We'd see genes for making females prefer long tails. Conversely, if we looked inside a male who actually has a short tail, we should see genes for making females prefer short tails. This is really a key point in the argument. The rationale for it is as follows.

If I am a male with a long tail, my father is more likely than not to have had a long tail too. This is just ordinary heredity. But also, since my father was chosen as a mate by my mother, my mother is more likely than not to have preferred long-tailed males. Therefore, if I have inherited genes for a long tail from my father, I am also likely to have inherited genes for preferring long tails from my mother. By the same reasoning, if you have inherited the genes for a short tail, the chances are that you have also inherited the genes for making females prefer a short tail.

We can follow the same kind of reasoning for females. If I am a female who prefers long-tailed males, the chances are that my mother also preferred long-tailed males. Therefore the chances are that my father had a long tail, since he was chosen by my mother. Therefore if I have inherited genes for preferring long tails, the chances are that I have also inherited genes for having a long tail, whether or not those genes are actually expressed in my female body. And if I have inherited genes for preferring short tails, the chances are that I have also inherited genes for *having* a short tail. The general conclusion is this. Any individual, of either sex, is likely to contain *both* genes for making males *have* a certain quality, *and* genes for making females *prefer* that selfsame quality, whatever that quality might be.

So, the genes for male qualities, and the genes for making females prefer those qualities, will not be randomly shuffled around the population, but will tend to be shuffled around *together*. This 'togetherness', which goes under the daunting technical name of linkage disequilibrium, plays curious tricks with the equations of mathematical geneticists. It has strange and wonderful consequences, not the least of which in practice, if Fisher and Lande are right, is the explosive evolution of peacocks' and widow birds' tails, and a host of other organs of attraction. These consequences can only be proved mathematically, but it is possible to say in words what they are, and we can try to gain some flavour of the mathematical argument in nonmathematical language. We still need our mental running shoes, although actually climbing boots is a better analogy. Each step in the argument is simple enough, but there is a long series of steps up the mountain of understanding, and if you miss any of the earlier steps you unfortunately can't take the later ones.

So far we have recognized the possibility of a complete range of female preferences, from females with a taste for long-tailed males through to females with the opposite taste, for short-tailed males. But if we actually did a poll of the females in a particular population, we would probably find that a majority of females shared the same general tastes in males. We can express the *range* of female tastes in the population in the same units — inches — as we express the range of male tail lengths. And we can express the *average* female preference in the same units of inches. It could turn out that the average female preference was exactly the same as the average male tail length, 3 inches in both cases. In this case female choice will not be an evolutionary force tending to change male tail length. Or it could turn out that the average female preference was for a tail rather longer than the average tail that actually exists, say 4 inches rather than 3. Leaving open, for the moment, why there might be such a discrepancy, just accept that there is one and ask the next obvious question. Why, if most females prefer males with 4-inch tails, do the majority of males actually have 3-inch tails? Why doesn't the average tail length in the population shift to 4 inches under the influence of female sexual selection? How can there be a discrepancy of 1 inch between the average preferred tail length and the actual average tail length?

The answer is that female taste is not the only kind of selection that bears upon male tail length. Tails have an important job to perform in flight, and a tail that is too long or too short will decrease the efficiency of flight. Moreover, a long tail costs more energy to carry around, and more to make it in the first place. Males with 4-inch tails might well pull the female birds, but the price the males would pay is their less-efficient flight, greater energy costs and greater vulnerability to predators. We can express this by saying that there is a *utilitarian optimum* tail length, which is different from the sexually selected optimum: an ideal tail length from the point of view of ordinary useful criteria; a tail length that is ideal from all points of view apart from attracting females.

Should we expect that the actual average tail length of males, 3 inches in our hypothetical example, will be the same as the utilitarian optimum? No, we should expect the utilitarian optimum to be less, say 2 inches. The reason is that the actual average tail length of 3 inches is the result of a compromise between utilitarian selection tending to make tails shorter, and sexual selection tending to make them longer. We may surmise that, if there were no need to attract females, average tail length would shrink towards 2 inches. If there were no need to worry about flying efficiency and energy costs, average tail length would shoot out towards 4 inches. The actual average of 3 inches is a compromise.

We left on one side the question of *why* females might agree in preferring a tail that departed from the utilitarian optimum. At first sight the very idea seems silly. Fashion-conscious females, with a taste for tails that are longer than they should be on good design criteria, are going to have poorly designed, inefficient, clumsily flying sons. Any mutant female who happened to have an unfashionable taste for shorter-tailed males, in particular a mutant female whose taste in tails happened to coincide with the utilitarian optimum, would produce efficient sons, well designed for flying, who would surely outcompete the sons of her more fashion-conscious rivals. Ah, but here is the rub. It is implicit in my metaphor of 'fashion'. The mutant female's sons may be efficient flyers, but they are not seen as attractive by the majority of females in the population. They will attract only minority females, fashion-defying females; and minority females, by definition, are harder to find than majority females, for the simple reason that they are thinner on the ground. In a society where only one in six males mates at all and the fortunate males have large harems, pandering to the majority tastes of females will have enormous benefits, benefits that are well capable of outweighing the utilitarian costs in energy and flight efficiency.

But even so, the reader may complain, the whole argument is based upon an arbitrary assumption. Given that most females prefer nonutilitarian long tails, the reader will admit, everything else follows. But *why* did this majority female taste come about in the first place? Why didn't the majority of females prefer tails that are *smaller* than the utilitarian optimum, or exactly the same length as the utilitarian optimum? Why shouldn't fashion coincide with utility? The answer is that any of these things might have happened, and in many species it probably did. My hypothetical case of females preferring long tails was, indeed arbitrary. But *whatever* the majority female taste had happened to be, and no matter how arbitrary, there would have been a tendency for that majority to be maintained by selection or even, under some conditions, actually increased – exaggerated. It is at this point in the argument that the lack of mathematical justification in my account becomes really noticeable. I could invite the reader simply to accept that the mathematical reasoning of Lande proves the point, and leave it at that. This might be the wisest course for me to pursue, but I shall have one try at explaining part of the idea in words.

The key to the argument lies in the point we established earlier about 'linkage disequilibrium', the 'togetherness' of genes for tails of a given length – any length – and corresponding genes for preferring tails of that self-same length. We can think about the 'togetherness factor' as a measurable number. If the togetherness factor is very high, this means that knowledge about an individual's genes for tail length enables us to predict, with great accuracy, his/her genes for preference, and vice versa. Conversely, if the togetherness factor is low, this means that knowledge about an individual's genes in one of the two departments – preference or tail length – gives us only a slight hint about his/her genes in the other department.

The kind of thing that affects the magnitude of the togetherness factor is the strength of the females' preference – how tolerant they are of what they see as imperfect males; how much of the variation in male tail length is governed by genes as opposed to environmental factors; and so on. If, as a result of all these effects, the togetherness factor – the tightness of binding of genes for tail length and genes for tail-length preference – is very strong, we can deduce the following consequence. Every time a male is chosen because of his long tail, not only are genes for long tails being chosen. At the same time, because of the 'togetherness' coupling, genes for *preferring* long tails are also being chosen. What this means is that genes that make females choose male tails of a particular length are, in effect, *choosing copies of themselves*. This is the essential ingredient of a self-reinforcing process: it has its own self-sustaining momentum. Evolution having started in a particular direction, this can, in itself, tend to make it persist in the same direction.

Another way to see this is in terms of what has become known as the 'green-beard effect'. The green-beard effect is a kind of academic biological joke. It is purely hypothetical, but it is instructive nevertheless. It was originally proposed as a way of explaining the fundamental principle underlying W. D. Hamilton's important theory of kin selection, which I discussed at length in *The Selfish Gene*. Hamilton, now my colleague at Oxford, showed that natural selection would favour genes for behaving altruistically towards close kin, simply because copies of those selfsame genes had a high probability of being in the bodies of kin. The 'green-beard' hypothesis puts the same point more generally, if less practically. Kinship, the argument runs, is only one possible way in which genes can, in effect, locate copies of themselves in other bodies. Theoretically, a gene could locate copies of itself by more direct means. Suppose a gene happened to arise that had the following two effects (genes with two or more effects are common): it makes its possessors have a conspicuous 'badge' such as a green beard, and it also affects their brains in such a way that they behave altruistically towards green-bearded individuals. A pretty improbable coincidence, admittedly, but if it ever did arise the evolutionary consequence is clear. The green-beard altruism gene would tend to be favoured by natural selection, for exactly the same kinds of reason as genes for altruism towards offspring or brothers. Every time a green-bearded individual helped another, the gene for giving this discriminating altruism would be favouring a copy of itself. The spread of the green-beard gene would be automatic and inevitable.

Nobody really believes, not even I, that the green-beard effect, in this ultra-simple form, will ever be found in nature. In nature, genes discriminate in favour of copies of themselves by means of less specific but more plausible labels than green beards. Kinship is just such a label. 'Brother' or, in practice, something like 'he who has just hatched in the nest from which I have just fledged', is a statistical label. Any gene that makes individuals behave altruistically towards bearers of such a label has a good statistical chance of aiding copies of itself: for brothers have a good statistical chance of sharing genes. Hamilton's theory of kin selection can be seen as one way in which a green-beard type of effect can be made plausible. Remember, by the way, that there is no suggestion here that genes 'want' to help copies of themselves. It is just that any gene that happens to have the *effect* of helping copies of itself will tend, willy nilly, to become more numerous in the population.

Kinship, then, can be seen as a way in which something like the green-beard effect can be made plausible. The Fisher theory of sexual selection can be explained as yet another way in which the green-beard can be made plausible. When the females of a population have strong preferences for male characteristics, it follows, by the reasoning we have been through, that each male body will tend to contain copies of genes that make females prefer his own characteristics. If a male has inherited a long tail from his father, the chances are that he has also inherited from his mother the genes that made her choose the long tail of his father. If he has a short tail, the chances are that he contains genes for making females prefer short tails. So, when a female exercises her choice of male, whichever way her preference lies, the chances are that the genes that bias her choice *are choosing copies of themselves* in the males. They are choosing copies of themselves using male tail length as a label, in a more complicated version of the way the hypothetical green-beard gene uses the green beard as a label.

If half the females in the population preferred long-tailed males, and the other half short-tailed males, genes for female choice would still be choosing copies of themselves, but there would be no tendency for one or other tail type to be favoured in general. There might be a tendency for the population to split into two — a long-tailed, long-preferring faction, and a short-tailed, short-preferring faction. But any such two-way split in female 'opinion' is an unstable state of affairs. The moment a majority, *however slight*, started to accrue among females for one type of preference rather than the other, that majority would be reinforced in subsequent generations. This is because males preferred by females of the minority school of thought would have a harder time finding mates; and females of the minority school of thought would have sons who had a relatively hard time finding mates, so minority females would have fewer grandchildren. Whenever small minorities tend to become even smaller minorities, and small majorities tend to become bigger majorities, we have a recipe for positive feedback: 'Unto every one that hath shall be given, and he shall have abundance: but from him that hath not shall be taken away even that which he hath.' Whenever we have an unstable balance, arbitrary, random beginnings are self-reinforcing. Just so, when we cut through a tree trunk, we may be uncertain whether the tree will fall to the north or the south; but, after remaining poised for a time, once it starts to fall in one direction or the other, nothing can bring it back.

Lacing our climbing boots even more securely we prepare to hammer in another piton. Remember that selection by females is pulling male tails in one direction, while 'utilitarian' selection is pulling them in the other ('pulling' in the evolutionary sense, of course), the actual average tail length being a compromise between the two pulls. Let us now recognize a quantity called the 'choice discrepancy'. This is the difference between the actual average tail length of males in the population, and the 'ideal' tail length that the average female in the population would really prefer. The units in which the choice discrepancy is measured are arbitrary, just as the Fahrenheit and Centigrade scales of temperature are arbitrary. Just as the Centigrade scale finds it convenient to fix its zero point at the freezing point of water, we shall find it convenient to fix our zero at the point where the pull of sexual selection exactly balances the opposite pull of utilitarian selection. In other words, a choice discrepancy of zero means that evolutionary change comes to a halt because the two opposite kinds of selection exactly cancel each other out.

Obviously, the larger the choice discrepancy, the stronger the evolutionary 'pull' exerted by females against the counteracting pull of utilitarian natural selection. What we are interested in is not the absolute value of the choice discrepancy at any particular time, but how the choice discrepancy *changes* in successive generations. As a result of a given choice discrepancy, tails get longer, and at the same time (remember that genes for choosing long tails are being selected in concert with genes for having long tails) the females' ideal preferred tail gets longer too. After a generation of this dual selection, both average tail length and average preferred tail length have become longer, but which has increased the most? This is another way of asking what will happen to the choice discrepancy.

The choice discrepancy could have stayed the same (if average tail length and average preferred tail length both increased by the same amount). It could have become smaller (if average tail length increased more than preferred tail length did). Or, finally, it could have become larger (if average tail length increased somewhat, but average preferred tail length increased even more). You can begin to see that, if the choice discrepancy gets smaller as tails get larger, tail length will evolve towards a stable equilibrium length. But if the choice discrepancy gets *larger* as tails get larger, future generations should theoretically see tails shooting out at ever increasing speed. This is, without any doubt, what Fisher must have calculated before 1930, although his brief published words were not clearly understood by others at the time.

Let us first take the case where the choice discrepancy becomes ever smaller as the generations go by. It will eventually become so small that the pull of female preference in one direction is exactly balanced by the pull of utilitarian selection in the other. Evolutionary change will then come to a halt, and the system is said to be in a state of equilibrium. The interesting thing Lande proved about this is that, at least under some conditions, there is not just one point of equilibrium, but many (theoretically an infinite number arranged in a straight line on a graph, but there's mathematics for you!). There is not just one balance point but many: for any strength of utilitarian selection pulling in one direction, the strength of female preference evolves in such a way as to reach a point where it balances it exactly.

So, if conditions are such that the choice discrepancy tends to become smaller as the generations go by, the population will come to rest at the 'nearest' point of equilibrium. Here utilitarian selection pulling in one direction will be exactly counteracted by female selection pulling in the other, and the tails of the males will stay the same length, regardless of how long that is. The reader may recognize that we have here a negative-feedback system, but it is a slightly weird kind of negative-feedback system. You can always tell a negative-feedback system by what happens if you 'perturb' it away from its ideal, 'set point'. If you perturb a room's temperature by opening the window, for instance, the thermostat responds by turning on the heater to compensate.

How might the system of sexual selection be perturbed? Remember that we are talking about the evolutionary timescale here, so it is difficult for us to do experiments – the equivalent of opening the window – and live to see the results. But no doubt, in nature, the system is perturbed often, for instance by spontaneous, random fluctuations in numbers of males due to chance, lucky or unlucky, events. Whenever this happens, given the conditions we have so far discussed, a combination of utilitarian selection and sexual selection will return the population to the nearest one of the set of equilibrium points. This probably will *not* be the same equilibrium point as before, but will be another point a little bit higher, or lower, along the line of equilibrium points. So, as time goes by, the population can drift up or down the line of equilibrium points. Drifting up the line means that tails get longer – theoretically there is no limit to how long. Drifting down the line means that tails get shorter – theoretically all the way down to a length of zero.

The analogy of a thermostat is often used to explain the idea of a point of equilibrium. We can develop the analogy to explain the more difficult idea of a *line* of equilibria. Suppose that a room has both a heating device and a cooling device, each with its own thermostat. Both thermostats are set to keep the room at the same fixed temperature, 70 degrees F. If the temperature drops below 70, the heater turns itself on and the refrigerator turns itself off. If the temperature rises above 70, the refrigerator turns itself on and the heater turns itself off. The analogue of the widow bird's tail length is not the temperature (which remains approximately constant at 70°) but the total rate of consumption of electricity. The point is that there are lots of different ways in which the desired temperature can be achieved. It can be achieved by both devices working very hard, the heater belting out hot air and the refrigerator working flat out to neutralize the heat. Or it can be achieved by the heater putting out a bit less heat, and the cooler working correspondingly less hard to neutralize it. Or it can be achieved by both devices working scarcely at all. Obviously, the latter is the most desirable solution from the point of view of the electricity bill but, as far as the object of maintaining the fixed temperature of 70 degrees is concerned, every one of a large series of working rates is equally satisfactory. We have a *line* of equilibrium points, rather than a single point. Depending upon details of how the system was set up, upon delays in the system and other things of the kind that preoccupy engineers, it is theoretically possible for the room's electricity-consumption rate to drift up and down the line of equilibrium points, while the temperature remains the same. If the room's temperature is perturbed a little below 70 degrees it will return, but it won't necessarily return to the same combination of work rates of the heater and the cooler. It may return to a different point along the line of equilibria.

In real practical engineering terms, it would be pretty difficult to set up a room so that a true line of equilibria existed. The line is in practice likely to 'collapse to a point'. Russell Lande's argument too, about a line of equilibria in sexual selection, rests upon assumptions that may well not be true in nature. It assumes, for instance, that there will be a steady supply of new mutations. It assumes that the act of choosing, by a female, is entirely cost-free. If this assumption is violated, as it well may be, the 'line' of equilibria collapses into a single point of equilibrium. But in any case, so far we have only discussed the case where the choice discrepancy becomes *smaller* as the successive generations of selection go by. Under other conditions the choice discrepancy may become larger.

It is a while since we discussed the matter, so let us remind ourselves of what this means. We have a population whose males are undergoing evolution of some characteristic such as tail length in widow birds, under the influence of female preference tending to make the tails longer and utilitarian selection tending to make the tails shorter. The reason there is any momentum in the evolution towards longer tails is that, whenever a female chooses a male of the type she 'likes', she is, because of the non-random association of genes, choosing copies of the very genes that made her do the choosing. So, in the next generation, not only will the males tend to have longer tails, but the females will tend to have a stronger preference for long tails. It is not obvious which of these two incremental processes will have the highest rate, generation by generation. We have so far considered the case where tail length increases faster, per generation, than preference. Now we come to consider the other possible case, where preference increases at an even higher rate, per generation, than tail length itself does. In other words, we are now going to discuss the case where the choice discrepancy gets bigger as the generations go by, not smaller as in the previous paragraphs.

Here the theoretical consequences are even more bizarre than before.

Instead of negative feedback, we have positive feedback. As the generations go by, tails get longer, but the female desire for long tails increases at a higher rate. This means that, theoretically, tails will get even longer still, and at an ever-accelerating rate as the generations go by. Theoretically, tails will go on expanding even after they are 10 miles long. In practice, of course, the rules of the game will have been changed long before these absurd lengths are reached, just as our steam engine with its reversed Watt governor would not *really* have gone on accelerating to a million revolutions per second. But although we have to water down the conclusions of the mathematical model when we come to the extremes, the model's conclusions may well hold true over a range of practically plausible conditions.

Now, 50 years late, we can understand what Fisher meant, when he baldly asserted that 'it is easy to see that the speed of development will be proportional to the development already attained, which will therefore increase with time exponentially, or in geometric progression'. His rationale was clearly the same as Lande's, when he said: 'The two characteristics affected by such a process, namely plumage development in the male, and sexual preference for such developments in the female, must thus advance together, and so long as the process is unchecked by severe counterselection, will advance with ever-increasing speed.'

The fact that Fisher and Lande both arrived by mathematical reasoning at the same intriguing conclusion does not mean that their theory is a correct reflection of what goes on in nature. It could be, as the Cambridge University geneticist Peter O'Donald, one of the leading authorities on the theory of sexual selection, has said, that the runaway property of the Lande model is 'built into' its starting assumptions, in such a way that it couldn't help emerging in a rather boring way at the other end of the mathematical reasoning. Some theorists, including Alan Grafen and W. D. Hamilton, prefer alternative kinds of theory in which choice made by a female really does have a beneficial effect on her progeny, in a utilitarian, eugenic sense. The theory they are together working on is that female birds act as diagnostic doctors, picking out those males who are least susceptible to parasites. Bright plumage, according to this characteristically ingenious theory of Hamilton, is a male's way of conspicuously advertising his health.

The theoretical importance of parasites would take too long to explain fully. Briefly, the problem with all 'eugenic' theories of female choice has always been as follows. If females really could successfully choose males with the best genes, their very success would reduce the range of choice available in the future: eventually, if there were only good genes around, there would be no point in choosing. Parasites remove this theoretical objection. The reason is that, according to Hamilton, parasites and hosts are running a never-ceasing *cyclical* arms race against one another. This in turn means that the 'best' genes in any one generation of birds are not the same as the best genes in future generations. What it takes to beat the current generation of parasites is no good against the next generation of evolving parasites. Therefore there will always be some males that happen to be genetically better equipped than others to beat the current crop of parasites. Females, therefore, can always benefit their offspring by choosing the healthiest of the current generation of males. The only *general* criteria that successive generations of females can use are the indicators that any vet might use – bright eyes, glossy plumage, and so on. Only genuinely healthy males can display these symptoms of health, so selection favours those males that display them to the full, and even exaggerate them into long tails and spreading fans.

But the parasite theory, though it may well be right, is off the point of my 'explosions' chapter. Returning to the Fisher/Lande runaway theory, what is needed now is evidence from real animals. How should we go about looking for such evidence? What methods might be used? A promising approach was made by Malte Andersson, from Sweden. As it happens, he worked on the very bird that I am using here to discuss the theoretical ideas, the long-tailed widow bird, and he studied it in its natural surroundings in Kenya. Andersson's experiments were made possible by a recent advance in technology: superglue. He reasoned as follows. If it is true that the actual tail length of males is a compromise between a utilitarian optimum on the one hand, and what females really want on the other, it should be possible to make a male super-attractive by giving him an extra long tail. This is where the superglue came in. I'll describe Andersson's experiment briefly, as it is a neat example of experimental design.

Andersson caught 36 male widow birds, and divided them into nine groups of four. Each group of four was treated alike. One member of each group of four (scrupulously chosen at random to avoid any unconscious bias) had his tail feathers trimmed to 14 centimetres (about 5½ inches). The portion removed was stuck, with quick-setting superglue, to the end of the tail of the second member of the group of four. So, the first one had an artificially shortened tail, the second one an artificially lengthened tail. The third bird was left with his tail untouched, for comparison. The fourth bird was also left with his tail the same length, but it wasn't untouched. Instead, the ends of the feathers were cut off and then glued back on again. This might seem a pointless exercise, but it is a good example of how careful you have to be in designing experiments. It could have been that the fact of having his tail feathers manipulated, or the fact of being caught and handled by a human, affected a bird, rather than the actual change in length itself. Group 4 was a 'control' for such effects.

The idea was to compare the mating success of each bird with its differently treated colleagues in its own group of four. After being treated in one of the four ways, every male was allowed to take up its former residence on its own territory. Here it resumed its normal business of trying to attract females into its territory, there to mate, build a nest and lay eggs. The question was, which member of each group of four would have the most success in pulling in females? Andersson measured this, not by literally watching females, but by waiting and then counting the number of nests containing eggs in each male's territory. What he found was that males with artificially elongated tails attracted nearly four times as many females as males with artificially shortened tails. Those with tails of normal, natural length had intermediate success.

The results were analysed statistically, in case they had resulted from chance alone. The conclusion was that if attracting females were the only criterion, males would be better off with longer tails than they actually have. In other words, sexual selection is constantly pulling tails (in the evolutionary sense) in the direction of getting longer. The fact that real tails are shorter than females would prefer suggests that there must be some other selection pressure keeping them shorter. This is 'utilitarian' selection. Presumably males with especially long tails are more likely to die than males with average tails. Unfortunately, Andersson did not have time to follow the subsequent fates of his doctored males. If he had, the prediction would have been that the males with extra tail-feathers glued on should, on average, have died younger than normal males, probably because of greater vulnerability to predators. Males with artificially shortened tails, on the other hand, should probably be expected to live longer than normal males. This is because the normal length is supposed to be a compromise between the sexual selection optimum and the utilitarian optimum. Presumably the birds with artificially shortened tails are closer to the utilitarian optimum, and therefore should live longer. There's a lot of supposition in all this, however. If the main utilitarian disadvantage of a long tail turned out to be the economic cost of growing it in the first place, rather than increased dangers of dying after it has grown, males who are handed an extra-long tail on a plate, as a free gift from Andersson, would not be expected to die particularly young as a result.

I have written as though female preference will tend to drag tails and other ornaments in the direction of getting larger. In theory, as we saw earlier, there is no reason why female preference should not pull in exactly the opposite direction, for instance in the direction of ever shortening, rather than lengthening, tails. The common wren has a tail so short and stubby that one is tempted to wonder whether it is, perhaps, shorter than it 'ought' to be for strictly utilitarian purposes. Competition between male wrens is intense, as you might guess from the disproportionate loudness of their song. Such singing is bound to be costly, and a male wren has even been known to sing himself, literally, to death. Successful males have more than one female in their territory, like widow birds. In such a competitive climate, we might expect positive feedbacks to get going. Could the wren's short tail represent the end product of a runaway process of evolutionary shrinkage?

Setting wrens on one side, peacock fans, and widow bird and bird of paradise tails, in their gaudy extravagance, are very plausibly seen as end-products of explosive, spiralling evolution by positive feedback. Fisher and his modern successors have shown us how this might have come about. Is this idea essentially tied to sexual selection, or can we find convincing analogies in other kinds of evolution? It is worth asking this question, if only because there are aspects of our own evolution that have more than a suggestion of the explosive about them, notably the extremely rapid swelling of our brains during the last few million years. It has been suggested that this is due to sexual selection itself, braininess being a sexually desirable character (or some manifestation of braininess, such as ability to remember the steps of a long and complicated ritual dance). But it could also be that brain size has exploded under the influence of a different kind of selection, analogous but not identical to sexual selection. I think it is helpful to distinguish two levels of possible analogy to sexual selection, a weak analogy and a strong analogy.

The weak analogy simply says the following. Any evolutionary process in which the end-product of one step in evolution sets the stage for the next step in evolution is potentially progressive, sometimes explosively so. We have already met this idea in the previous chapter, in the form of 'arms races'. Each evolutionary improvement in predator design changes the pressures on prey, thereby making prey become better at avoiding the predators. This in turn puts pressure on the predators to improve, so we have an ever-rising spiral. As we saw, it is likely that neither predators nor prey will necessarily enjoy a higher success rate as a result, because their enemies are improving at the same time. But nevertheless, both prey and predators are becoming progressively better *equipped*. This, then, is the weak analogy with sexual selection. The strong analogy with sexual selection notes that the essence of the Fisher/Lande theory is the 'green beard'-like phenomenon whereby genes for female choice automatically tend to choose copies of *themselves*, a process with an automatic tendency to go explosive. It is not clear that there are examples of this kind of phenomenon other than sexual selection itself.

I suspect that a good place to look for analogies to explosive evolution of the sexual selection kind is in human cultural evolution. This is because, here again, choice by whim matters, and such choice may be subject to the 'fashion' or 'majority always wins' effect. Once again, the warning with which I began this chapter should be heeded. Cultural 'evolution' is not really evolution at all if we are being fussy and purist about our use of words, but there may be enough in common between them to justify some comparison of principles. In doing this we must not make light of the differences. Let us get these matters out of the way before returning to the particular issue of explosive spirals.

It has frequently been pointed out – indeed any fool can see – that there is something quasi-evolutionary about many aspects of human history. If you sample a particular aspect of human life at regular intervals, say you sample the state of scientific knowledge, the kind of music being played, dress fashions, or vehicles of transport, at intervals of one century or perhaps one decade, you will find *trends*. If we have three samplings, at successive times A, B and C, then, to say that there is a trend is to say that the measurement taken at time B will be intermediate between the measurements taken at times A and C. Although there are exceptions, everyone will agree that trends of this kind characterize many aspects of civilized life. Admittedly the directions of trends sometimes reverse (for example, skirt lengths), but this is true of genetic evolution too.

Many trends, particularly trends in useful technology as opposed to frivolous fashions, can, without much argument over value-judgements, be identified as *improvements*. There can be no doubt, for instance, that vehicles for getting about the world have improved steadily and without reversal, over the past 200 years, passing from horse-drawn through steam-drawn vehicles, and culminating today in supersonic jet planes. I am using the word improvement in a neutral way. I don't mean to say that everybody would agree that the quality of life has improved as a result of these changes; personally I often doubt

it. Nor do I mean to deny the popular view that standards of workman-
ship have gone *down* as mass production has replaced skilled
craftsmen. But looking at means of transport purely from the point of
view of *transport*, which means getting from one part of the world to
another, there can be no disputing the historical trend towards some
kind of improvement, even if it is only an improvement in speed.
Similarly, over a timescale of decades or even years, there is a progressive improvement in the quality of high-fidelity sound
amplification equipment that is undeniable, even if you agree with me
sometimes that the world would be a more agreeable place if the
amplifier had never been invented. It is not that tastes have changed; it
is an objective, measurable fact that fidelity of reproduction is now
better than it was in 1950, and in 1950 it was better than in 1920. The
quality of picture reproduction is undeniably better in modern television sets than in earlier ones, although of course the same may not be
true of the quality of the entertainment transmitted. The quality of
machines for killing in war shows a dramatic trend towards improvement – they are capable of killing more people faster as the years go by.
The sense in which this is not an improvement is too obvious to
labour.

There is no doubt about it, in the narrow technical sense things do
get better as time goes by. But this is only obviously true of technically
useful things such as aeroplanes and computers. There are many other
aspects of human life that show true trends without these trends being,
in any obvious sense, improvements. Languages clearly evolve in that
they show trends, they diverge, and as the centuries go by after their
divergence they become more and more mutually unintelligible. The
numerous islands of the Pacific provide a beautiful workshop for the
study of language evolution. The languages of different islands clearly
resemble each other, and their differences can be measured precisely
by the numbers of words that differ between them, a measure that is
closely analogous to the molecular taxonomic measures that we shall
discuss in Chapter 10. Difference between languages, measured in
numbers of divergent words, can be plotted on a graph against distance
between islands, measured in miles, and it turns out that the points on
the graph fall on a curve whose precise mathematical shape tells us
something about rates of diffusion from island to island. Words
travelled by canoe, island-hopping at intervals proportional to the
degree of remoteness of the islands concerned. Within any one island
words change at a steady rate, in very much the same way as genes
occasionally mutate. Any island, if completely isolated, would exhibit
some evolutionary change in its language as time went by, and hence

some divergence from the languages of other islands. Islands that are near each other obviously have a higher rate of word flow between them, via canoe, than islands that are far from each other. Their languages also have a more recent common ancestor than the languages of islands that are far apart. These phenomena, which explain the observed pattern of resemblances between near and distant islands, are closely analogous to the facts about finches on different islands of the Galápagos Archipelago which originally inspired Charles Darwin. Genes island-hop in the bodies of birds, just as words island-hop in canoes.

Languages, then, evolve. But although modern English has evolved from Chaucerian English, I don't think many people would wish to claim that modern English is an improvement on Chaucerian English. Ideas of improvement or quality do not normally enter our heads when we speak of language. Indeed, to the extent that they do, we often see change as deterioration, as degeneration. We tend to see earlier usages as correct, recent changes as corruptions. But we can still detect evolution-like trends that are progressive in a purely abstract, value-free sense. And we can even find evidence of positive feedbacks, in the form of escalations (or, looking at it from the other direction, degenerations) in meaning. For instance, the word 'star' was used to mean a film actor of quite exceptional celebrity. Then it degenerated to mean any ordinary actor who was playing one of the principal roles in a film. Therefore, in order to recapture the original meaning of exceptional celebrity, the word had to escalate to 'superstar'. Later, studio publicity began to use 'superstar' for actors that many people had never heard of, so there was a further escalation to 'megastar'. Now there are quite a few advertised 'megastars' whom I, at least, have never heard of before, so perhaps we are due for yet another escalation. Shall we soon hear tell of 'hyperstars'? A similar positive feedback has driven down the currency of the word 'chef'. It comes, of course, from the French *chef de cuisine*, meaning *chief* or head of the kitchen. This is the meaning given in the Oxford Dictionary. By definition, then, there can be only one chef per kitchen. But, perhaps to satisfy their dignity, ordinary (male) cooks, even junior hamburger-slingers, started referring to themselves as 'chefs'. The result is that now the tautological phrase 'head chef' is frequently heard!

But if this is an analogy for sexual selection, it is so, at best, only in what I have called the 'weak' sense. Let me jump now straight to the nearest approach I can think of to a 'strong' analogy: to the world of 'pop' records. If you listen to discussion among aficionados of pop records, or switch on the mid-Atlantic mouthings of disc jockeys on the radio, you will discover a very curious thing. Whereas other genres of art criticism betray some preoccupation with style or skill of performance, with mood, emotional impact, with the qualities and properties of the art-form, the 'pop' music sub-culture is almost exclusively preoccupied with *popularity itself*. It is quite clear that the important thing about a record is not what it sounds like, but *how many people are buying it*. The whole sub-culture is obsessed with a rank ordering of records, called the Top 20 or Top 40, which is based only upon sales figures. The thing that really matters about a record is where it lies in the Top 20. This, when you think about it, is a very singular fact, and a very interesting one if we are thinking about R. A. Fisher's theory of runaway evolution. It is probably also significant that a disc jockey seldom mentions the current position of a record in the charts, without at the same time telling us its position in the previous week. This allows the listener to assess, not just the present popularity of a record, but also its rate and direction of *change* in popularity.

It appears to be a fact that many people will buy a record for no better reason than that large numbers of other people are buying the same record, or are likely to do so. Striking evidence comes from the fact that record companies have been known to send representatives into key shops to buy up large numbers of their own records, in order to push sales figures up into the region where they may 'take off'. (This is not so difficult to do as it sounds, because the Top 20 figures are based upon sales returns from a small sample of record shops. If you know which these key shops are, you don't have to buy all that many records from them in order to make a significant impact on nationwide estimates of sales. There are also well-authenticated stories of shop assistants in these key shops being bribed.)

To a lesser extent, the same phenomenon of popularity being popular for its own sake is well known in the worlds of book publishing, womens' fashions, and advertising generally. One of the best things an advertiser can say about a product is that it is the best-selling product of its kind. Best-seller lists of books are published weekly, and it is undoubtedly true that as soon as a book sells enough copies to appear in one of these lists, its sales increase even more, simply by virtue of that fact. Publishers speak of a book 'taking off', and those publishers with some knowledge of science even speak of a 'critical mass for take-off'. The analogy here is to an atomic bomb. Uranium-235 is stable as long as you don't have too much of it in one place. There is a critical mass which, once exceeded, permits a chain reaction or runaway process to get going, with devastating results. An atom bomb contains two lumps of uranium-235, both smaller than the critical mass. When the bomb is detonated the two lumps are thrust together, the critical mass is exceeded, and that is the end of a medium-sized city. When a book's sales 'go critical', the numbers reach the point where word-of-mouth recommendations et cetera cause its sales suddenly to take-off in a runaway fashion. Rates of sales suddenly become dramatically larger than they were before critical mass was reached, and there may be a period of exponential growth before the inevitable levelling out and subsequent decline.

The underlying phenomena are not difficult to understand. Basically we have here yet more examples of positive feedback. A book's, or even a pop record's, real qualities are not negligible in determining its sales but, nevertheless, wherever there are positive feedbacks lurking, there is bound to be a strong arbitrary element determining which book or record succeeds, and which fails. If critical mass and take-off are important elements in any success story, there is bound to be a lot of luck, and there is also plenty of scope for manipulation and exploitation by people that understand the system. It is, for instance, worth laying out a considerable sum of money to promote a book or record up to the point where it just 'goes critical', because you then don't need to spend so much money on promoting it thereafter: the positive feedbacks take over and do the work of publicity for you.

The positive feedbacks here have something in common with those of sexual selection according to the Fisher/Lande theory, but there are differences too. Peahens that prefer long-tailed peacocks are favoured solely because *other* females have the same preference. The male's qualities themselves are arbitrary and irrelevant. In this respect, the record enthusiast who wants a particular record just because it is in the Top 20 is behaving just like a peahen. But the precise mechanisms by which the positive feedbacks work in the two cases are different. And this, I suppose, brings us back to where we began in this chapter, with a warning that analogies should be taken so far, and no farther.

CHAPTER 9

PUNCTURING PUNCTUATIONISM

The children of Israel, according to the Exodus story, took 40 years to migrate across the Sinai desert to the promised land. That is a distance of some 200 miles. Their average speed was, therefore, approximately 24 yards per day, or 1 yard per hour; say 3 yards per hour if we allow for night stops. However we do the calculation, we are dealing with an absurdly slow average speed, much slower than the proverbially slow snail's pace (an incredible 55 yards per hour is the speed of the world record snail according to the *Guinness Book of Records*). But of course nobody really believes that the average speed was continuously and uniformly maintained. Obviously the Israelites travelled in fits and starts, perhaps camping for long periods in one spot before moving on. Probably many of them had no very clear idea that they were *travelling* in any particularly consistent direction, and they meandered round and round from oasis to oasis as nomadic desert herdsmen are wont to do. Nobody, I repeat, really believes that the average speed was continuously and uniformly maintained.

But now suppose that two eloquent young historians burst upon the scene. Biblical history so far, they tell us, has been dominated by the 'gradualistic' school of thought. 'Gradualist' historians, we are told, literally believe that the Israelites travelled 24 yards per day; they folded their tents every morning, crawled 24 yards in an east-northeasterly direction, and then pitched camp again. The only alternative to 'gradualism', we are told, is the dynamic new 'punctuationist' school of history. According to the radical young punctuationists, the Israelites spent most of their time in 'stasis', not moving at all but camped, often for years at a time, in one place. Then they would move on, rather fast, to a new encampment, where they again stayed for several years. Their progress towards the promised land, instead of being gradual and continuous, was jerky: long periods of stasis punctuated by brief periods of rapid movement. Moreover, their bursts of movement were not always in the direction of the promised land, but were in almost random directions. It is only when we look, with hindsight, at the large scale *macromigrational* pattern, that we can see a trend in the direction of the promised land.

Such is the eloquence of the punctuationist biblical historians that they become a media sensation. Their portraits adorn the front covers of mass circulation news magazines. No television documentary about biblical history is complete without an interview with at least one leading punctuationist. People who know nothing else of biblical scholarship remember just the one fact: that in the dark days before the punctuationists burst upon the scene, everybody else got it wrong. Note that the publicity value of the punctuationists has nothing to do with the fact that they may be right. It has everything to do with the allegation that earlier authorities were 'gradualist' and wrong. It is because the punctuationists sell themselves as revolutionaries that they are listened to, not because they are right.

My story about the punctuationist biblical historians is, of course, not really true. It is a parable about an analogous alleged controversy among students of biological evolution. In some respects it is an unfair parable, but it is not totally unfair and it has enough truth in it to justify its telling at the beginning of this chapter. There is a highly advertised school of thought among evolutionary biologists whose proponents call themselves punctuationists, and they did invent the term 'gradualist' for their most influential predecessors. They have enjoyed enormous publicity, among a public that knows almost nothing else about evolution, and this is largely because their position has been represented, by secondary reporters more than by themselves, as radically different from the positions of previous evolutionists, especially Charles Darwin. So far, my biblical analogy is a fair one.

The respect in which the analogy is unfair is that in the story of the biblical historians 'the gradualists' were *obviously* non-existent straw men, fabricated by the punctuationists. In the case of the evolutionary 'gradualists', the fact that they are non-existent straw men is not quite so obvious. It needs to be demonstrated. It is possible to interpret the words of Darwin and many other evolutionists as gradualist in intent, but it then becomes important to realize that the word gradualist can be interpreted in different ways to mean different things. Indeed, I shall develop an interpretation of the word 'gradualist' according to which just about everybody is a gradualist. In the evolutionary case, unlike in the parable of the Israelites, there is genuine controversy lurking, but that genuine controversy is about little details which are nowhere near important enough to justify all the media hype.

Among evolutionists, the 'punctuationists' were originally drawn from the ranks of palaeontology. Palaeontology is the study of fossils. It is a very important branch of biology, because evolutionary ancestors all died long ago and fossils provide us with our only direct evidence of the animals and plants of the distant past. If we want to know what our evolutionary ancestors looked like, fossils are our main hope. As soon as people realized what fossils really were – previous schools of thought had held that they were creations of the devil, or that they were the bones of poor sinners drowned in the flood – it became clear that any theory of evolution must have certain expectations about the fossil record. But there has been some discussion of exactly what those expectations are, and this is partly what the punctuationism argument is about.

We are lucky to have fossils at all. It is a remarkably fortunate fact of geology that bones, shells and other hard parts of animals, before they decay, can occasionally leave an imprint which later acts as a mould, which shapes hardening rock into a permanent memory of the animal. We don't know what proportion of animals are fossilized after their death – I personally would consider it an honour to be fossilized – but it is certainly very small indeed. Nevertheless, however small the proportion fossilized, there are certain things about the fossil record that any evolutionist should expect to be true. We should be very surprised, for example, to find fossil humans appearing in the record before mammals are supposed to have evolved! If a single, well-verified mammal skull were to turn up in 500 million year-old rocks, our whole modern theory of evolution would be utterly destroyed. Incidentally, this is a sufficient answer to the canard, put about by creationists and their journalistic fellow travellers, that the whole theory of evolution is an 'unfalsifiable' tautology. Ironically, it is also the reason why creationists are so keen on the fake human footprints, which were carved during the depression to fool tourists, in the dinosaur beds of Texas.

Anyway, if we arrange our genuine fossils in order, from oldest to youngest, the theory of evolution expects to see some sort of orderly sequence rather than a higgledy-piggledy jumble. More to the point in this chapter, different versions of the theory of evolution, for instance 'gradualism' and 'punctuationism', might expect to see different kinds of pattern. Such expectations can be tested only if we have some means of *dating* fossils, or at least of knowing the order in which they were laid down. The problems of dating fossils, and the solutions of these problems, require a brief digression, the first of several for which the reader's indulgence is asked. They are necessary for the explanation of the main theme of the chapter.

We have long known how to arrange fossils in the order in which they were laid down. The method is inherent in the very phrase 'laid down'. More recent fossils are obviously laid down on top of older fossils rather than underneath them, and they therefore lie above them in rock sediments. Occasionally volcanic upheavals can turn a chunk of rock right over and then, of course, the order in which we find fossils as we dig downwards will be exactly reversed; but this is rare enough to be obvious when it occurs. Even though we seldom find a complete historical record as we dig down through the rocks of any one area, a good record can be pieced together from overlapping portions in different areas (actually, although I use the image of 'digging down', palaeontologists seldom literally dig downwards through strata; they are more likely to find fossils exposed by erosion at various depths). Long before they knew how to date fossils in actual millions of years, palaeontologists had worked out a reliable scheme of geological eras, and they knew in great detail which era came before which. Certain kinds of shells are such reliable indicators of the ages of rocks that they are among the main indicators used by oil prospectors in the field. By themselves, however, they can tell us only about the relative ages of rock strata, never their absolute ages.

More recently, advances in physics have given us methods to put absolute dates, in millions of years, on rocks and the fossils that they contain. These methods depend upon the fact that particular radioactive elements decay at precisely known rates. It is as though precision-made miniature stopwatches had been conveniently buried in the rocks. Each stopwatch was started at the moment that it was laid down. All that the palaeontologist has to do is dig it up and read off the time on the dial. Different kinds of radioactive decay-based geological stopwatches run at different rates. The radiocarbon stopwatch buzzes round at a great rate, so fast that, after some thousands of years, its spring is almost wound down and the watch is no longer reliable. It is useful for dating organic material on the archaeological/historical timescale where we are dealing in hundreds or a few thousands of years, but it is no good for the evolutionary timescale where we are dealing in millions of years.

For the evolutionary timescale other kinds of watch, such as the potassium–argon watch, are suitable. The potassium–argon watch is so slow that it would be unsuitable for the archaeological/historical timescale. That would be like trying to use the hour hand on an ordinary watch to time an athlete sprinting a hundred yards. For timing the megamarathon that is evolution, on the other hand, something like the potassium–argon watch is just what is needed. Other radioactive 'stopwatches', each with its own characteristic rate of slowing down, are the rubidium–strontium, and the uranium–thorium–lead watches. So, this digression has told us that if a palaeontologist is presented with a fossil, he can usually know when the animal lived, on an absolute timescale of millions of years. We got into this discussion of dating and timing in the first place, you will remember, because we were interested in the expectations about the fossil record that various kinds of evolutionary theory – 'punctuationist', 'gradualist', etc. – should have. It is now time to discuss what those various expectations are.

Suppose, first, that nature had been extraordinarily kind to palaeontologists (or perhaps unkind, when you think of the extra work involved), and given them a fossil of every animal that ever lived. If we could indeed look at such a complete fossil record, carefully arranged in chronological order, what should we, as evolutionists, expect to see? Well, if we are 'gradualists', in the sense caricatured in the parable of the Israelites, we should expect something like the following. Chronological sequences of fossils will always exhibit smooth evolutionary trends with fixed rates of change. In other words, if we have three fossils, A, B and C, A being ancestral to B, which is ancestral to C, we should expect B to be proportionately intermediate in form between A and C. For instance, if A had a leg length of 20 inches and C had a leg length of 40 inches, B's legs should be intermediate, their exact length being proportional to the time that elapsed between A's existence and B's.

If we carry the caricature of gradualism to its logical conclusion, just as we calculated the average speed of the Israelites as 24 yards per day, so we can calculate the average rate of lengthening of the legs in the evolutionary line of descent from A to C. If, say, A lived 20 million years earlier than C (to fit this vaguely into reality, the earliest known member of the horse family, *Hyracotherium*, lived about 50 million years ago, and was the size of a terrier), we have an evolutionary growth rate of 20 leg-inches per 20 million years, or one-millionth of an inch per year. Now the caricature of a gradualist is supposed to believe that the legs steadily grew, over the generations, at this very slow rate: say four-millionths of an inch per generation, if we assume a horse-like generation-time of about 4 years. The gradualist is supposed to believe that, through all those millions of generations, individuals with legs four-millionths of an inch longer than the average had an advantage over individuals with legs of average length. To believe this is like believing that the Israelites travelled 24 yards every day across the desert.

The same is true even of one of the fastest known evolutionary changes, the swelling of the human skull from an *Australopithecus*-like ancestor, with a brain volume of about 500 cubic centimetres (cc), to the modern *Homo sapiens*'s average brain volume of about 1,400 cc. This increase of about 900 cc, nearly a tripling in brain volume, has been accomplished in no more than three million years. By evolutionary standards this is a rapid rate of change: the brain seems to swell like a balloon and indeed, seen from some angles, the modern human skull does rather resemble a bulbous, spherical balloon in comparison to the flatter, sloping-browed skull of *Australopithecus*. But if we count up the number of generations in three million years (say about four per century), the average rate of evolution is less than a hundredth of a cubic centimetre per generation. The caricature of a gradualist is supposed to believe that there was a slow and inexorable change, generation by generation, such that in all generations sons were slightly brainier than their fathers, brainier by 0.01 cc. Presumably the extra hundredth of a cubic centimetre is supposed to provide each succeeding generation with a significant survival advantage compared with the previous generation.

But a hundredth of a cubic centimetre is a tiny quantity in comparison to the range of brain sizes that we find among modern humans. It is an often-quoted fact, for instance, that the writer Anatole France – no fool, and a Nobel prizewinner – had a brain size of less than 1,000 cc, while at the other end of the range, brains of 2,000 cc are not unknown: Oliver Cromwell is frequently cited as an example, though I do not know with what authenticity. The average per-generation increment of 0.01 cc, then, which is supposed by the caricature of a gradualist to give a significant survival advantage, is a mere hundred-thousandth part of the *difference* between the brains of Anatole France and Oliver Cromwell! It is fortunate that the caricature of a gradualist does not really exist.

Well, if this kind of gradualist is a non-existent caricature – a windmill for punctuationist lances – is there some other kind of gradualist who really exists and who holds tenable beliefs? I shall show that the answer is yes, and that the ranks of gradualists, in this second sense, include all sensible evolutionists, among them, when you look carefully at their beliefs, those that call themselves punctuationists. But we have to understand why the punctuationists *thought* that their views were revolutionary and exciting. The starting point for discussing these matters is the apparent existence of 'gaps' in the fossil record, and it is to these gaps that we now turn.

From Darwin onwards evolutionists have realized that, if we arrange all our available fossils in chronological order, they do *not* form a smooth sequence of scarcely perceptible change. We can, to be sure, discern long-term trends of change – legs get progressively longer, skulls get progressively more bulbous, and so on – but the trends as seen in the fossil record are usually jerky, not smooth. Darwin, and most others following him, have assumed that this is mainly because the fossil record is imperfect. Darwin's view was that a complete fossil record, if only we had one, *would* show gentle rather than jerky change. But since fossilization is such a chancy business, and finding such fossils as there are is scarcely less chancy, it is as though we had a cine film with most of the frames missing. We can, to be sure, see movement of a kind when we project our film of fossils, but it is more jerky than Charlie Chaplin, for even the oldest and scratchiest Charlie Chaplin film hasn't completely lost nine-tenths of its frames.

The American palaeontologists Niles Eldredge and Stephen Jay Gould, when they first proposed their theory of punctuated equilibria in 1972, made what has since been represented as a very different suggestion. They suggested that, actually, the fossil record may not be as imperfect as we thought. Maybe the 'gaps' are a true reflection of what really happened, rather than being the annoying but inevitable consequences of an imperfect fossil record. Maybe, they suggested, evolution really did in some sense go in sudden bursts, punctuating long periods of 'stasis', when no evolutionary change took place in a given lineage.

Before we come to the sort of sudden bursts that they had in mind, there are some conceivable meanings of 'sudden bursts' that they most definitely did not have in mind. These must be cleared out of the way because they have been the subject of serious misunderstandings. Eldredge and Gould certainly would agree that some very important gaps really are due to imperfections in the fossil record. Very big gaps, too. For example the Cambrian strata of rocks, vintage about 600 million years, are the oldest ones in which we find most of the major invertebrate groups. And we find many of them already in an advanced state of evolution, the very first time they appear. It is as though they were just planted there, without any evolutionary history. Needless to say, this appearance of sudden planting has delighted creationists. Evolutionists of all stripes believe, however, that this really does represent a very large gap in the fossil record, a gap that is simply due to the fact that, for some reason, very few fossils have lasted from periods before about 600 million years ago. One good reason might be that many of these animals had only soft parts to their bodies: no shells or bones to fossilize. If you are a creationist you may think that this is special pleading. My point here is that, when we are talking about gaps of this magnitude, there is no difference whatever in the interpretations of 'punctuationists' and 'gradualists'. Both schools of thought despise so-called scientific creationists equally, and both agree that the *major* gaps are real, that they are true imperfections in the fossil record. Both schools of thought agree that the only alternative explanation of the sudden appearance of so many complex animal types in the Cambrian era is divine creation, and both would reject this alternative.

There is another conceivable sense in which evolution might be said to go in sudden jerks, but which is also not the sense being proposed by Eldredge and Gould, at least in most of their writings. It is conceivable that some of the apparent 'gaps' in the fossil record really do reflect sudden change in a single generation. It is conceivable that there really never were any intermediates; conceivable that large evolutionary changes took place in a single generation. A son might be born so different from his father that he properly belongs in a different species from his father. He would be a mutant individual, and the mutation would be such a large one that we should refer to it as a macromutation. Theories of evolution that depend upon macromutation are called 'saltation' theories, from *saltus*, the Latin for 'jump'. Since the theory of punctuated equilibria frequently is confused with true saltation, it is important here to discuss saltation, and show why it cannot be a significant factor in evolution.

Macromutations – mutations of large effect – undoubtedly occur. What is at issue is not whether they occur but whether they play a role in evolution; whether, in other words, they are incorporated into the gene pool of a species, or whether, on the contrary, they are always eliminated by natural selection. A famous example of a macromutation is 'antennapaedia' in fruitflies. In a normal insect the antennae have something in common with the legs, and they develop in the embryo in a similar way. But the differences are striking as well, and the two sorts of limb are used for very different purposes: the legs for walking; the antennae for feeling, smelling and otherwise sensing things. Antennapaedic flies are freaks in which the antennae develop just like legs. Or, another way of putting it, they are flies that have no antennae but an extra pair of legs, growing out of the sockets where the antennae ought to be. This is a true mutation in that it results from an error in the copying of DNA. And it breeds true if antennapaedic flies are cosseted in the laboratory so that they survive long enough to breed at all. They would not survive long in the wild, as their movements are clumsy and their vital senses are impaired.

So, macromutations do happen. But do they play a role in evolution? People called saltationists believe that macromutations are a means by which major jumps in evolution could take place in a single generation. Richard Goldschmidt, whom we met in Chapter 4, was a true saltationist. If saltationism were true, apparent 'gaps' in the fossil record needn't be gaps at all. For example, a saltationist might believe that the transition from sloping-browed *Australopithecus* to dome-browed *Homo sapiens* took place in a single macromutational step, in a single generation. The difference in form between the two species is probably less than the difference between a normal and an antennapaedic fruitfly, and it is theoretically conceivable that the first *Homo sapiens* was a freak child – probably an ostracized and persecuted one – of two normal *Australopithecus* parents.

There are very good reasons for rejecting all such saltationist theories of evolution. One rather boring reason is that if a new species really did arise in a single mutational step, members of the new species might have a hard time finding mates. But I find this reason less telling and interesting than two others which have already been foreshadowed in our discussion of why major jumps across Biomorph Land are to be ruled out. The first of these points was put by the great statistician and biologist R. A. Fisher, whom we met in other connections in previous chapters. Fisher was a stalwart opponent of all forms of saltationism, at a time when saltationism was much more fashionable than it is today, and he used the following analogy. Think, he said, of a microscope which is almost, but not quite perfectly, in focus and otherwise well adjusted for distinct vision. What are the odds that, if we make some random change to the state of the microscope (corresponding to a mutation), we shall improve the focus and general quality of the image? Fisher said:

> It is sufficiently obvious that any large derangement will have a very small probability of improving the adjustment, while in the case of alterations much less than the smallest of those intentionally effected by the maker or the operator, the chance of improvement should be almost exactly one half.

I have already remarked that what Fisher found 'easy to see' could place formidable demands on the mental powers of ordinary scientists, and the same is true of what Fisher thought was 'sufficiently obvious'. Nevertheless, further cogitation almost always shows him to have been right, and in this case we can prove it to our own satisfaction without too much difficulty. Remember that we are assuming the microscope to be almost in correct focus before we start. Suppose that the lens is slightly lower than it ought to be for perfect focus, say a tenth of an inch too close to the slide. Now if we move it a small amount, say a hundredth of an inch, in a random direction, what are the odds that the focus will improve? Well, if we happen to move it *down* a hundredth of an inch, the focus will get worse. If we happen to move it *up* a hundredth of an inch, the focus will get better. Since we are moving it in a random direction, the chance of each of these two eventualities is one half. The smaller the movement of adjustment, in relation to the initial error, the closer will the chance of improvement approach one half. That completes the justification of the second part of Fisher's statement.

But now, suppose we move the microscope tube a large distance – equivalent to a macromutation – also in a random direction; suppose we move it a full inch. Now it doesn't matter which direction we move it in, up or down, we shall still make the focus worse than it was before. If we chance to move it down, it will now be one and one-tenth inches away from its ideal position (and will probably have crunched through the slide). If we chance to move it up, it will now be nine-tenths of an inch away from its ideal position. Before the move, it was only one-tenth of an inch away from its ideal position so, either way, our 'macromutational' big move has been a bad thing. We have done the calculation for a very big move ('macromutation') and a very small move ('micromutation'). We can obviously do the same calculation for a range of intermediate sizes of move, but there is no point in doing so. I think it really will now be sufficiently obvious that the smaller we make the move, the closer we shall approach the extreme case in which the odds of an improvement are one-half; and the larger we make the move, the closer we shall approach the extreme case in which the odds of an improvement are zero.

The reader will have noticed that this argument depends upon the initial assumption that the microscope was already pretty close to being in focus before we even started making random adjustments. If the microscope starts 2 inches out of focus, then a random change of 1 inch has a 50 per cent chance of being an improvement, just as a random change of one-hundredth of an inch has. In this case the 'macromutation' appears to have the advantage that it moves the microscope into focus more quickly. Fisher's argument will, of course, apply here to 'megamutations' of, say, 6 inches movement in a random direction.

Why, then, was Fisher allowed to make his initial assumption that the microscope was nearly in focus at the start? The assumption flows from the role of the microscope in the analogy. The microscope after its random adjustment stands for a mutant animal. The microscope before its random adjustment stands for the normal, unmutated parent of the supposed mutant animal. Since it is a parent, it must have survived long enough to reproduce, and therefore it cannot be all that far from being well-adjusted. By the same token, the microscope before the random jolt cannot be all that far from being in focus, or the animal that it stands for in the analogy couldn't have survived at all. It is only an analogy, and there is no point in arguing over whether 'all that far' means an inch or a tenth of an inch or a thousandth of an inch. The important point is that if we consider mutations of ever-increasing magnitude, there will come a point when, the larger the mutation is, the less likely it is to be beneficial; while if we consider mutations of ever-decreasing magnitude, there will come a point when the chance of a mutation's being beneficial is 50 per cent.

The argument over whether macromutations such as antennapaedia could ever be beneficial (or at least could avoid being harmful), and therefore whether they could give rise to evolutionary change, therefore turns on *how* 'macro' the mutation is that we are considering. The more 'macro' it is, the more likely it is to be deleterious, and the less likely it is to be incorporated in the evolution of a species. As a matter of fact, virtually all the mutations studied in genetics laboratories – which are pretty macro because otherwise geneticists wouldn't notice them – are deleterious to the animals possessing them (ironically I've met people who think that this is an argument *against* Darwinism!). Fisher's microscope argument, then, provides one reason for scepticism about 'saltation' theories of evolution, at least in their extreme form.

The other general reason for not believing in true saltation is also a statistical one, and its force also depends quantitatively on *how* macro is the macromutation we are postulating. In this case it is concerned with the complexity of evolutionary changes. Many, though not all, of the evolutionary changes we are interested in are advances in complexity of design. The extreme example of the eye, discussed in earlier chapters, makes the point clear. Animals with eyes like ours evolved from ancestors with no eyes at all. An extreme saltationist might postulate that the evolution took place in a single mutational step. A parent had no eye at all, just bare skin where the eye might be. He had a freak offspring with a fully developed eye, complete with variable focus lens, iris diaphragm for 'stopping down', retina with millions of three-colour photocells, all with nerves correctly connected up in the brain to provide him with correct, binocular, stereoscopic colour vision.

In the biomorph model we assumed that this kind of multi-dimensional improvement could not occur. To recapitulate on why that was a reasonable assumption, to make an eye from nothing you need not just one improvement but a large number of improvements. Any one of these improvements is pretty improbable by itself, but not so improbable as to be impossible. The greater the number of simultaneous improvements we consider, the more improbable is their simultaneous occurrence. The coincidence of their simultaneous occurrence is equivalent to leaping a large distance across Biomorph Land, and happening to land on one particular, predesignated spot. If we choose to consider a sufficiently large number of improvements, their joint occurrence becomes so improbable as to be, to all intents and purposes, impossible. The argument has already been sufficiently made, but it may be helpful to draw a distinction between two kinds of hypothetical macromutation, both of which *appear* to be ruled out by the complexity argument but only one of which, in fact, *is* ruled out by the complexity argument. I label them, for reasons that will become clear, Boeing 747 macromutations and Stretched DC8 macromutations.

Boeing 747 macromutations are the ones that really are ruled out by the complexity argument just given. They get their name from the astronomer Sir Fred Hoyle's memorable misunderstanding of the theory of natural selection. He compared natural selection, in its alleged improbability, to a hurricane blowing through a junkyard and chancing to assemble a Boeing 747. As we saw in Chapter 1, this is an entirely false analogy to apply to natural selection, but it is a very good analogy for the idea of certain kinds of macromutation giving rise to evolutionary change. Indeed, Hoyle's fundamental error was that he, in effect, thought (without realizing it) that the theory of natural selection *did* depend upon macromutation. The idea of a single macromutation's giving rise to a fully functioning eye with the properties listed above, where there was only bare skin before, is, indeed, just about as improbable as a hurricane assembling a Boeing 747. This is why I refer to this kind of hypothetical macromutation as a Boeing 747 macromutation.

Stretched DC8 macromutations are mutations that, although they may be large in the magnitude of their effects, turn out not to be large in terms of their complexity. The Stretched DC8 is an airliner that was made by modifying an earlier airliner, the DC8. It is like a DC8, but with an elongated fuselage. It was an improvement at least from one point of view, in that it could carry more passengers than the original DC8. The stretching is a large increase in length, and in that sense is analogous to a macromutation. More interestingly, the increase in length is, at first sight, a complex one. To elongate the fuselage of an airliner, it is not enough just to insert an extra length of cabin tube. You also have to elongate countless ducts, cables, air tubes and electric wires. You have to put in lots more seats, ashtrays, reading lights, 12-channel music selectors and fresh-air nozzles. At first sight there seems to be much more complexity in a Stretched DC8 than there is in an ordinary DC8, but is there really? The answer is no, at least to the extent that the 'new' things in the stretched plane are just 'more of the same'. The biomorphs of Chapter 3 frequently show macromutations of the Stretched DC8 variety.

What has this to do with mutations in real animals? The answer is that some real mutations cause large changes that are very like the change from DC8 to Stretched DC8, and some of these, although in a sense 'macro' mutations, have definitely been incorporated in evolution. Snakes, for instance, all have many more vertebrae than their ancestors. We could be sure of this even if we didn't have any fossils, because snakes have many more vertebrae than their surviving relatives. Moreover, different species of snakes have different numbers of vertebrae, which means that vertebral number must have changed in evolution since their common ancestor, and quite often at that.

Now, to change the number of vertebrae in an animal, you need to do more than just shove in an extra bone. Each vertebra has, associated with it, a set of nerves, a set of blood vessels, a set of muscles etc., just as each row of seats in an airliner has a set of cushions, a set of head rests, a set of headphone sockets, a set of reading-lights with their associated cables etc. The middle part of the body of a snake, like the middle part of the body of an airliner, is composed of a number of *segments*, many of which are exactly like each other, however complex they all individually may be. Therefore, in order to add new segments, all that has to be done is a simple process of duplication. Since there already exists genetic machinery for making one snake segment – genetic machinery of great complexity, which took many generations of step-by-step, gradual evolution to build up – new identical segments may easily be added by a single mutational step. If we think of genes as 'instructions to a developing embryo', a gene for inserting extra segments may read, simply, 'more of the same here'. I imagine that the instructions for building the first Stretched DC8 were somewhat similar.

We can be sure that, in the evolution of snakes, numbers of vertebrae changed in whole numbers rather than in fractions. We cannot imagine a snake with 26.3 vertebrae. It either had 26 or 27, and it is obvious that there must have been cases when an offspring snake had at least one whole vertebra more than its parents did. This means that it had a whole extra set of nerves, blood vessels, muscle blocks, etc. In a sense, then, this snake was a *macro*-mutant, but only in the weak 'Stretched DC8' sense. It is easy to believe that individual snakes with half a dozen more vertebrae than their parents could have arisen in a single mutational step. The 'complexity argument' against saltatory evolution does not apply to Stretched DC8 macromutations because, if we look in detail at the nature of the change involved, they are in a real sense not true macromutations at all. They are only macromutations if we look, naïvely, at the finished product, the adult. If we look at the *processes* of embryonic development they turn out to be micromutations, in the sense that only a small change in the embryonic *instructions* had a large apparent effect in the adult. The same goes for antennapaedia in fruitflies and the many other so-called 'homeotic mutations'.

This concludes my digression on macromutation and saltatory evolution. It was necessary, because the theory of punctuated equilibria is frequently confused with saltatory evolution. But it *was* a digression, because the theory of punctuated equilibria is the main topic of this chapter, and that theory in truth has no connection with macromutation and true saltation.

The 'gaps' that Eldredge and Gould and the other 'punctuationists' are talking about, then, have nothing to do with true saltation, and they are much much smaller gaps than the ones that excite creationists. Moreover, Eldredge and Gould originally introduced their theory, *not* as radically and revolutionarily antipathetic to ordinary, 'conventional' Darwinism – which is how it later came to be sold – but as something that *followed* from long-accepted conventional Darwinism, properly understood. To gain this proper understanding, I'm afraid we need another digression, this time into the question of how new species originate, the process known as 'speciation'.

Darwin's answer to the question of the origin of species was, in a general sense, that species were descended from other species. Moreover, the family tree of life is a branching one, which means that more than one modern species can be traced back to one ancestral one. For instance, lions and tigers are now members of different species, but they have both sprung from a single ancestral species, probably not very long ago. This ancestral species may have been the same as one of the two modern species; or it may have been a third modern species; or maybe it is now extinct. Similarly, humans and chimps now clearly belong to different species, but their ancestors of a few million years ago belonged to one single species. Speciation is the process by which a single species becomes two species, one of which may be the same as the original single one.

The reason speciation is thought to be a difficult problem is this. All the members of the single would-be ancestral species are capable of interbreeding with one another: indeed, to many people, this is what is *meant* by the phrase 'single species'. Therefore, every time a new daughter species begins to be 'budded off', the budding off is in danger of being frustrated by interbreeding. We can imagine the would-be ancestors of the lions and the would-be ancestors of the tigers failing to split apart because they keep interbreeding with one another and therefore staying similar to one another. Don't, incidentally, read too much into my use of words like 'frustrated', as though the ancestral lions and tigers, in some sense, 'wanted' to separate from each other. It is simply that, as a matter of fact, species obviously *have* diverged from one another in evolution, and at first sight the fact of interbreeding makes it hard for us to see how this divergence came about.

It seems almost certain that the principal correct answer to this problem is the obvious one. There will be no problem of interbreeding if the ancestral lions and the ancestral tigers happen to be in different parts of the world, where they can't interbreed with each other. Of course, they didn't go to different continents in order to allow them-selves to diverge from one another: they didn't think of themselves as ancestral lions or ancestral tigers! But, given that the single ancestral species spread to different continents anyway, say Africa and Asia, the ones that happened to be in Africa could no longer interbreed with the ones that happened to be in Asia because they never met them. If there was any tendency for the animals on the two continents to evolve in different directions, either under the influence of natural selection or under the influence of chance, interbreeding no longer constituted a barrier to their diverging and eventually becoming two distinct species.

I have spoken of different continents to make it clear, but the principle of geographical separation as a barrier to interbreeding can apply to animals on different sides of a desert, a mountain range, a river, or even a motorway. It can also apply to animals separated by no barrier other than sheer distance. Shrews in Spain cannot interbreed with shrews in Mongolia, and they can diverge, evolutionarily speaking, from shrews in Mongolia, even if there is an unbroken chain of interbreeding shrews connecting Spain to Mongolia. Nevertheless the idea of geographical separation as the key to speciation is clearer if we think in terms of an actual physical barrier, such as the sea or a mountain range. Chains of islands, indeed, are probably fertile nurseries for new species.

Here, then, is our orthodox neo-Darwinian picture of how a typical species is 'born', by divergence from an ancestral species. We start with the ancestral species, a large population of rather uniform, mutually interbreeding animals, spread over a large land mass. They could be any sort of animal, but let's carry on thinking of shrews. The landmass is cut in two by a mountain range. This is hostile country and the shrews are unlikely to cross it, but it is not quite impossible and very occasionally one or two do end up in the lowlands on the other side. Here they can flourish, and they give rise to an outlying population of the species, effectively cut off from the main population. Now the two populations breed and breed separately, mixing their genes on each side of the mountains but not across the mountains. As time goes by, any changes in the genetic composition of one population are spread by breeding throughout that population but *not* across to the other population. Some of these changes may be brought about by natural selection, which may be different on the two sides of the mountain range: we should hardly expect weather conditions, and predators and parasites, to be exactly the same on the two sides. Some of the changes may be due to chance alone. Whatever the genetic changes are due to, breeding tends to spread them *within* each of the two populations, but not *between* the two populations. So the two populations diverge genetically: they become progressively more unlike each other.

They become so unlike each other that, after a while, naturalists would see them as belonging to different 'races'. After a longer time, they will have diverged so far that we should classify them as different species. Now imagine that the climate warms up so that travel through the mountain passes becomes easier and some of the new species start trickling back to their ancestral homelands. When they meet the descendants of their long-lost cousins, it turns out that they have diverged so far in their genetic makeup that they can no longer successfully interbreed with them. If they do hybridize with them the resulting offspring are sickly, or sterile like mules. So natural selection penalizes any predilection, on the part of individuals on either side, towards hybridizing with the other species or even race. Natural selection thereby finishes off the process of 'reproductive isolation' that began with the chance intervention of a mountain range. 'Speciation' is complete. We now have two species where previously there was one, and the two species can coexist in the same area without interbreeding with one another.

Actually, the likelihood is that the two species would not coexist for very long. This is not because they would interbreed but because they would compete. It is a widely accepted principle of ecology that two species with the same way of life will not coexist for long in one place, because they will compete and one or other will be driven extinct. Of course our two populations of shrews might no longer have the same way of life; for instance, the new species, during its period of evolution on the other side of the mountains, might have come to specialize on a different kind of insect prey. But if there is significant competition between the two species, most ecologists would expect one or other species to go extinct in the area of overlap. If it happened to be the original, ancestral species that was driven extinct, we should say that it had been replaced by the new, immigrant species.

The theory of speciation resulting from initial geographical separation has long been a cornerstone of mainstream, orthodox neo-Darwinism, and it is still accepted on all sides as the main process by which new species come into existence (some people think there are others as well). Its incorporation into modern Darwinism was largely due to the influence of the distinguished zoologist Ernst Mayr. What the 'punctuationists' did, when they first proposed their theory, was to ask themselves: Given that, like most neo-Darwinians, we accept the orthodox theory that speciation starts with geographical isolation, what should we expect to see in the fossil record?

Recall the hypothetical population of shrews, with a new species diverging on the far side of a mountain range, then eventually returning to the ancestral homelands and, quite possibly, driving the ancestral species extinct. Suppose that these shrews had left fossils; suppose even that the fossil record was *perfect*, with no gaps due to the unfortunate omission of key stages. What should we expect these fossils to show us? A smooth transition from ancestral species to daughter species? Certainly not, at least if we are digging in the main landmass where the original ancestral shrews lived, and to which the new species returned. Think of the history of what actually happened in the main landmass. There were the ancestral shrews, living and breeding happily away, with no particular reason to change. Admittedly their cousins the other side of the mountains were busy evolving, but their fossils are all on the other side of the mountain so we don't find them in the main landmass where we are digging. Then, suddenly (suddenly by geological standards, that is), the new species returns, competes with the main species and, perhaps, replaces the main species. Suddenly the fossils that we find as we move up through the strata of the main landmass change. Previously they were all of the ancestral species. Now, abruptly and without visible transitions, fossils of the new species appear, and fossils of the old species disappear.

The 'gaps', far from being annoying imperfections or awkward embarrassments, turn out to be exactly what we should positively *expect*, if we take seriously our orthodox neo-Darwinian theory of speciation. The reason the 'transition' from ancestral species to descendant species appears to be abrupt and jerky is simply that, when we look at a series of fossils from any one place, we are probably not looking at an *evolutionary* event at all: we are looking at a *migrational* event, the arrival of a new species from another geographical area. Certainly there were evolutionary events, and one species really did evolve, probably gradually, from another. But in order to see the evolutionary transition documented in the fossils we should have to dig elsewhere – in this case on the other side of the mountains.

The point that Eldredge and Gould were making, then, could have been modestly presented as a helpful rescuing of Darwin and his successors from what had seemed to them an awkward difficulty. Indeed that is, at least in part, how it *was* presented – initially. Darwinians had always been bothered by the apparent gappiness of the fossil record, and had seemed forced to resort to special pleading about imperfect evidence. Darwin himself had written:

> The geological record is extremely imperfect and this fact will to a large extent explain why we do not find interminable varieties, connecting together all the extinct and existing forms of life by the finest graduated steps. He who rejects these views on the nature of the geological record, will rightly reject my whole theory.

Eldredge and Gould could have made this their main message: Don't worry Darwin, even if the fossil record *were* perfect you shouldn't expect to see a finely graduated progression if you only dig in one place, for the simple reason that most of the evolutionary change took place somewhere else! They could have gone further and said:

> Darwin, when you said that the fossil record was imperfect, you were understating it. Not only is it imperfect, there are good reasons for expecting it to be *particularly* imperfect just when it gets interesting, just when evolutionary change is taking place; this is partly because evolution usually occurred in a different place from where we find most of our fossils; and it is partly because, even if we are fortunate enough to dig in one of the small outlying areas where most evolutionary change went on, that

evolutionary change (though still gradual) occupies such a short time that we should need an extra *rich* fossil record in order to track it!

But no, instead they chose, especially in their later writings in which they were eagerly followed by journalists, to sell their ideas as being radically *opposed* to Darwin's and opposed to the neo-Darwinian synthesis. They did this by emphasizing the 'gradualism' of the Darwinian view of evolution as opposed to the sudden, jerky, sporadic 'punctuationism' of their own. They even, especially Gould, saw analogies between themselves and the old schools of 'catastrophism' and 'saltationism'. Saltationism we have already discussed. Catastrophism was an eighteenth- and nineteenth-century attempt to reconcile some form of creationism with the uncomfortable facts of the fossil record. Catastrophists believed that the apparent progression of the fossil record really reflected a series of discrete creations, each one terminated by a catastrophic mass extinction. The latest of these catastrophes was Noah's flood.

Comparisons between modern punctuationism on the one hand, and catastrophism or saltationism on the other, have a purely poetic force. They are, if I may coin a paradox, deeply superficial. They sound impressive in an artsy, literary way, but they do nothing to aid serious understanding, and they can give spurious aid and comfort to modern creationists in their disturbingly successful fight to subvert American education and textbook publishing. The fact is that, in the fullest and most serious sense, Eldredge and Gould are really just as gradualist as Darwin or any of his followers. It is just that they would compress all the gradual change into brief bursts, rather than having it go on all the time; and they emphasize that most of the gradual change goes on in geographical areas away from the areas where most fossils are dug up.

So, it is not really the *gradualism* of Darwin that the punctuationists oppose: gradualism means that each generation is only slightly different from the previous generation; you would have to be a saltationist to oppose that, and Eldredge and Gould are not saltationists. Rather, it turns out to be Darwin's alleged belief in the constancy of rates of evolution that they and the other punctuationists object to. They object to it because they think that evolution (still undeniably gradualistic evolution) occurs rapidly during relatively brief bursts of activity (speciation events, which provide a kind of crisis atmosphere in which the alleged normal resistance to evolutionary change is broken); and that evolution occurs very slowly or not at all during long intervening periods of stasis. When we say 'relatively' brief we mean, of course, brief relative to the geological timescale in general.

Even the evolutionary jerks of the punctuationists, though they may be instantaneous by geological standards, still have a duration that is measured in tens or hundreds of thousands of years.

A thought of the famous American evolutionist G. Ledyard Stebbins is illuminating at this point. He isn't specifically concerned with jerky evolution, but is just seeking to dramatize the speed with which evolutionary change can happen, when seen against the timescale of available geological time. He imagines a species of animal, of about the size of a mouse. He then supposes that natural selection starts to favour an increase in body size, but only very very slightly. Perhaps larger males enjoy a slight advantage in the competition for females. At any time, males of average size are slightly less successful than males that are a tiny bit bigger than average. Stebbins put an exact figure on the mathematical advantage enjoyed by larger individuals in his hypothetical example. He set it at a value so very very tiny that it wouldn't be measurable by human observers. And the rate of evolutionary change that it brings about is consequently so slow that it wouldn't be noticed during an ordinary human lifetime. As far as the scientist studying evolution on the ground is concerned, then, these animals are not evolving at all. Nevertheless they are evolving, very slowly at a rate given by Stebbins's mathematical assumption, and, even at this slow rate, they would eventually reach the size of elephants. How long would this take? Obviously a long time by human standards, but human standards aren't relevant. We are talking about geological time. Stebbins calculates that at his assumed very slow rate of evolution, it would take about 12,000 generations for the animals to evolve from an average weight of 40 grams (mouse size) to an average weight of over 6,000,000 grams (elephant size). Assuming a generation-time of 5 years, which is longer than that of a mouse but shorter than that of an elephant, 12,000 generations would occupy about 60,000 years. 60,000 years is too *short* to be measured by ordinary geological methods of dating the fossil record. As Stebbins says, 'The origin of a new kind of animal in 100,000 years or less is regarded by paleontologists as "sudden" or "instantaneous".'

The punctuationists aren't talking about jumps in evolution, they are talking about episodes of relatively rapid evolution. And even these episodes don't have to be rapid by human standards, in order to appear instantaneous by geological standards. Whatever we may think of the theory of punctuated equilibria itself, it is all too easy to confuse gradualism (the belief, held by modern punctuationists as well as Darwin, that there are no sudden leaps between one generation and the next) with 'constant evolutionary speedism' (opposed by punctuationists and allegedly, though not actually, held by Darwin). They are not the same thing at all. The proper way to characterize the beliefs of punctuationists is: 'gradualistic, but with long periods of "stasis" (evolutionary stagnation) punctuating brief episodes of rapid gradual change'. The emphasis is then thrown onto the long periods of *stasis* as being the previously overlooked phenomenon that really needs explaining. It is the emphasis on stasis that is the punctuationists' real contribution, not their claimed opposition to gradualism, for they are truly as gradualist as anybody else.

Even the emphasis on stasis can be found, in less-exaggerated form, in Mayr's theory of speciation. He believed that, of the two geographically separated races, the original large ancestral population is less likely to change than the new, 'daughter' population (on the other side of the mountains in the case of our shrew example). This is not just because the daughter population is the one that has moved to new pastures, where conditions are likely to be different and natural selection pressures changed. It is also because there are some theoretical reasons (which Mayr emphasized but whose importance can be disputed) for thinking that large breeding populations have an inherent tendency to *resist* evolutionary change. A suitable analogy is the inertia of a large heavy object; it is hard to shift. Small, outlying populations, by virtue of being small, are inherently more likely, so the theory goes, to change, to evolve. Therefore, although I spoke of the two populations or races of shrews as diverging from each other, Mayr would prefer to see the original, ancestral population as relatively static, and the new population as diverging from it. The branch of the evolutionary tree does not fork into two equal twigs: rather, there is a main stem with a side twig sprouting from it.

The proponents of punctuated equilibrium took this suggestion of Mayr, and exaggerated it into a strong belief that 'stasis', or lack of evolutionary change, is the norm for a species. They believe that there are genetic forces in large populations that actively *resist* evolutionary change. Evolutionary change, for them, is a rare event, coinciding with speciation. It coincides with speciation in the sense that, in their view, the conditions under which new species are formed – geographical separation of small, isolated subpopulations – are the very conditions under which the forces that normally resist evolutionary change are relaxed or overthrown. Speciation is a time of upheaval, or revolution. And it is during these times of upheaval that evolutionary change is concentrated. For most of the history of a lineage it stagnates.

It isn't true that Darwin believed that evolution proceeded at a constant rate. He certainly didn't believe it in the ludicrously extreme sense that I satirized in my parable of the children of Israel, and I don't think he really believed it in any important sense. Quotation of the following well-known passage from the fourth edition (and later editions) of *The Origin of Species* annoys Gould because he thinks it is unrepresentative of Darwin's general thought:

> Many species once formed never undergo any further change . . . ; and the periods, during which species have undergone modification, though long as measured by years, have probably been short in comparison with the periods during which they retain the same form.

Gould wants to shrug off this sentence and others like it, saying:

> You cannot do history by selective quotation and search for qualifying footnotes. General tenor and historical impact are the proper criteria. Did his contemporaries or descendants ever read Darwin as a saltationist?

Gould is right, of course, about general tenor and historical impact, but the final sentence of this quotation from him is a highly revealing *faux pas*. *Of course*, nobody has ever read Darwin as a saltationist and, of course, Darwin was consistently hostile to saltationism, but the whole point is that saltationism is not the issue when we are discussing punctuated equilibrium. As I have stressed, the theory of punctuated equilibrium, by Eldredge and Gould's own account, is not a saltationist theory. The jumps that it postulates are not real, single-generation jumps. They are spread out over large numbers of generations over periods of, by Gould's own estimation, perhaps tens of thousands of years. The theory of punctuated equilibrium is a gradualist theory, albeit it emphasizes long periods of stasis intervening between *relatively* short bursts of gradualistic evolution. Gould has misled himself by his own rhetorical emphasis on the purely poetic or literary resemblance between punctuationism, on the one hand, and true saltationism on the other.

I think it would clarify matters if, at this point, I summarized a range of possible points of view about rates of evolution. Out on a limb we have true saltationism, which I have already discussed sufficiently. True saltationists don't exist among modern biologists. Everyone that is not a saltationist is a gradualist, and this includes Eldredge and Gould, however they may choose to describe themselves. Within gradualism, we may distinguish various beliefs about rates of (gradual) evolution. Some of these beliefs, as we have seen, bear a purely superficial ('literary' or 'poetic') resemblance to true, anti-gradualist saltationism, which is why they are sometimes confused with it.

At another extreme we have the sort of 'constant speedism' that I caricatured in the Exodus parable with which I began this chapter. An extreme constant speedist believes that evolution is plodding along steadily and inexorably all the time, whether or not there is any branching or speciation going on. He believes that quantity of evolutionary change is strictly proportional to time elapsed. Ironically, a form of constant speedism has recently become highly favoured among modern molecular geneticists. A good case can be made for believing that evolutionary change at the level of protein molecules really does plod along at a constant rate exactly like the hypothetical children of Israel; and this *even if* externally visible characteristics like arms and legs are evolving in a highly punctuated manner. We have already met this topic in Chapter 5, and I shall mention it again in the next chapter. But as far as adaptive evolution of large-scale structures and behaviour patterns are concerned, just about all evolutionists would reject constant speedism, and Darwin certainly would have rejected it. Everyone that is not a constant speedist is a variable speedist.

Within variable speedism we may distinguish two kinds of belief, labelled, 'discrete variable speedism' and 'continuously variable speedism'. An extreme 'discretist' not only believes that evolution varies in speed. He thinks that the speed flips abruptly from one discrete level to another, like a car's gearbox. He might believe, for instance, that evolution has only two speeds: very fast and stop (I cannot help being reminded here of the humiliation of my first school report, written by the Matron about my performance as a seven-year-old in folding clothes, taking cold baths, and other daily routines of boarding-school life: 'Dawkins has only three speeds: slow, very slow, and stop'). 'Stopped' evolution is the 'stasis' that is thought by punctuationists to characterize large populations. Top-gear evolution is the evolution that goes on during speciation, in small isolated populations round the edge of large, evolutionarily static populations. According to this view, evolution is always in one or other of the two gears, never in between. Eldredge and Gould tend in the direction of discretism, and in this respect they are genuinely radical. They may be called 'discrete variable speedists'. Incidentally, there is no *particular* reason why a discrete variable speedist should necessarily emphasize speciation as the time of high-gear evolution. In practice, however, most of them do.

'Continuously variable speedists', on the other hand, believe that evolutionary rates fluctuate continuously from very fast to very slow and stop, with all intermediates. They see no particular reason to emphasize certain speeds more than others. In particular, stasis, to them, is just an extreme case of ultra-slow evolution. To a punctuationist, there is something very special about stasis. Stasis, to him, is not just evolution that is so slow as to have a rate of zero: stasis is not just passive lack of evolution because there is no driving force in favour of change. Rather, stasis represents a positive *resistance* to evolutionary change. It is almost as though species are thought to take active steps *not* to evolve, *in spite of* driving forces in favour of evolution.

More biologists agree that stasis is a real phenomenon than agree about its causes. Take, as an extreme example, the coelacanth *Latimeria*. The coelacanths were a large group of 'fish' (actually, although they are called fish they are more closely related to us than they are to trout and herrings) that flourished more than 250 million years ago and apparently died out at about the same time as the dinosaurs. I say 'apparently' died out because in 1938, much to the zoological world's astonishment, a weird fish, a yard and a half long and with unusual leg-like fins, appeared in the catch of a deep-sea fishing boat off the South African coast. Though almost destroyed before its priceless worth was recognized, its decaying remains were fortunately brought to the attention of a qualified South African zoologist just in time. Scarcely able to believe his eyes, he identified it as a living coelacanth, and named it *Latimeria*. Since then, a few other specimens have been fished up in the same area, and the species has now been properly studied and described. It is a 'living fossil', in the sense that it has changed hardly at all since the time of its fossil ancestors, hundreds of millions of years ago.

So, we have stasis. What are we to make of it? How do we explain it? Some of us would say that the lineage leading to *Latimeria* stood still because natural selection did not move it. In a sense it had no 'need' to evolve because these animals had found a successful way of life deep in the sea where conditions did not change much. Perhaps they never participated in any arms races. Their cousins that emerged onto the land did evolve because natural selection, under a variety of hostile conditions including arms races, forced them to. Other biologists, including some of those that call themselves punctuationists, might say that the lineage leading to modern *Latimeria* actively resisted change, *in spite of* what natural selection pressures there might have been. Who is right? In the particular case of *Latimeria* it is hard to know, but there is one way in which, in principle, we might go about finding out.

Let us, to be fair, stop thinking in terms of *Latimeria* in particular. It is a striking example but a very extreme one, and it is not one on which the punctuationists particularly want to rely. Their belief is that less extreme, and shorter-term, examples of stasis are commonplace; are, indeed, the norm, because species have genetic mechanisms that actively resist change, even if there are forces of natural selection urging change. Now here is the very simple experiment which, in principle at least, we can do to test this hypothesis. We can take wild populations and impose our own forces of selection upon them. According to the hypothesis that species actively resist change, we should find that, if we try to breed for some quality, the species should dig in its heels, so to speak, and refuse to budge, at least for a while. If we take cattle and attempt to breed selectively for high milk yield, for instance, we should fail. The genetic mechanisms of the species should mobilize their anti-evolution forces and fight off the pressure to change. If we try to make chickens evolve higher egg-laying rates we should fail. If bullfighters, in pursuit of their contemptible 'sport', try to increase the courage of their bulls by selective breeding, they should fail. These failures should only be temporary, of course. Eventually, like a dam bursting under pressure, the alleged anti-evolution forces will be overcome, and the lineage can then move rapidly to a new equilibrium. But we should experience at least some resistance when we first initiate a new program of selective breeding.

The fact is, of course, that we do not fail when we try to shape evolution by selectively breeding animals and plants in captivity, nor do we experience a period of initial difficulty. Animal and plant species are usually immediately amenable to selective breeding, and breeders detect no evidence of any intrinsic, anti-evolution forces. If anything, selective breeders experience difficulty *after* a number of generations of successful selective breeding. This is because after some generations of selective breeding the available genetic variation runs out, and we have to wait for new mutations. It is conceivable that coelacanths stopped evolving because they stopped mutating – perhaps because they were protected from cosmic rays at the bottom of the sea! – but nobody, as far as I know, has seriously suggested this, and in any case this is not what punctuationists mean when they talk of species having built-in resistance to evolutionary change.

They mean something more like the point I was making in Chapter 7 about 'cooperating' genes: the idea that groups of genes are so well adapted to each other that they resist invasion by new mutant genes which are not members of the club. This is quite a sophisticated idea, and it can be made to sound plausible. Indeed, it was one of the theoretical props of Mayr's inertia idea, already referred to. Nevertheless, the fact that, whenever we try selective breeding, we encounter no initial resistance to it, suggests to me that, if lineages go for many generations in the wild without changing, this is not because they resist change but because there is no natural selection pressure in favour of changing. They don't change because individuals that stay the same survive better than individuals that change.

Punctuationists, then, are really just as gradualist as Darwin or any other Darwinian; they just insert long periods of stasis between spurts of gradual evolution. As I said, the one respect in which punctuationists do differ from other schools of Darwinism is in their strong emphasis on stasis as something positive: as an active resistance to evolutionary change rather than as, simply, absence of evolutionary change. And this is the one respect in which they are quite probably wrong. It remains for me to clear up the mystery of why they *thought* they were so far from Darwin and neo-Darwinism.

The answer lies in a confusion of two meanings of the word 'gradual', coupled with the confusion, which I have been at pains to dispel here but which lies at the back of many peoples' minds, between punctuationism and saltationism. Darwin was a passionate anti-saltationist, and this led him to stress, over and over again, the extreme gradualness of the evolutionary changes that he was proposing. The reason is that saltation, to him, meant what I have called Boeing 747 macromutation. It meant the sudden calling into existence, like Pallas Athene from the head of Zeus, of brand-new complex organs at a single stroke of the genetic wand. It meant fully formed, complex working eyes springing up from bare skin, in a single generation. The reason it meant these things to Darwin is that that is exactly what it meant to some of his most influential opponents, and they really believed in it as a major factor in evolution.

The Duke of Argyll, for instance, accepted the evidence that evolution had happened, but he wanted to smuggle divine creation in by the back door. He wasn't alone. Instead of a single, once and for all creation in the Garden of Eden, many Victorians thought that the deity had intervened repeatedly, at crucial points in evolution. Complex organs like eyes, instead of evolving from simpler ones by slow degrees as Darwin had it, were thought to have sprung into existence in a single instant. Such people rightly perceived that such instant 'evolution', if it occurred, would imply supernatural intervention: that is what they believed in. The reasons are the statistical ones I have discussed in connection with hurricanes and Boeing 747s. 747 saltationism is, indeed, just a watered-down form of creationism. Putting it the other way around, divine creation is the ultimate in saltation. It is the ultimate leap from inanimate clay to fully formed man. Darwin perceived this too. He wrote in a letter to Sir Charles Lyell, the leading geologist of his day:

> If I were convinced that I required such additions to the theory of natural selection, I would reject it as rubbish . . . I would give nothing for the theory of Natural selection, if it requires miraculous additions at any one stage of descent.

This is no petty matter. In Darwin's view, the whole *point* of the theory of evolution by natural selection was that it provided a *non* - miraculous account of the existence of complex adaptations. For what it is worth, it is also the whole point of this book. For Darwin, any evolution that had to be helped over the jumps by God was not evolution at all. It made a nonsense of the central point of evolution. In the light of this, it is easy to see why Darwin constantly reiterated the *gradualness* of evolution. It is easy to see why he wrote that sentence quoted in Chapter 4:

> If it could be demonstrated that any complex organ existed, which could not possibly have been formed by numerous, successive, slight modifications, my theory would absolutely break down.

There is another way of looking at the fundamental importance of gradualness for Darwin. His contemporaries, like many people still today, had a hard time believing that the human body and other such complex entities could conceivably have come into being through evolutionary means. If you think of the single-celled *Amoeba* as our remote ancestor – as, until quite recently, it was fashionable to do – many people found it hard in their minds to bridge the gap between *Amoeba* and man. They found it inconceivable that from such simple beginnings something so complex could emerge. Darwin appealed to the idea of a gradual series of small steps as a means of overcoming this kind of incredulity. You may find it hard to imagine an *Amoeba* turning into a man, the argument runs; but you do not find it hard to imagine an *Amoeba* turning into a slightly different kind of *Amoeba*. From this it is not hard to imagine it turning into a slightly different kind of slightly different kind of . . . , and so on. As we saw in Chapter 3, this argument overcomes our incredulity only if we stress that there was an extremely large number of steps along the way, and only if each step is very tiny. Darwin was constantly battling against this source of incredulity, and he constantly made use of the same weapon: the emphasis on gradual, almost imperceptible change, spread out over countless generations.

Incidentally, it is worth quoting J. B. S. Haldane's characteristic piece of lateral thinking in combating the same source of incredulity. Something like the transition from *Amoeba* to man, he pointed out, goes on in every mother's womb in a mere nine months. Development is admittedly a very different process from evolution but, nevertheless, anyone sceptical of the very *possibility* of a transition from single cell to man has only to contemplate his own foetal beginnings to have his doubts allayed. I hope I shall not be thought a pedant if I stress, by the way, that the choice of *Amoeba* for the title of honorary ancestor is simply following a whimsical tradition. A bacterium would be a better choice, but even bacteria, as we know them, are modern organisms.

To resume the argument, Darwin laid great stress on the gradualness of evolution because of what he was arguing *against*: the misconceptions about evolution that were prevalent in the nineteenth century. The *meaning* of 'gradual', in the context of those times, was 'opposite of saltation'. Eldredge and Gould, in the context of the late twentieth century, use 'gradual' in a very different sense. They in effect, though not explicitly, use it to mean 'at a constant speed', and they oppose to it their own notion of 'punctuation'. They criticize gradualism in this sense of 'constant speedism'. No doubt they are right to do so: in its extreme form it is as absurd as my Exodus parable.

But to couple this justifiable criticism with a criticism of Darwin is simply to confuse two quite separate meanings of the word 'gradual'. In the sense in which Eldredge and Gould are opposed to gradualism, there is no particular reason to doubt that Darwin would have agreed with them. In the sense of the word in which Darwin was a passionate gradualist, Eldredge and Gould are also gradualists. The theory of punctuated equilibrium is a minor gloss on Darwinism, one which Darwin himself might well have approved if the issue had been discussed in his time. As a minor gloss, it does not deserve a particularly large measure of publicity. The reason it has in fact received such publicity, and why I have felt obliged to devote a whole chapter of this book to it, is simply that the theory has been sold – oversold by some journalists – as if it were radically opposed to the views of Darwin and his successors. Why has this happened?

There are people in the world who desperately want not to have to believe in Darwinism. They seem to fall into three main classes. First, there are those who, for religious reasons, want evolution itself to be untrue. Second, there are those who have no reason to deny that evolution has happened but who, often for political or ideological reasons, find Darwin's theory of its *mechanism* distasteful. Of these, some find the idea of natural selection unacceptably harsh and ruthless; others confuse natural selection with randomness, and hence 'meaninglessness', which offends their dignity; yet others confuse Darwinism with Social Darwinism, which has racist and other disagreeable overtones. Third, there are people, including many working in what they call (often as a singular noun) 'the media', who just like seeing applecarts upset, perhaps because it makes good journalistic copy; and Darwinism has become sufficiently established and respectable to be a tempting applecart.

Whatever the motive, the consequence is that if a reputable scholar breathes so much as a hint of criticism of some detail of current Darwinian theory, the fact is eagerly seized on and blown up out of all proportion. So strong is this eagerness, it is as though there were a powerful amplifier, with a finely tuned microphone selectively listening out for anything that sounds the tiniest bit like opposition to Darwinism. This is most unfortunate, for serious argument and criticism is a vitally important part of any science, and it would be tragic if scholars felt the need to muzzle themselves because of the microphones. Needless to say the amplifier, though powerful, is not hi-fi: there is plenty of distortion! A scientist who cautiously whispers some slight misgiving about a current nuance of Darwinism is liable to hear his distorted and barely recognizable words booming and echoing out through the eagerly waiting loudspeakers.

Eldredge and Gould don't whisper, they shout, with eloquence and power! What they shout is often pretty subtle, but the message that gets across is that something is wrong with Darwinism. Hallelujah, 'the scientists' said it themselves! The editor of *Biblical Creation* has written:

> it is undeniable that the credibility of our religious and scientific position has been greatly strengthened by the recent lapse in neo-Darwinian morale. And this is something we must exploit to the full.

Eldredge and Gould have both been doughty champions in the fight against redneck creationism. They have shouted their complaints at the misuse of their own words, only to find that, for *this* part of their message, the microphones suddenly went dead on them. I can sympathize, for I have had a similar experience with a different set of microphones, in this case politically rather than religiously tuned.

What needs to be said now, loud and clear, is the truth: that the theory of punctuated equilibrium lies firmly within the neo-Darwinian synthesis. It always did. It will take time to undo the damage wrought by the overblown rhetoric, but it will be undone. The theory of punctuated equilibrium will come to be seen in proportion, as an interesting but minor wrinkle on the surface of neo-Darwinian theory. It certainly provides no basis for any 'lapse in neo-Darwinian morale', and no basis whatever for Gould to claim that the synthetic theory (another name for neo-Darwinism) 'is effectively dead'. It is as if the discovery that the Earth is not a perfect sphere but a slightly flattened spheroid were given banner treatment under the headline:

COPERNICUS WRONG. FLAT EARTH THEORY VINDICATED.

But, to be fair, Gould's remark was aimed not so much at the alleged 'gradualism' of the Darwinian synthesis as at another of its claims. This is the claim, which Eldredge and Gould dispute, that all evolution, even on the grandest geological timescale, is an extrapolation of events that take place within populations or species. They believe that there is a higher form of selection which they call 'species selection'. I am deferring this topic to the next chapter. The next chapter is also the place to deal with another school of biologists who, on equally flimsy grounds, have in some cases been passed off as anti-Darwinian, the so-called 'transformed cladists'. These belong within the general field of taxonomy, the science of classification.

CHAPTER 10

THE ONE TRUE TREE
OF LIFE

This book is mainly about evolution as the solution of the complex 'design' problem; evolution as the true explanation for the phenomena that Paley thought proved the existence of a divine watchmaker. This is why I keep going on about eyes and echolocation. But there is another whole range of things that the theory of evolution explains. These are the phenomena of diversity: the pattern of different animal and plant types distributed around the world, and the distribution of characteristics among them. Although I am mainly concerned with eyes and other pieces of complex machinery, I mustn't neglect this other aspect of evolution's role in helping us to understand nature. So this chapter is about taxonomy.

Taxonomy is the science of classification. For some people it has an undeservedly dull reputation, a subconscious association with dusty museums and the smell of preserving fluid, almost as though it were being confused with taxidermy. In fact it is anything but dull. It is, for reasons that I do not fully understand, one of the most acrimoniously controversial fields in all of biology. It is of interest to philosophers and historians. It has an important role to play in any discussion of evolution. And from within the ranks of taxonomists have come some of the most outspoken of those modern biologists who pretend to be anti-Darwinian.

Although taxonomists mostly study animals or plants, all sorts of other things can be classified: rocks, warships, books in a library, stars, languages. Orderly classification is often represented as a measure of convenience, a practical necessity, and this is indeed a part of the truth. The books in a large library are nearly useless unless they are organized in some nonrandom way so that books on a particular subject can be found when you want them. The science, or it may be an art, of librarianship is an exercise in applied taxonomy. For the same kind of reason, biologists find their life made easier if they can pigeonhole animals and plants in agreed categories with names. But to say that this is the only reason for animal and plant taxonomy would be to miss most of the point. For evolutionary biologists there is something very special about the classification of living organisms, something that is not true of any other kind of taxonomy. It follows from the idea of evolution that there is one uniquely correct branching family tree of all living things, and we can base our taxonomy upon it. In addition to its uniqueness, this taxonomy has the singular property that I shall call *perfect nesting*. What this means, and why it is so important, is a major theme of this chapter.

Let us use the library as an example of nonbiological taxonomy. There is no single, unique, correct solution to the problem of how the books in a library or a bookshop should be classified. One librarian might divide his collection up into the following major categories: science, history, literature, other arts, foreign works, etc. Each of these major departments of the library would be subdivided. The science wing of the library might have subdivisions into biology, geology, chemistry, physics, and so on. The books in the biology section of the science wing might be subdivided into shelves devoted to physiology, anatomy, biochemistry, entomology, and so on. Finally, within each shelf, the books might be housed in alphabetical order. Other major wings of the library, the history wing, the literature wing, the foreign-language wing, and so on, would be subdivided in similar ways. The library is, therefore, hierarchically divided in a way that makes it possible for a reader to home in on the book that he wants. Hierarchical classification is convenient because it enables the borrower to find his way around the collection of books quickly. It is for the same kind of reason that the words in dictionaries are arranged in alphabetical order.

But there is no unique hierarchy by which the books in a library must be arranged. A different librarian might choose to organize the same collection of books in a different, but still hierarchical, way. He might not, for instance, have a separate foreign-language wing, but might prefer to house books, regardless of language, in their appropriate subject areas: German biology books in the biology section, German history books in the history section, and so on. A third librarian might adopt the radical policy of housing all books, on whatever subject, in chronological order of publication, relying on card indexes (or computer equivalents) to find books on desired topics.

These three library plans are quite different from each other, but they would probably all work adequately and would be thought acceptable by many readers, though not, incidentally, by the choleric, elderly London clubman whom I once heard on the radio berating his club's committee for employing a librarian. The library had got along for a hundred years without organization, and he didn't see why it needed organizing now. The interviewer mildly asked him how he thought the books ought to be arranged. 'Tallest on the left, shortest on the right!', he roared without hesitation. Popular bookshops classify their books into major sections that reflect popular demand. Instead of science, history, literature, geography, and so on, their major departments are gardening, cookery, 'TV titles', the occult, and I once saw a shelf prominently labelled 'RELIGION AND UFOs'.

So, there is no *correct* solution to the problem of how to classify books. Librarians can have sensible disagreements with one another about classification policy, but the criteria by which arguments are won or lost will not include the 'truth' or 'correctness' of one classification system relative to another. Rather, the criteria that are bandied about in argument will be 'convenience for library users', 'speed of finding books', and so on. In this sense the taxonomy of books in a library can be said to be arbitrary. This doesn't imply that it is unimportant to devise a good classification system; far from it. What it does mean is that there is no single classification system which, in a world of perfect information, would be universally agreed as the only correct classification. The taxonomy of living creatures on the other hand, as we shall see, does have this strong property that the taxonomy of books lacks; at least it does if we take up an evolutionary standpoint.

It is, of course, possible to devise any number of systems for classifying living creatures, but I shall show that all but one of these are just as arbitrary as any librarian's taxonomy. If it is simply convenience that is required, a museum keeper might classify his specimens according to size and keeping conditions: large stuffed specimens; small dried specimens pinned on cork boards in trays; pickled ones in bottles; microscopic ones on slides, and so on. Such groupings of convenience are common in zoos. In the London Zoo rhinos are housed in the 'Elephant House' for no better reason than that they need the same kind of stoutly fortified cages as elephants. An applied biologist might classify animals into harmful (subdivided into medical pests, agricultural pests and directly dangerous biters or stingers), beneficial (subdivided in similar ways) and neutral. A nutritionist might classify animals according to the food value of their meat to humans, again with elaborate subdivision of categories. My grandmother once embroidered a cloth book about animals for children, which classified them by their feet. Anthropologists have documented numerous elaborate systems of animal taxonomy used by tribes around the world.

But of all the systems of classification that could be dreamed up, there is one unique system, unique in the sense that words like 'correct' and 'incorrect', 'true' and 'false' can be applied to it with perfect agreement given perfect information. That unique system is the system based on evolutionary relationships. To avoid confusion I shall give this system the name that biologists give to its strictest form: cladistic taxonomy.

In cladistic taxonomy, the ultimate criterion for grouping organisms together is closeness of cousinship or, in other words, relative recency of common ancestry. Birds, for instance, are distinguished from non-birds by the fact that they are all descended from a common ancestor, which is not an ancestor of any non-bird. Mammals are all descended from a common ancestor, which is not an ancestor of any non-mammal. Birds and mammals have a more remote common ancestor, which they share with lots of other animals like snakes and lizards and tuataras. The animals descended from this common ancestor are all called amniotes. So, birds and mammals are amniotes. 'Reptiles' is not a true taxonomic term, according to cladists, because it is defined by exception: all amniotes except birds and mammals. In other words, the most recent common ancestor of all 'reptiles' (snakes, turtles, etc.) is also ancestral to some non-'reptiles', namely birds and mammals.

Within mammals, rats and mice share a recent common ancestor with each other; leopards and lions share a recent common ancestor with each other; so do chimpanzees and humans with each other. Closely related animals are animals that share a recent common ancestor. More distantly related animals share an earlier common ancestor. Very distantly related animals, like people and slugs, share a very early common ancestor. Organisms can never be totally *un*related to one another, since it is all but certain that life as we know it originated only once on earth.

True cladistic taxonomy is strictly hierarchical, an expression which I shall use to mean that it can be represented as a tree whose branches always diverge and never converge again. In my view (some schools of taxonomists, that we shall discuss later, would disagree), it is strictly hierarchical *not* because hierarchical classification is convenient, like a librarian's classification, nor because everything in the world falls naturally into a hierarchical pattern, but simply because the pattern of evolutionary descent is hierarchical. Once the tree of life has branched beyond a certain minimum distance (basically the bounds of the species), the branches never ever come together again (there may be very rare exceptions, as in the origin of the eukaryotic cell mentioned in Chapter 7). Birds and mammals are descended from a common ancestor, but they are now separate branches of the evolutionary tree, and they will never come together again: there will never be a hybrid between a bird and a mammal. A group of organisms that has this property of all being descended from a common ancestor, which is not an ancestor of any non-member of the group, is called a *clade*, after the Greek for a tree branch.

Another way of representing this idea of strict hierarchy is in terms of 'perfect nesting'. We write the names of any set of animals on a large sheet of paper and draw rings round related sets. For example, rat and mouse would be united in a small ring indicating that they are close cousins, with a recent common ancestor. Guinea-pig and capybara would be united with each other in another small ring. The rat/mouse ring and the guinea-pig/capybara ring would, in turn, be united with each other (and beavers and porcupines and squirrels and lots of other animals) in a larger ring labelled with its own name, rodents. Inner rings are said to be 'nested' inside larger, outer rings. Somewhere else on the paper, lion and tiger would be united with one another in a small ring. This ring would be included with others in a ring labelled cats. Cats, dogs, weasels, bears, etc. would all be united, in a series of rings within rings, in a single large ring labelled carnivores. The rodent ring and the carnivore ring would then take part in a more global series of rings within rings in a very large ring labelled mammals.

The important thing about this system of rings within rings is that it is *perfectly nested*. Never, not on a single solitary occasion, will the rings that we draw intersect each other. Take any two overlapping rings, and it will always be true to say that one lies wholly inside the other. The area enclosed by the inner one is always totally enclosed by the outer one: there are never any partial overlaps. This property of perfect taxonomic nesting is not exhibited by books, languages, soil types, or schools of thought in philosophy. If a librarian draws a ring round the biology books and another ring round the theology books, he will find that the two rings overlap. In the zone of overlap are books with titles like 'Biology and Christian Belief'.

On the face of it, we might expect the classification of languages to exhibit the property of perfect nesting. Languages, as we saw in Chapter 8, evolve in a rather animal-like way. Languages that have recently diverged from a common ancestor, like Swedish, Norwegian and Danish, are much more similar to each other than they are to languages that diverged longer ago, like Icelandic. But languages don't only diverge, they also merge. Modern English is a hybrid between Germanic and Romance languages that had diverged much earlier, and English would therefore not fit neatly in any hierarchical nesting diagram. The rings that enclosed English would be found to intersect, to overlap partially. Biological classificatory rings never intersect in this way, because biological evolution above the species level is always divergent.

Returning to the library example, no librarian can entirely avoid the problem of intermediates or overlaps. It is no use housing the biology and theology sections next door to each other and putting intermediate books in the corridor between them; for what then do we do with books that are intermediate between biology and chemistry, between physics and theology, history and theology, history and biology? I think I am right in saying that the problem of intermediates is inescapably, inherently a part of all taxonomic systems other than that which springs from evolutionary biology. Speaking personally, it is a problem that gives me almost physical discomfort when I am attempting the modest filing tasks that arise in my professional life: shelving my own books, and reprints of scientific papers that colleagues (with the kindest of intentions) send me; filing administrative papers; old letters, and so on. Whatever categories one adopts for a filing system, there are always awkward items that don't fit, and the uncomfortable indecision leads me, I am sorry to say, to leave odd papers out on the table, sometimes for years at a time until it is safe to throw them away. Often one has unsatisfactory recourse to a 'miscellaneous' category, a category which, once initiated, has a menacing tendency to grow. I sometimes wonder whether librarians, and keepers of all museums except biological museums, are particularly prone to ulcers.

In the taxonomy of living creatures these filing problems do not arise. There are no 'miscellaneous' animals. As long as we stay above the level of the species, and as long as we study only modern animals (or animals in any given time slice: see below) there are no awkward intermediates. If an animal appears to be an awkward intermediate, say it seems to be exactly intermediate between a mammal and a bird, an evolutionist can be confident that it *must* definitely be one or the other. The appearance of intermediacy must be an illusion. The unlucky librarian can take no such reassurance. It is perfectly possible for a book to belong simultaneously in both the history and the biology departments. Cladistically inclined biologists never indulge in any librarians' arguments over whether it is 'convenient' to classify whales as mammals or as fish, or as intermediate between mammals and fish. The only argument we have is a factual one. In this case, as it happens, the facts lead all modern biologists to the same conclusion. Whales are mammals and not fish, and they are not, even to a tiny degree, intermediate. They are no closer to fish than humans are, or duck-billed platypuses, or any other mammals.

Indeed, it is important to understand that all mammals – humans, whales, duck-billed platypuses, and the rest – are *exactly equally* close to fish, since all mammals are linked to fish via the same common ancestor. The myth that mammals, for instance, form a ladder or 'scale', with 'lower' ones being closer to fish than 'higher' ones, is a piece of snobbery that owes nothing to evolution. It is an ancient, pre-evolutionary notion, sometimes called the 'great chain of being', which should have been destroyed by evolution but which was, mysteriously, absorbed into the way many people thought about evolution.

At this point I cannot resist drawing attention to the irony in the challenge that creationists are fond of hurling at evolutionists: 'Produce your intermediates. If evolution were true, there should be animals that are half way between a cat and a dog, or between a frog and an elephant. But has anyone ever seen a frelephant?' I have been sent creationist pamphlets that attempt to ridicule evolution with drawings of grotesque chimeras, horse hindquarters grafted to a dog's front end, for instance. The authors seem to imagine that evolutionists should expect such intermediate animals to exist. This not only misses the point, it is the precise antithesis of the point. One of the strongest expectations the theory of evolution gives us is that intermediates of this kind should *not* exist. This is the burden of my comparison between animals and library books.

The taxonomy of evolved living beings, then, has the unique property of providing perfect agreement in a world of perfect information. That is what I meant by saying that words like 'true' and 'false' could be applied to claims in cladistic taxonomy, though not to claims in any librarian's taxonomy. We must make two qualifications. First, in the real world we don't have perfect information. Biologists may disagree with one another over the facts of ancestry, and the disputes may be difficult to settle because of imperfect information – not enough fossils, say. I shall return to this. Second, a different kind of problem arises if we have too *many* fossils. The neat and clear-cut discreteness of classification is liable to evaporate if we try to include all animals that have ever lived, rather than just modern animals. This is because, however distant from each other two modern animals may be – say they are a bird and a mammal – they did, once upon a time, have a common ancestor. If we are faced with trying to fit that ancestor into our modern classification, we may have problems.

The moment we start to consider extinct animals, it is no longer true that there are no intermediates. On the contrary, we now have to contend with potentially continuous series of intermediates. The distinction between modern birds, and modern non-birds like mammals, is a clear-cut one only because the intermediates converging backwards on the common ancestor are all dead. To make the point most forcibly, think again of a hypothetically 'kind' nature, providing us with a complete fossil record; with a fossil of every animal that ever lived. When I introduced this fantasy in the previous chapter, I mentioned that from one point of view nature would actually be being *un*kind. I was thinking then of the toil of studying and describing all the fossils, but we now come to another aspect of that paradoxical unkindness. A complete fossil record would make it very difficult to classify animals into discrete nameable groups. If we had a complete fossil record, we should have to give up discrete names and resort to some mathematical or graphical notation of sliding scales. The human mind far prefers discrete names, so in one sense it is just as well that the fossil record is poor.

If we consider all animals that have ever lived instead of just modern animals, such words as 'human' and 'bird' become just as blurred and unclear at the edges as words like 'tall' and 'fat'. Zoologists can argue unresolvably over whether a particular fossil is, or is not, a bird. Indeed they often do argue this very question over the famous fossil *Archaeopteryx*. It turns out that if 'bird/non-bird' is a clearer distinction than 'tall/short', it is only because in the bird/non-bird case the awkward intermediates are all dead. If a curiously selective plague came along and killed all people of intermediate height, 'tall' and 'short' would come to have just as precise a meaning as 'bird' or 'mammal'.

It isn't just zoological classification that is saved from awkward ambiguity only by the convenient fact that most intermediates are now extinct. The same is true of human ethics and law. Our legal and moral systems are deeply species-bound. The director of a zoo is legally entitled to 'put down' a chimpanzee that is surplus to requirements, while any suggestion that he might 'put down' a redundant keeper or ticket-seller would be greeted with howls of incredulous outrage. The chimpanzee is the property of the zoo. Humans are nowadays not supposed to be anybody's property, yet the rationale for discriminating against chimpanzees in this way is seldom spelled out, and I doubt if there is a defensible rationale at all. Such is the breathtaking speciesism of our Christian-inspired attitudes, the abortion of a single human zygote (most of them are destined to be spontaneously aborted anyway) can arouse more moral solicitude and righteous indignation than the vivisection of any number of intelligent adult chimpanzees! I have heard decent, liberal scientists, who had no intention of actually cutting up live chimpanzees, nevertheless passionately defending their *right* to do so if they chose, without interference from the law. Such people are often the first to bristle at the smallest infringement of *human* rights. The only reason we can be comfortable with such a double standard is that the intermediates between humans and chimps are all dead.

The last common ancestor of humans and chimps lived perhaps as recently as five million years ago, definitely more recently than the common ancestor of chimps and orang-utans, and perhaps 30 million years more recently than the common ancestor of chimps and monkeys. Chimpanzees and we share more than 99 per cent of our genes. If, in various forgotten islands around the world, survivors of all intermediates back to the chimp/human common ancestor were discovered, who can doubt that our laws and our moral conventions would be profoundly affected, especially as there would presumably be some interbreeding along the spectrum? Either the whole spectrum would have to be granted full human rights (Votes for Chimps), or there would have to be an elaborate apartheid-like system of discriminatory laws, with courts deciding whether particular individuals were legally 'chimps' or legally 'humans'; and people would fret about their daughter's desire to marry one of 'them'. I suppose the world is already too well explored for us to hope that this chastening fantasy will ever come true. But anybody who thinks that there is something obvious and self-evident about human 'rights' should reflect that it is just sheer luck that these embarrassing intermediates happen not to have survived. Alternatively, maybe if chimpanzees hadn't been discovered until today they would now be seen as the embarrassing intermediates.

Readers of the previous chapter may remark that the whole argument, that categories become blurred if we don't stick to contemporary animals, assumes that evolution goes at a constant speed, rather than being punctuated. The more our view of evolution approaches the extreme of smooth, continuous change, the more pessimistic shall we be about the very possibility of applying such words as bird or non-bird, human or non-human, to all animals that ever lived. An extreme saltationist could believe that there really was a first human, whose mutant brain was twice the size of his father's brain and that of his chimp-like brother.

The advocates of punctuated equilibrium are for the most part not, as we have seen, true saltationists. Nevertheless, to them the problem of the ambiguity of names is bound to seem less severe than it will on a more continuous view. The naming problem would arise even for punctuationists if literally every animal that had ever lived was preserved as a fossil, because the punctuationists are really gradualists when we come right down to detail. But, since they assume that we are particularly unlikely to find fossils documenting short periods of rapid transition, while being particularly likely to find fossils documenting the long periods of stasis, the 'naming problem' will be less severe on a punctuationist view than on a nonpunctuationist view of evolution.

It is for this reason that the punctuationists, especially Niles Eldredge, make a big point of treating 'the species' as a real 'entity'. To a non-punctuationist, 'the species' is definable only because the awkward intermediates are dead. An extreme anti-punctuationist, taking a long view of the entirety of evolutionary history, cannot see 'the species' as a discrete entity at all. He can see only a smeary continuum. On his view a species never has a clearly defined beginning, and it only sometimes has a clearly defined end (extinction); often a species does not end decisively but turns gradually into a new species. A punctuationist, on the other hand, sees a species as coming into existence at a particular time (strictly there is a transition period with a duration of tens of thousands of years, but this duration is short by geological standards). Moreover, he sees a species as having a definite, or at least rapidly accomplished, end, not a gradual fading into a new species. Since most of the life of a species, on the punctuationist view, is spent in unchanging stasis, and since a species has a discrete beginning and end, it follows that, to a punctuationist, a species can be said to have a definite, measurable 'life span'. The non-punctuationist would not see a species as having a 'life span' like an individual organism. The extreme punctuationist sees 'the species' as a discrete entity that really deserves its own name. The extreme anti-punctuationist sees 'the species' as an arbitrary stretch of a continuously flowing river, with no particular reason to draw lines delimiting its beginning and end.

In a punctuationist book on the history of a group of animals, say the history of the horses over the past 30 million years, the characters in the drama may all be species rather than individual organisms, because the punctuationist author thinks of species as real 'things', with their own discrete identity. Species will suddenly arrive on the scene, and as suddenly they will disappear, replaced by successor species. It will be a history of successions, as one species gives way to another. But if an anti-punctuationist writes the same history, he will use species names only as a vague convenience. When he looks longitudinally through time, he ceases to see species as discrete entities. The real actors in his drama will be individual organisms in shifting populations. In his book it will be individual animals that give way to descendant individual animals, not species that give way to species. It is not surprising, then, that punctuationists tend to believe in a kind of natural selection at the species level, which they regard as analogous to Darwinian selection at the ordinary individual level. Non-punctuationists, on the other hand, are likely to see natural selection as working at no higher level than the individual organism. The idea of 'species selection' has less appeal for them, because they do not think of species as entities with a discrete existence through geological time.

This is a convenient moment to deal with the hypothesis of species selection, which is left over, in a sense, from the previous chapter. I shan't spend very much time on it, as I have spelled out in *The Extended Phenotype* my doubts about its alleged importance in evolution. It is true that the vast majority of species that have ever lived have gone extinct. It is also true that new species come into existence at a rate that at least balances the extinction rate, so that there is a kind of 'species pool' whose composition is changing all the time. Nonrandom recruitment to the species pool and nonrandom removal of species from it could, it is true, theoretically constitute a kind of higher-level natural selection. It is possible that certain characteristics of species bias their probability of going extinct, or of budding off new species. The species that we see in the world will tend to have whatever it takes to come into the world in the first place – to 'be speciated' – and whatever it takes not to go extinct. You can call that a form of natural selection if you wish, although I suspect that it is closer to single-step selection than to cumulative selection. What I am sceptical about is the suggestion that this kind of selection has any great importance in explaining evolution.

This may just reflect my biased view of what is important. As I said at the beginning of this chapter, what I mainly want a theory of evolution to do is explain complex, well-designed mechanisms like hearts, hands, eyes and echolocation. Nobody, not even the most ardent species selectionist, thinks that species selection can do this. Some people do think that species selection can explain certain long-term trends in the fossil record, such as the rather commonly observed trend towards larger body size as the ages go by. Modern horses, as we have seen, are bigger than their ancestors of 30 million years ago. Species selectionists object to the idea that this came about through consistent individual advantage: they don't see the fossil trend as indicating that large individual horses were consistently more successful than small individual horses within their species. What they think happened is this. There were lots of species, a species pool. In some of these species, average body size was large, in others it was small (perhaps because in some species large individuals did best, in other species small individuals did best). The species with large body size were less likely to go extinct (or more likely to bud off new species like themselves) than the species with small body size. Whatever went on within species, according to the species selectionist, the fossil trend towards larger body size was due to a succession of *species* with progressively larger average body size. It is even possible that in the majority of species *smaller* individuals were favoured, yet the fossil trend could still be towards larger body size. In other words the selection of *species* could favour that minority of species in which larger individuals were favoured. Exactly this point was made, admittedly in a spirit of devil's advocacy, by the great neo-Darwinian theorist George C. Williams, long before modern species selectionism came on the scene.

It could be said that we have here, and maybe in all alleged examples of species selection, not so much an evolutionary trend, more a *successional* trend, like the trend towards larger and larger plants as a piece of waste ground becomes colonized successively by small weeds, larger herbs, shrubs and, finally, the mature 'climax' forest trees. Anyway, whether you call it a successional trend or an evolutionary trend, the species selectionists may well be right to believe that it is this kind of trend that they, as palaeontologists, are often dealing with in successive strata of the fossil record. But, as I said, nobody wants to say that species selection is an important explanation for the evolution of complex adaptations. Here is why.

Complex adaptations are in most cases not properties of species, they are properties of individuals. Species don't have eyes and hearts; the individuals in them do. If a species goes extinct because of poor eyesight, this presumably means that every individual in that species died because of poor eyesight. Quality of eyesight is a property of individual animals. What kinds of traits can *species* be said to have? The answer must be traits that affect the survival and reproduction of the species, in ways that cannot be reduced to the sum of their effects on individual survival and reproduction. In the hypothetical example of the horses, I suggested that that minority of species in which larger individuals were favoured were less likely to go extinct than the majority of species in which smaller individuals were favoured. But this is pretty unconvincing. It is hard to think of reasons why species survivability should be decoupled from the sum of the survivabilities of the individual members of the species.

A better example of a species level trait is the following hypothetical one. Suppose that in some species all individuals make their living in the same way. All koalas, for instance, live in eucalyptus trees and eat only eucalyptus leaves. Species like this can be called uniform. Another species might contain a diversity of individuals that make their living in different ways. Each individual could be just as specialized as an individual koala, but the species as a whole contains a variety of dietary habits. Some members of the species eat nothing but eucalyptus leaves; others nothing but wheat; others nothing but yams; others nothing but lime peel, and so on. Call this second kind of species variegated species. Now I think it is easy to imagine circumstances in which uniform species would be more likely to go extinct than variegated species. Koalas rely totally on a supply of eucalyptus, and a eucalyptus plague analogous to Dutch elm disease would finish them. In the variegated species, on the other hand, *some* members of the species would survive any particular plague of food plants, and the species could go on. It is also easy to believe that variegated species are more likely to bud off new, daughter species than uniform species. Here, perhaps would be examples of true species-level selection. Unlike shortsightedness or long-leggedness, 'uniformness' and 'variegatedness' are true species-level traits. The trouble is that examples of such species-level traits are few and far between.

There is an interesting theory by the American evolutionist Egbert Leigh which can be interpreted as a possible candidate example of true species-level selection, although it was suggested before the phrase 'species selection' came into vogue. Leigh was interested in that perennial problem, the evolution of 'altruistic' behaviour in individuals. He correctly recognized that if individual interests conflict with those of the species, the individual interests – short-term interests – must prevail. Nothing, it seems, can prevent the march of selfish genes. But Leigh made the following interesting suggestion. There must be some groups or species in which, as it happens, what is best for the individual pretty much coincides with what is best for the species. And there must be other species in which the interests of the individual happen to depart especially strongly from the interests of the species. Other things being equal, the second type of species could well be more likely to go extinct. A form of species selection, then, could favour, not individual self-sacrifice, but those species in which individuals are not *asked* to sacrifice their own welfare. We could then see apparently unselfish individual behaviour evolving, because species selection has favoured those species in which individual self-interest is best served by their own apparent altruism.

Perhaps the most dramatic example of a truly species-level trait concerns the mode of reproduction, sexual versus asexual. For reasons that I haven't the space to go into, the existence of sexual reproduction poses a big theoretical puzzle for Darwinians. Many years ago R. A. Fisher, usually hostile to any idea of selection at levels higher than the individual organism, was prepared to make an exception for the special case of sexuality itself. Sexually reproducing species, he argued, for reasons that, again, I shan't go into (they aren't as obvious as one might think), are capable of evolving faster than asexually reproducing species. Evolving is something that species do, not something that individual organisms do: you can't talk of an organism as evolving. Fisher was suggesting, then, that species-level selection is partly responsible for the fact that sexual reproduction is so common among modern animals. But, if so, we are dealing with a case of single-step selection, not cumulative selection.

Asexual species when they occur tend, according to the argument, to go extinct because they don't evolve fast enough to keep up with the changing environment. Sexual species tend not to go extinct, because they can evolve fast enough to keep up. So what we see around us are mostly sexual species. But the 'evolution' whose rate varies between the two systems is, of course, ordinary Darwinian evolution by cumulative selection at the individual level. The species selection, such as it is, is simple single-step selection, chosing between only two traits, asexuality versus sexuality, slow evolution versus fast evolution. The machinery of sexuality, sex organs, sexual behaviour, the cellular machinery of sexual cell division, all these must have been put together by standard, low-level Darwinian cumulative selection, *not* by species selection. In any case, as it happens, the modern consensus is against the old theory that sexuality is maintained by some kind of group level or species level selection.

To conclude the discussion of species selection, it could account for the pattern of species existing in the world at any particular time. It follows that it could also account for changing patterns of species as geological ages give way to later ages, that is, for changing patterns in the fossil record. But it is not a significant force in the evolution of the complex machinery of life. The most it can do is to choose between various alternative complex machineries, given that those complex machineries have already been put together by true Darwinian selection. As I have put it before, species selection may occur but it doesn't seem to *do* anything much! I now return to the subject of taxonomy and its methods.

I said that cladistic taxonomy has the advantage over librarians' types of taxonomy that there is one unique, true hierarchical nesting pattern in nature, waiting to be discovered. All that we have to do is develop methods of discovering it. Unfortunately there are practical difficulties. The most interesting bugbear of the taxonomist is evolutionary convergence. This is such an important phenomenon that I have already devoted half a chapter to it. In Chapter 4 we saw how, over and over again, animals have been found to resemble unrelated animals in other parts of the world, because they have similar ways of life. New World army ants resemble Old World driver ants. Uncanny resemblances have evolved between the quite unrelated electric fish of Africa and South America; and between true wolves and the marsupial 'wolf' *Thylacinus* of Tasmania. In all these cases I simply asserted without justification that these resemblances were convergent: that they had evolved independently in unrelated animals. But how do we know that they are unrelated? If taxonomists use resemblances to measure closeness of cousinship, why weren't taxonomists fooled by the uncannily close resemblances that seem to unite these pairs of animals? Or, to twist the question round into a more worrying form, when taxonomists tell us that two animals really *are* closely related – say rabbits and hares – how do we know that the taxonomists haven't been fooled by massive convergence?

This question really is worrying, because the history of taxonomy is replete with cases where later taxonomists have declared their predecessors wrong for precisely this reason. In Chapter 4 we saw that an Argentinian taxonomist had pronounced the litopterns ancestral to true horses, whereas they are now thought to be convergent on true horses. The African porcupine was long believed to be closely related to the American porcupines, but the two groups are now thought to have evolved their prickly coats independently. Presumably prickles were useful to both for similar reasons in the two continents. Who is to say that future generations of taxonomists won't change their minds yet again? What confidence can we vest in taxonomy, if convergent evolution is such a powerful faker of deceptive resemblances? The main reason why I personally feel optimistic is the arrival on the scene of powerful new techniques based on molecular biology.

To recapitulate from earlier chapters, all animals and plants and bacteria, however different they appear to be from one another, are astonishingly uniform when we get down to molecular basics. This is most dramatically seen in the genetic code itself. The genetic dictionary has 64 DNA words of three letters each. Every one of these words has a precise translation into protein language (either a particular amino acid or a punctuation mark). The language appears to be arbitrary in the same sense as a human language is arbitrary (there is nothing intrinsic in the sound of the word 'house', for instance, which suggests to the listener any attribute of a dwelling). Given this, it is a fact of great significance that every living thing, no matter how different from others in external appearance it may be, 'speaks' almost exactly the same language at the level of the genes. The genetic code is universal. I regard this as near-conclusive proof that all organisms are descended from a single common ancestor. The odds of the same dictionary of arbitrary 'meanings' arising twice are almost unimaginably small. As we saw in Chapter 6, there may once have been other organisms that used a different genetic language, but they are no longer with us. All surviving organisms are descended from a single ancestor from which they have inherited a nearly identical, though arbitrary, genetic dictionary, identical in almost every one of its 64 DNA words.

Just think of the impact of this fact on taxonomy. Before the age of molecular biology, zoologists could be confident of the cousinship of only animals that shared a very large number of anatomical features. Molecular biology suddenly opened a new treasure chest of resemblances to add to the meagre list offered by anatomy and embryology. The 64 identities (resemblances is too weak a word) of the shared genetic dictionary is only the start. Taxonomy has been transformed. What were once vague guesses of cousinship have become statistical near-certainties.

The almost complete word-for-word universality of the genetic dictionary is, for the taxonomist, too much of a good thing. Once it has told us that all living things are cousins, it cannot tell us which pairs are closer cousins than others. But other molecular information can, because here we find variable degrees of resemblance rather than total identity. The product of the genetic translating machinery, remember, is protein molecules. Each protein molecule is a sentence, a chain of amino acid words from the dictionary. We can read these sentences, either in their translated protein form or in their original DNA form. Though all living things share the same dictionary, they don't all make the same sentences with their shared dictionary. This offers us the opportunity to work out varying degrees of cousinship. The protein sentences, though different in detail, are often similar in overall pattern. For any pair of organisms, we can always find sentences that are sufficiently similar to be obviously slightly 'garbled' versions of the same ancestral sentence. We have already seen this in the example of the minor differences between the histone sequences of cows and peas.

Taxonomists can now compare molecular sentences exactly as they might compare skulls or leg bones. Closely similar protein or DNA sentences can be assumed to come from close cousins; more different sentences from more distant cousins. These sentences are all constructed from the universal dictionary of no more than 64 words. The beauty of modern molecular biology is that we can measure the difference between two animals exactly, as the precise number of words by which their versions of a particular sentence differ. In the terms of the genetic hyperspace of Chapter 3, we can measure exactly how many steps separate one animal from another, at least with respect to a particular protein molecule.

An added advantage of using molecular sequences in taxonomy is that, according to one influential school of geneticists, the 'neutralists' (we shall meet them again in the next chapter), most of the evolutionary change that goes on at the molecular level is *neutral*. This means that it is not due to natural selection but is effectively random, and therefore that, except through occasional bad luck, the bugbear of convergence is not there to mislead the taxonomist. A related fact is that, as we have already seen, any one kind of molecule seems to evolve at a roughly constant rate in widely different animal groups. This means that the number of differences between comparable molecules in two animals, say between human cytochrome and warthog cytochrome, is a good measure of the time that has elapsed since their common ancestor lived. We have a pretty accurate 'molecular clock'. The molecular clock allows us to estimate, not just which pairs of animals have the most recent common ancestors, but also approximately *when* those common ancestors lived.

The reader may be puzzled, at this point, by an apparent inconsistency. This whole book emphasizes the overriding importance of natural selection. How then can we now emphasize the randomness of evolutionary change at the molecular level? To anticipate Chapter 11, there really is no quarrel with respect to the evolution of adaptations, which are the main subject of this book. Not even the most ardent neutralist thinks that complex working organs like eyes and hands have evolved by random drift. Every sane biologist agrees that these can only have evolved by natural selection. It is just that the neutralists think – rightly, in my opinion – that such adaptations are the tip of the iceberg: probably most evolutionary change, when seen at the molecular level, is non-functional.

So long as the molecular clock is a fact – and it does seem to be true that each kind of molecule changes at roughly its own characteristic rate per million years – we can use it to date branch points in the evolutionary tree. And if it is really true that most evolutionary change, at the molecular level, is neutral, this is a wonderful gift for the taxonomist. It means that the problem of convergence may be swept away by the weapon of statistics. Every animal has great volumes of genetic text written in its cells, text most of which, according to the neutralist theory, has nothing to do with fitting it to its peculiar way of life; text that is largely untouched by selection and largely not subject to convergent evolution except as a result of sheer chance. The chance that two large pieces of selectively neutral text could resemble each other by luck can be calculated, and it is very low indeed. Even better, the constant rate of molecular evolution actually lets us *date* branch points in evolutionary history.

It is hard to exaggerate the extra power that the new molecular sequence-reading techniques have added to the taxonomist's armoury. Not all molecular sentences in all animals have yet, of course, been deciphered, but already one can walk into the library and look up the exact word-for-word, letter-for-letter, phraseology of, say, the α-haemoglobin sentences of a dog, a kangaroo, a spiny anteater, a chicken, a viper, a newt, a carp and a human. Not all animals have haemoglobin, but there are other proteins, for instance histones, of which a version exists in every animal and plant, and again many of them can already be looked up in the library. These are not vague measurements of the kind which, like leg length or skull width, might vary with the age and health of the specimen, or even with the eyesight of the measurer. They are precisely worded alternative versions of the same sentence in the same language, which can be placed side by side and compared with each other as minutely and as exactly as a fastidious Greek scholar might compare two parchments of the same Gospel. DNA sequences are the gospel documents of all life, and we have learned to decipher them.

The basic taxonomists' assumption is that close cousins will have more similar versions of a particular molecular sentence than more distant cousins. This is called the 'parsimony principle'. Parsimony is another name for economic meanness. Given a set of animals whose sentences are known, say the eight animals listed in the previous paragraph, our task is to discover which of all the possible tree diagrams linking the eight animals is the most parsimonious. The most parsimonious tree is the tree which is 'economically meanest' with its assumptions, in the sense that it assumes the minimum number of word changes in evolution, and the minimum amount of convergence. We are entitled to assume the minimum amount of convergence on grounds of sheer improbability. It is unlikely, especially if much of molecular evolution is neutral, that two unrelated animals would have hit upon exactly the same sequence, word for word, letter for letter.

There are computational difficulties in trying to look at all possible trees. When there are only three animals to be classified, the number of possible trees is only three: A united with B excluding C; A with C excluding B; and B with C excluding A. You can do the same calculation for larger numbers of animals to be classified, and the number of possible trees rises steeply. When there are only four animals to be considered, the total number of possible trees of cousinship is still manageable at only 15. It doesn't take the computer long to work out which of the 15 is the most parsimonious. But if there are 20 animals to be considered, I reckon the number of possible trees to be 8,200,794,532,637,891,559,375 (see Figure 9). It has been calculated that the fastest of today's computers would take 10,000 million years, approximately the age of the universe, to discover the most parsimonious tree for a mere 20 animals. And taxonomists often want to construct trees of more than 20 animals.

Although molecular taxonomists have been the first to make much

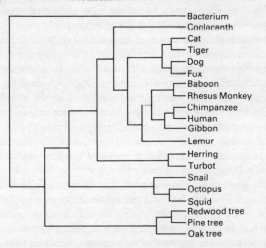

Figure 9 This family tree is correct. There are 8200794532637891559374 other ways of classifying these 20 organisms, and all of them are wrong.

of it, this problem of exploding large numbers has actually been lurking all along in non-molecular taxonomy. Nonmolecular taxonomists have simply evaded it by making intuitive guesses. Of all the possible family trees that might be tried, huge numbers of trees can be eliminated immediately — for instance, all those millions of conceivable family trees that place humans closer to earthworms than to chimps. Taxonomists don't even bother to consider such obviously absurd trees of cousinship, but instead home in on the relatively few trees that do not too drastically violate their preconceptions. This is probably fair, although there is always the danger that the truly most parsimonious tree is one of those that have been thrown out without consideration. Computers, too, can be programmed to take short cuts, and the problem of the exploding large numbers can be mercifully cut down.

Molecular information is so rich that we can do our taxonomy separately, over and over again, for different proteins. We can then use our conclusions, drawn from the study of one molecule, as a check on our conclusions based on the study of another molecule. If we are worried that the story told by one protein molecule is really confounded by convergence, we can immediately check it by looking at another protein molecule. Convergent evolution is really a special kind of coincidence. The thing about coincidences is that, even if they happen once, they are far less likely to happen twice. And even less likely to happen three times. By taking more and more separate protein molecules, we can all but eliminate coincidence.

For instance, in one study by a group of New Zealand biologists, 11 animals were classified, not once but five times independently, using five different protein molecules. The 11 animals were sheep, rhesus monkey, horse, kangaroo, rat, rabbit, dog, pig, human, cow and chimpanzee. The idea was first to work out a tree of relationships among the 11 animals using one protein. Then see whether you get the *same* tree of relationships using a different protein. Then do the same for a third, fourth and fifth protein. Theoretically, if evolution were not true for example, it is possible for each of the five proteins to give a completely different tree of 'relationships'.

The five protein sequences were all available to be looked up in the library, for all 11 animals. For 11 animals, there are 654,729,075 possible trees of relationships to be considered, and the usual short-cut methods had to be employed. For each of the five protein molecules, the computer printed out the most parsimonious tree of relationship. This gives five independent best guesses as to the true tree of relationships among these 11 animals. The neatest result that we could hope for is that all five estimated trees turn out to be identical. The probability of getting this result by sheer luck is very small indeed: the number has 31 noughts after the decimal point. We should not be surprised if we fail to get agreement quite as perfect as this: a certain amount of convergent evolution and coincidence is only to be expected. But we should be worried if there is not a substantial measure of agreement among the different trees. In fact the five trees turned out to be not quite identical, but they are very similar. All the five molecules agree in placing human, chimp and monkey close to each other, but there are some disagreements over which animal is the next closest to this cluster: haemoglobin B says the dog is; fibrinopeptide B says the rat is; fibrinopeptide A says that a cluster consisting of rat and rabbit is; haemoglobin A says that a cluster consisting of rat, rabbit and dog is.

We have a definite common ancestor with the dog, and another definite common ancestor with the rat. These two ancestors really existed, at a particular moment in history. One of them has to be more recent than the other, so either haemoglobin B or fibrinopeptide B must be wrong in its estimate of evolutionary relationships. Such minor discrepancies needn't worry us, as I have said. We expect a certain amount of convergence and coincidence. If we are truly closer to the dog, then this means that we and the rat have converged on one another with respect to our fibrinopeptide B. If we are truly closer to the rat, this means that we and the dog have converged on each other with respect to our haemoglobin B. We can get an idea of which of these two is the more likely, by looking at yet other molecules. But I shan't pursue the matter: the point has been made.

I said that taxonomy was one of the most rancorously ill-tempered of biological fields. Stephen Gould has well characterized it with the phrase 'names and nastiness'. Taxonomists seem to feel passionately about their schools of thought, in a way that we expect in political science or economics, but not usually in academic science. It is clear that members of a particular school of taxonomy think of themselves as a beleaguered band of brothers, like the early Christians. I first realized this when a taxonomist acquaintance told me, white-faced with dismay, the 'news' that So-and-so (the name doesn't matter) had '*gone over* to the cladists'.

The following brief account of taxonomic schools of thought will probably annoy some members of those schools, but no more than they habitually infuriate each other so no undue harm will be done. In terms of their fundamental philosophy, taxonomists fall into two main camps. On the one hand there are those that make no bones about the fact that their aim is openly to discover evolutionary relationships. To them (and to me) a good taxonomic tree *is* a family tree of evolutionary relationships. When you do taxonomy you are using all methods at your disposal to make the best guess you can about the closeness of cousinship of animals to one another. It is hard to find a name for these taxonomists because the obvious name, 'evolutionary taxonomists', has been usurped for one particular sub-school. They are sometimes called 'phyleticists'. I have written this chapter, so far, from a phyleticist's point of view.

But there are many taxonomists who proceed in a different way, and for quite sensible reasons. Although they are likely to agree that one ultimate aim of doing taxonomy is to make discoveries about evolutionary relationships, they insist on keeping the *practice* of taxonomy separate from the theory – presumably evolutionary theory – of what has led to the pattern of resemblances. These taxonomists study patterns of resemblances in their own right. They do not pre-judge the issue of whether the pattern of resemblances is caused by evolutionary history and whether close resemblance is due to close cousinship. They prefer to construct their taxonomy using the pattern of resemblances alone.

One advantage of doing this is that, if you have any doubts about the truth of evolution, you can use the pattern of resemblances to test it. If evolution is true, resemblances among animals should follow certain predictable patterns, notably the pattern of hierarchical nesting. If evolution is false, goodness knows *what* pattern we should expect, but there is no obvious reason to expect a nested hierarchical pattern. If you assume evolution throughout the *doing* of your taxonomy, this school insists, you can't then use the results of your taxonomic work to support the truth of evolution: the argument would be circular. This argument would have force if anybody was seriously in doubt about the truth of evolution. Once again, it is hard to find a suitable name for this second school of thought among taxonomists. I shall call them the 'pure-resemblance measurers'.

The phyleticists, the taxonomists that openly try to discover evolutionary relationships, further split into two schools of thought. These are the cladists, who follow the principles laid down in Willi Hennig's famous book *Phylogenetic Systematics*; and the 'traditional' evolutionary taxonomists. Cladists are obsessed with branches. For them, the goal of taxonomy is to discover the order in which lineages split from each other in evolutionary time. They don't care how much, or how little, those lineages have changed since the branch-point. 'Traditional' (*don't* think of it as a pejorative name) evolutionary taxonomists differ from cladists mainly in that they don't consider only the branching kind of evolution. They also take account of the total quantity of change that occurs during evolution, not just branching.

Cladists think in terms of branching trees, right from the outset of their work. They ideally begin by writing down all possible branching trees for the animals they are dealing with (two-way branching trees only, because there are limits to anyone's patience!). As we saw when discussing molecular taxonomy, this gets difficult if you are trying to classify lots of animals, because the number of possible trees becomes astronomically large. But as we also saw, there are fortunately short cuts and serviceable approximations which mean that this kind of taxonomy can, in practice, be done.

If, for the sake of argument, we were trying to classify just the three animals squid, herring and human, the only three possible two-way branching trees are the following:

1. Squid and herring are close to each other, human is the 'outgroup'.

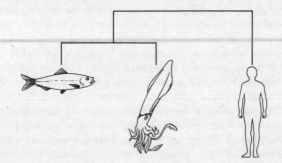

2. Human and herring are close to each other, squid is the outgroup.

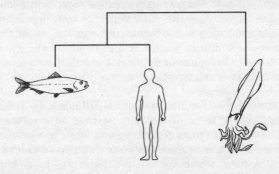

3. Squid and human are close to each other, herring is the outgroup.

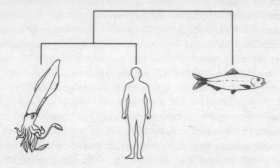

Cladists would look at each of the three possible trees in turn, and choose the best tree. How is the best tree recognized? Basically, it is the tree that unites the animals that have the most features in common. We label as the 'outgroup' the animal that has fewest features in common with the other two. Of the above list of trees, the second would be preferred, because human and herring share many more features in common with each other than squid and herring do or than squid and human do. Squid is the outgroup because it doesn't have many features in common with either human or herring.

Actually, it isn't quite as simple as just counting features in common, because some kinds of features are deliberately ignored. Cladists want to give special weight to features that are recently evolved. Ancient features that all mammals inherited from the first mammal, for instance, are useless for doing classifications within the mammals. The methods they use for deciding which features are ancient are interesting, but they would take us outside the scope of this book. The main thing to remember at this stage is that, at least in principle, the cladist thinks about all possible bifurcating trees that *might* unite the set of animals he is dealing with, and tries to choose the one correct tree. And the true cladist makes no bones about the fact that he thinks of the branching trees or 'cladograms' as family trees, trees of closeness of evolutionary cousinship.

If pushed to the extreme, the obsession with branchings alone could give strange results. It is theoretically possible for a species to be *identical* in every detail to its distant cousins, while being exceedingly different from its closer cousins. For instance, suppose that two very similar fish species, which we can call Jacob and Esau, lived 300 million years ago. Both these species founded dynasties of descendants,

which last to the present day. Esau's descendants stagnated. They went
on living in the deep sea but they didn't evolve. The result is that a
modern descendant of Esau is essentially the same as Esau, and is
therefore also very like Jacob. Jacob's descendants evolved and pro-
liferated. They eventually gave rise to all modern mammals. But one
lineage of Jacob's descendants also stagnated in the deep sea, and also
leaves modern descendants. These modern descendants are fish that
are so similar to Esau's modern descendants that they are hard to tell
apart.

Now, how shall we classify these animals? The traditional
evolutionary taxonomist would recognize the great similarity between
the primitive deep-sea descendants of Jacob and Esau, and would
classify them together. The strict cladist could not do that. The deep-sea
descendants of Jacob, for all that they look just like the deep-sea
descendants of Esau, are, nevertheless, closer cousins of mammals.
Their common ancestor with mammals lived more recently, even if
only slightly more recently, than their common ancestor with Esau's
descendants. Therefore they must be classified together with
mammals. This may seem strange, but personally I can treat it with
equanimity. It is at least utterly logical and clear. There are indeed
virtues in both cladism and traditional evolutionary taxonomy, and I
don't much mind how people classify animals so long as they tell me
clearly how they are doing it.

Turning now to the other major school of thought, the
pure-resemblance measurers, they again can be split into two sub-
schools. Both the sub-schools agree to banish evolution from their
day-to-day thoughts while they do taxonomy. But they disagree over
how to proceed in their day-to-day taxonomy. One sub-school among
these taxonomists are sometimes called 'pheneticists' and sometimes
called 'numerical taxonomists'. I shall call them the 'average-distance
measurers'. The other school of resemblance measurers call themselves 'transformed cladists'. This is a poor name, since the one thing
these people are *not* is cladists! When Julian Huxley invented the term
clade he defined it, clearly and unambiguously, in terms of
evolutionary branching and evolutionary ancestry. A clade is the set of
all organisms descended from a particular ancestor. Since the main
point of 'transformed cladists' is to avoid all notions of evolution and
of ancestry, they cannot sensibly call themselves cladists. The reason
they do so is one of history: they started out as true cladists, and kept
some of the methods of cladists while abandoning their fundamental
philosophy and rationale. I suppose I have no choice but to call them
transformed cladists, although I do so with reluctance.

The average-distance measurers not only refuse to use evolution in their taxonomy (although they all believe in evolution). They are consistent in that they don't even assume that the pattern of resemblance will necessarily be a simply branching hierarchy. They try to employ methods that will uncover a hierarchical pattern if one is really there, but not if it isn't. They try to ask Nature to tell them whether she is really organized hierarchically. This is not an easy task, and it is probably fair to say that methods are not really available for achieving this aim. Nevertheless the aim seems to me to be all of a piece with the laudable one of avoiding preconceptions. Their methods are often rather sophisticated and mathematical, and they are just as suitable for classifying nonliving things, for instance rocks or archaeological relics, as for classifying living organisms.

They usually begin by measuring everything they can about their animals. You have to be a bit clever about how you interpret these measurements, but I shan't go into that. The end result is that the measurements are all combined together to produce an index of resemblance (or, its opposite, an index of difference) between each animal and each other animal. If you wish, you can actually visualize the animals as clouds of points in space. Rats, mice, hamsters, etc. would all be found in one part of the space. Far away in another part of the space would be another little cloud, consisting of lions, tigers, leopards, cheetahs, etc. The distance between any two points in the space is a measure of how closely the two animals resemble each other, when large numbers of their attributes are combined together. The distance between lion and tiger is small. So is the distance between rat and mouse. But the distance between rat and tiger, or mouse and lion, is large. The combining together of attributes is usually done with the aid of a computer. The space that these animals are sitting in is superficially a bit like Biomorph Land, but the 'distances' reflect bodily resemblances rather than genetic resemblances.

Having calculated an index of average resemblance (or distance) between each animal and each other animal, the computer is next programmed to scan the set of distances/resemblances and to try to fit them into a hierarchical clustering pattern. Unfortunately there is a lot of controversy about exactly which calculation method should be used to look for clusters. There is no one obviously correct method, and the methods don't all give the same answer. Worse, it is possible that some of these computer methods are over-'eager' to 'see' hierarchically arranged clusters within clusters, even if they aren't really there. The school of distance measurers, or 'numerical taxonomists', has become a bit unfashionable lately. My view is that the unfashionableness is a temporary phase, as fashions often are, and that this kind of 'numerical taxonomy' is by no means easily to be written off. I expect a come-back.

The other school of pure-pattern measurers are the ones that call themselves transformed cladists, for reasons of history as we have seen. It is from within this group that the 'nastiness' mainly emanates. I shall not follow the usual practice of tracing their historical origins from within the ranks of true cladists. In their underlying philosophy, so-called transformed cladists have more in common with the other school of pure-pattern measurers, the ones often called 'pheneticists' or 'numerical taxonomists', whom I have just discussed under the title of average-distance measurers. What these share with each other is an antipathy to dragging evolution into the practice of taxonomy, although this does not *necessarily* betoken any hostility to the idea of evolution itself.

What the transformed cladists share with true cladists is many of their methods in practice. Both think, right from the start, in terms of bifurcating trees. And both pick out certain kinds of characteristics as taxonomically important, other kinds of characteristics as taxonomically worthless. They differ with respect to the rationale that they give to this discrimination. Like average-distance measurers, transformed cladists are not out to discover family trees. They are looking for trees of pure resemblance. They agree with the average-distance measurers to leave open the question of whether the pattern of resemblance reflects evolutionary history. But unlike the distance measurers, who, at least in theory, are prepared to let Nature tell them whether she is actually hierarchically organized, the transformed cladists *assume* that she is. It is an axiom, an article of faith with them, that things are to be classified into branching hierarchies (or, equivalently, into nested nests). Because the branching tree has nothing to do with evolution, it need not necessarily be applied to living things. The methods of transformed cladistics can, according to their advocates, be used for classifying not just animals and plants but stones, planets, library books and Bronze Age pots. In other words they would not subscribe to the point I made with my library comparison, that evolution is the only sound basis for a uniquely hierarchical classification.

The average-distance measurers, as we saw, measure how far each animal is from each other animal, where 'far' means 'does not re-semble' and 'near' means 'resembles'. Only then, after calculating a sort of summed average index of resemblance, do they start trying to inter-pret their results in terms of a branching, cluster-within-clustery hierarchy or 'tree' diagram. The transformed cladists, however, like the true cladists that they once were, bring in clustery, branchy thinking right at the outset. Like true cladists, they would begin, at least in principle, by writing down all possible bifurcating trees, and then choosing the best.

But what are they actually talking about when they consider each possible 'tree', and what do they mean by the best? What hypothetical state of the world does each tree correspond to? To a true cladist, a follower of W. Hennig, the answer is very clear. Each of the 15 possible trees uniting four animals represents a possible family tree. Of all the 15 conceivable family trees uniting four animals, one and only one must be the correct one. The history of the animals' ancestors really did happen, in the world. There are 15 possible histories, if we make the assumption that all branchings are two-way branchings. Fourteen of those possible histories must be wrong. Only one can be right; can correspond to the way the history actually happened. Of all the 135,135 possible family trees culminating in 8 animals, 135,134 must be wrong. Only one represents historical truth. It may not be easy to be sure *which* one is the correct one, but the true cladist can at least be sure *that* not more than one is correct.

But what do the 15 (or 135,135, or whatever it is) possible trees, and the one correct tree, correspond to in the nonevolutionary world of the transformed cladist? The answer, as my colleague and former student Mark Ridley has pointed out in his *Evolution and Classification*, is nothing very much. The transformed cladist refuses to allow the concept of *ancestry* to enter his considerations. 'Ancestor', to him, is a dirty word. But on the other hand he insists that classification must be a branching hierarchy. So, if the 15 (or 135,135) possible hierarchical trees are not trees of ancestral history, what on earth are they? There is nothing for it but to appeal to ancient philosophy for some woolly, idealistic notion that the world just is organized hierarchically; some notion that everything in the world has its 'opposite', its mystical ying or yang. It never gets much more concrete than that. It certainly is not possible, in the nonevolutionary world of the transformed cladist, to make strong and clear statements such as 'only one out of the 945 possible trees uniting 6 animals can be right; all the rest must be wrong'.

Why is ancestor a dirty word to cladists? It is not (I hope) that they think that there never were any ancestors. It is rather that they have decided that ancestors have no place in taxonomy. This is a defensible position as far as the day-to-day *practice* of taxonomy is concerned. No cladist actually draws flesh and blood ancestors on family trees, though traditional evolutionary taxonomists sometimes do. Cladists, of all stripes, treat all relationships between real, observed animals as *cousinships*, as a matter of form. This is perfectly sensible. What is not sensible is to carry this over into a taboo against the very *concept* of ancestors, against the use of the language of ancestry in providing the fundamental justification for adopting the hierarchically branching tree as the basis for your taxonomy.

I have left till last the oddest aspect of the transformed cladism school of taxonomy. Not content with a perfectly sensible belief that there is something to be said for leaving evolutionary and ancestral assumptions out of the *practice* of taxonomy, a belief that they share with pheneticist 'distance measurers', some transformed cladists have gone right over the top and concluded that there must be something wrong with evolution itself! The fact is almost too bizarre to credit, but some of the leading 'transformed cladists' profess an actual hostility to the idea of evolution itself, especially the Darwinian theory of evolution. Two of them, G. Nelson and N. Platnick from the American Museum of Natural History in New York, have gone so far as to write that 'Darwinism . . . is, in short, a theory that has been put to the test and found false'. I should love to know what this 'test' is and, even more, I should love to know by what alternative theory Nelson and Platnick would explain the phenomena that Darwinism explains, especially adaptive complexity.

It isn't that any transformed cladists are themselves fundamentalist creationists. My own interpretation is that they enjoy an exaggerated idea of the importance of taxonomy in biology. They have decided, perhaps rightly, that they can do taxonomy better if they forget about evolution, and especially if they never use the concept of the ancestor in thinking about taxonomy. In the same way, a student of, say, nerve cells, might decide that he is not aided by thinking about evolution. The nerve specialist agrees that his nerve cells are the products of evolution, but he does not need to use this fact in his research. He needs to know a lot about physics and chemistry, but he believes that Darwinism is irrelevant to his day-to-day research on nerve impulses. This is a defensible position. But you can't reasonably say that, because you don't need to use a particular theory in the day to day practice of your particular branch of science, therefore that theory is *false*. You will only say this if you have a remarkably grandiose estimation of the importance of your own branch of science.

Even then, it isn't logical. A physicist certainly doesn't need Darwinism in order to do physics. He might think that biology is a trivial subject compared with physics. It would follow from this that, in his opinion, Darwinism is of trivial importance to science. But he could not sensibly conclude from this that it is therefore *false*! Yet this is essentially what some of the leaders of the school of transformed cladistics seem to have done. 'False', note well, is precisely the word Nelson and Platnick used. Needless to say, their words have been picked up by the sensitive microphones that I mentioned in the previous chapter, and the result has been considerable publicity. They have earned themselves a place of honour in fundamentalist, creationist literature. When a leading transformed cladist came to give a guest lecture in my university recently, he drew a bigger crowd than any other guest lecturer that year! It isn't hard to see why.

There is no doubt at all that remarks like 'Darwinism . . . is a theory that has been put to the test and found false', coming from established biologists on the staff of a respected national museum, will be meat and drink to creationists and others who actively have an interest in perpetrating falsehoods. This is the only reason I have troubled my readers with the topic of transformed cladism at all. As Mark Ridley more mildly said, in a review of the book in which Nelson and Platnick made that remark about Darwinism being false, Who would have guessed that all that they really *meant* was that ancestral species are tricky to represent in cladistic classification? Of course it is difficult to pin down the precise identity of ancestors, and there is a good case for not even trying to do so. But to make statements that encourage others to conclude that there never *were* any ancestors is to debauch language and betray truth.

Now I'd better go out and dig the garden, or something.

CHAPTER 11

DOOMED RIVALS

No serious biologist doubts the fact that evolution has happened, nor that all living creatures are cousins of one another. Some biologists, however, have had doubts about Darwin's particular theory of *how* evolution happened. Sometimes this turns out to be just an argument about words. The theory of punctuated evolution, for instance, can be represented as anti-Darwinian. As I argued in Chapter 9, however, it is really a minor variety of Darwinism, and does not belong in any chapter about rival theories. But there are other theories that are most definitely *not* versions of Darwinism, theories that go flatly against the very spirit of Darwinism. These rival theories are the subject of this chapter. They include various versions of what is called Lamarckism; also other points of view such as 'neutralism', 'mutationism' and creationism which have, from time to time, been advanced as alternatives to Darwinian selection.

The obvious way to decide between rival theories is to examine the evidence. Lamarckian types of theory, for instance, are traditionally rejected – and rightly so – because no good evidence for them has ever been found (not for want of energetic trying, in some cases by zealots prepared to fake evidence). In this chapter I shall take a different tack, largely because so many other books have examined the evidence and concluded in favour of Darwinism. Instead of examining the evidence for and against rival theories, I shall adopt a more armchair approach. My argument will be that Darwinism is the only known theory that is in principle *capable* of explaining certain aspects of life. If I am right it means that, even if there were no actual evidence in favour of the Darwinian theory (there is, of course) we should still be justified in preferring it over all rival theories.

One way in which to dramatize this point is to make a prediction. I predict that, if a form of life is ever discovered in another part of the universe, however outlandish and weirdly alien that form of life may be in detail, it will be found to resemble life on Earth in one key respect: it will have evolved by some kind of Darwinian natural selection. Unfortunately, this is a prediction that we shall, in all probability, not be able to test in our lifetimes, but it remains a way of dramatizing an important truth about life on our own planet. The Darwinian theory is in principle capable of explaining life. No other theory that has ever been suggested is in principle capable of explaining life. I shall demonstrate this by discussing all known rival theories, not the evidence for or against them, but their adequacy, in principle, as explanations for life.

First, I must specify what it means to 'explain' life. There are, of course, many properties of living things that we could list, and some of them might be explicable by rival theories. Many facts about the distribution of protein molecules, as we have seen, may be due to neutral genetic mutations rather than Darwinian selection. There is one particular property of living things, however, that I want to single out as explicable *only* by Darwinian selection. This property is the one that has been the recurring topic of this book: adaptive complexity. Living organisms are well fitted to survive and reproduce in their environments, in ways too numerous and statistically improbable to have come about in a single chance blow. Following Paley, I have used the example of the eye. Two or three of an eye's well-'designed' features could, conceivably, have come about in a single lucky accident. It is the sheer number of interlocking parts, all well adapted to seeing and well adapted to each other, that demands a special kind of explanation beyond mere chance. The Darwinian explanation, of course, involves chance too, in the form of mutation. But the chance is filtered cumulatively by selection, step by step, over many generations. Other chapters have shown that this theory is capable of providing a satisfying explanation for adaptive complexity. In this chapter I shall argue that all other known theories are *not* capable of so doing.

First, let us take Darwinism's most prominent historical rival, Lamarckism. When Lamarckism was first proposed in the early nineteenth century, it was not as a rival to Darwinism, because Darwinism had not yet been thought of. The Chevalier de Lamarck was ahead of his time. He was one of those eighteenth-century intellectuals who argued in favour of evolution. In this he was right, and he would deserve to be honoured for this alone, along with Charles Darwin's grandfather Erasmus and others. Lamarck also offered the best theory of the mechanism of evolution that anyone could come up with at the time, but there is no reason to suppose that, if the Darwinian theory of mechanism had been around at the time, he would have rejected it. It was not around, and it is Lamarck's misfortune that, at least in the English-speaking world, his name has become a label for an error – his theory of the *mechanism* of evolution – rather than for his correct belief in the *fact* that evolution has occurred. This is not a history book, and I shall not attempt a scholarly dissection of exactly what Lamarck himself said. There was a dose of mysticism in Lamarck's actual words – for instance, he had a strong belief in progress up what many people, even today, think of as the ladder of life; and he spoke of animals striving as if they, in some sense, consciously *wanted* to evolve. I shall extract from Lamarckism those non-mystical elements which, at least at first sight, seem to have a sporting chance of offering a real alternative to Darwinism. These elements, the only ones adopted by modern 'neo-Lamarckians', are basically two: the inheritance of acquired characteristics, and the principle of use and disuse.

The principle of use and disuse states that those parts of an organism's body that are used grow larger. Those parts that are not used tend to wither away. It is an observed fact that when you exercise particular muscles they grow; muscles that are never used shrink. By examining a man's body we can tell which muscles he uses and which he does not. We may even be able to guess his profession or his recreation. Enthusiasts of the 'body-building' cult make use of the principle of use and disuse to 'build' their bodies, almost like a piece of sculpture, into whatever unnatural shape is demanded by fashion in this peculiar minority culture. Muscles are not the only parts of the body that respond to use in this kind of way. Walk barefoot, and you acquire tougher skin on your soles. It is easy to tell a farmer from a bank clerk by looking at their hands alone. The farmer's hands are horny, toughened by long exposure to rough work. If the clerk's hands are horny at all, it amounts only to a little callus on the writing finger.

The principle of use and disuse enables animals to become better at the job of surviving in their world, progressively better during their own lifetime as a result of living in that world. Humans, through direct exposure to sunlight, or lack of it, develop a skin colour which equips them better to survive in the particular local conditions. Too much sunlight is dangerous. Enthusiastic sunbathers with very fair skins are susceptible to skin cancer. Too little sunlight, on the other hand, leads to vitamin-D deficiency and rickets, sometimes seen in hereditarily black children living in Scandinavia. The brown pigment melanin, which is synthesized under the influence of sunlight, makes a screen to protect the underlying tissues from the harmful effects of further sunlight. If a suntanned person moves to a less sunny climate the melanin disappears, and the body is able to benefit from what little sun there is. This can be represented as an instance of the principle of use and disuse: skin goes brown when it is 'used', and fades to white when it is not 'used'. Some tropical races, of course, inherit a thick screen of melanin whether or not they are exposed to sunlight as individuals.

Now turn to the other main Lamarckian principle, the idea that such acquired characteristics are then inherited by future generations. All evidence suggests that this idea is simply false, but through most of history it has been believed to be true. Lamarck did not invent it, but simply incorporated the folk wisdom of his time. In some circles it is still believed. My mother had a dog who occasionally affected a limp, holding up one hind leg and hobbling on the other three. A neighbour had an older dog who had unfortunately lost one hind leg in a car accident. She was convinced that her dog must be the father of my mother's dog, the evidence being that he had obviously inherited his limp. Folk wisdom and fairy tales are filled with similar legends. Many people either believe, or would like to believe, in the inheritance of acquired characteristics. Until this century it was the dominant theory of heredity among serious biologists too. Darwin himself believed in it, but it was not a part of his theory of evolution so his name is not linked to it in our minds.

If you put the inheritance of acquired characteristics together with the principle of use and disuse, you have what looks like a good recipe for evolutionary improvement. It is this recipe that is commonly labelled the Lamarckian theory of evolution. If successive generations toughen their feet by walking barefoot over rough ground, each generation, so the theory goes, will have a slightly tougher skin than the generation before. Each generation gets an advantage over its predecessor. In the end, babies will be born with already toughened feet (which as a matter of fact they are, though for a different reason as we shall see). If successive generations bask in the tropical sun, they will go browner and browner as, according to the Lamarckian theory, each generation will inherit some of the previous generation's tan. In time they will be born black (again as a matter of fact they are, but not for the Lamarckian reason).

The legendary examples are the blacksmith's arms and the giraffe's neck. In villages where the blacksmith inherited his trade from his father, grandfather and great grandfather before him, he was thought to inherit the well-trained muscles of his ancestors too. Not just inherit them but add to them through his own exercise, and pass on the improvements to his son. Ancestral giraffes with short necks desperately needed to reach high leaves on trees. They strove mightily upwards, thereby stretching neck muscles and bones. Each generation ended up with a slightly longer neck than its predecessor, and it passed its head start on to the next generation. All evolutionary advancement, according to the pure Lamarckian theory, follows this pattern. The animal strives for something that it needs. As a result the parts of the body used in the striving grow larger, or otherwise change in an appropriate direction. The change is inherited by the next generation and so the process goes on. This theory has the advantage that it is cumulative – an essential ingredient of any theory of evolution if it is to fulfil its role in our world view, as we have seen.

The Lamarckian theory seems to have great emotional appeal, for certain types of intellectual as well as for laymen. I was once approached by a colleague, a celebrated Marxist historian and a most cultivated and well-read man. He understood, he said, that the facts all seemed to be against the Lamarckian theory, but was there really no hope that it might be true? I told him that in my opinion there was none, and he accepted this with sincere regret, saying that for ideological reasons he had wanted Lamarckism to be true. It seemed to offer such positive hopes for the betterment of humanity. George Bernard Shaw devoted one of his enormous Prefaces (in *Back to Methuselah*) to a passionate advocacy of the inheritance of acquired characteristics. His case was based not upon biological knowledge, of which he would cheerfully have admitted he had none. It was based upon an emotional loathing of the implications of Darwinism, that 'chapter of accidents':

> it seems simple, because you do not at first realize all that it involves. But when its whole significance dawns on you, your heart sinks into a heap of sand within you. There is a hideous fatalism about it, a ghastly and damnable reduction of beauty and intelligence, of strength and purpose, of honor and aspiration.

Arthur Koestler was another distinguished man of letters who could not abide what he saw as the implications of Darwinism. As Stephen Gould has sardonically but correctly put it, throughout his last six books Koestler conducted 'a campaign against his own misunderstanding of Darwinism'. He sought refuge in an alternative which was never entirely clear to me but which can be interpreted as an obscure version of Lamarckism.

Koestler and Shaw were individualists who thought for themselves. Their eccentric views on evolution have probably not been very influential although I do remember, to my shame, that my own appreciation of Darwinism as a teenager was held back for at least a year by Shaw's bewitching rhetoric in *Back to Methuselah*. The emotional appeal of Lamarckism, and the accompanying emotional hostility to Darwinism, has at times had a more sinister impact, via powerful ideologies used as a substitute for thought. T. D. Lysenko was a second-rate agricultural plant breeder of no distinction other than in the field of politics. His anti-Mendelian fanaticism, and his fervent, dogmatic belief in the inheritance of acquired characteristics, would have been harmlessly ignored in most civilized countries. Unfortunately he happened to live in a country where ideology mattered more than scientific truth. In 1940 he was appointed director of the Institute of Genetics of the Soviet Union, and he became immensely influential. His ignorant views on genetics became the only ones allowed to be taught in Soviet schools for a generation. Incalculable damage was done to Soviet agriculture. Many distinguished Soviet geneticists were banished, exiled or imprisoned. For example, N. I. Vavilov, a geneticist of worldwide reputation, died of malnutrition in a windowless prison cell after a prolonged trial on ludicrously trumped up charges such as 'spying for the British'.

It is not possible to prove that acquired characteristics are never inherited. For the same reason we can never prove that fairies do not exist. All we can say is that no sightings of fairies have ever been confirmed, and that such alleged photographs of them as have been produced are palpable fakes. The same is true of alleged human footprints in Texan dinosaur beds. Any categorical statement I make that fairies don't exist is vulnerable to the possibility that, one day, I may see a gossamer-winged little person at the bottom of my garden. The status of the theory of the inheritance of acquired characteristics is similar. Almost all attempts to demonstrate the effect have simply failed. Of those that have apparently succeeded, some have turned out to be fakes; for example, the notorious injection of Indian ink under the skin of the midwife toad, recounted by Arthur Koestler in his book of that name. The rest have failed to be replicated by other workers. Nevertheless, just as somebody may one day see a fairy at the bottom of the garden when sober and in possession of a camera, somebody may one day prove that acquired characteristics can be inherited.

There is a little more that can be said, however. Some things that have never been reliably seen are, nevertheless, believable insofar as they do not call in question everything else that we know. I have seen no good evidence for the theory that plesiosaurs live today in Loch Ness, but my world view would not be shattered if one were found. I should just be surprised (and delighted), because no plesiosaur fossils are known for the last 60 million years and that seems a long time for a small relict population to survive. But no great scientific principles are at stake. It is simply a matter of fact. On the other hand, science has amassed a good understanding of how the universe ticks, an understanding that works well for an enormous range of phenomena, and certain allegations would be incompatible, or at least very hard to reconcile, with this understanding. For example, this is true of the allegation, sometimes made on spurious biblical grounds, that the universe was created only about 6,000 years ago. This theory is not just unauthenticated. It is incompatible, not only with orthodox biology and geology, but with the physical theory of radioactivity and with cosmology (heavenly bodies more than 6,000 light-years away shouldn't be visible if nothing older than 6,000 years exists; the Milky Way shouldn't be detectable, nor should any of the 100,000 million other galaxies whose existence modern cosmology acknowledges).

There have been times in the history of science when the whole of orthodox science has been rightly thrown over because of a single awkward fact. It would be arrogant to assert that such overthrows will never happen again. But we naturally, and rightly, demand a higher standard of authentication before accepting a fact that would turn a major and successful scientific edifice upside down, than before accepting a fact which, even if surprising, is readily accommodated by existing science. For a plesiosaur in Loch Ness, I would accept the evidence of my own eyes. If I saw a man levitating himself, before rejecting the whole of physics I would suspect that I was the victim of a hallucination or a conjuring trick. There is a continuum, from theories that probably are not true but easily could be, to theories that could only be true at the cost of overthrowing large edifices of successful orthodox science.

Now, where does Lamarckism stand in this continuum? It is usually presented as well over on the 'not true but easily could be' end of the continuum. I want to make a case that, while not in the same class as levitation by the power of prayer, Lamarckism, or more specifically the inheritance of acquired characteristics, is closer to the 'levitation' end of the continuum than to the 'Loch Ness monster' end. The inheritance of acquired characteristics is not one of those things that easily could be true but probably isn't. I shall argue that it could only be true if one of our most cherished and successful principles of embryology is overthrown. Lamarckism therefore needs to be subjected to more than the usual 'Loch Ness monster' level of scepticism. What, then, is this widely accepted and successful embryological principle that would have to be overthrown before Lamarckism could be accepted? That is going to take a little explaining. The explanation will seem like a digression, but its relevance will become clear eventually. And remember that this is all before we start the argument that Lamarckism, even if it *were* true, would still be incapable of explaining the evolution of adaptive complexity.

The field of discourse, then, is embryology. There has traditionally been a deep divide between two different attitudes to the way single cells turn into adult creatures. The official names for them are preformationism and epigenesis, but in their modern forms I shall call them the blueprint theory and the recipe theory. The early preformationists believed that the adult body was *preformed* in the single cell from which it was to develop. One of them imagined that he could see in his microscope a little miniature human – a 'homunculus' – curled up inside a sperm (not egg!). Embryonic development, for him, was simply a process of growth. All the bits of the adult body were already there, preformed. Presumably each male homunculus had his own ultra-miniature sperms in which his own children were coiled up, and each of them contained his coiled up grandchildren . . . Quite apart from this problem of infinite regress, naive preformationism neglects the fact, which was hardly less obvious in the seventeenth century than now, that children inherit attributes from the mother as well as the father. To be fair, there were other preformationists called ovists, rather more numerous than the 'spermists', who believed that the adult was preformed in the egg rather than the sperm. But ovism suffers from the same two problems as spermism.

Modern preformationism does not suffer from either of these problems, but it is still wrong. Modern preformationism – the blueprint theory – holds that the DNA in a fertilized egg is equivalent to a blueprint of the adult body. A blueprint is a scaled-down miniature of the real thing. The real thing – house, car, or whatever it is – is a three-dimensional object, while a blueprint is two-dimensional. You can represent a three-dimensional object such as a building by means of a set of two-dimensional slices: a ground plan of every floor, various elevation views, and so on. This reduction in dimensions is a matter of convenience. Architects could provide builders with matchstick and balsa-wood scale models of buildings in three dimensions, but a set of two-dimensional models on flat paper – blueprints – is easier to carry around in a briefcase, easier to amend, and easier to work from.

A further reduction to *one* dimension is necessary if blueprints are to be stored in a computer's pulse code and, for example, transmitted by telephone line to another part of the country. This is easily done by recoding each two dimensional blueprint as a one-dimensional 'scan'. Television pictures are coded in this way for transmission over the airwaves. Again, the dimensional compression is an essentially trivial coding device. The important point is that there is still one-to-one correspondence between blueprint and building. Each bit of the blueprint corresponds to a matching bit of the building. There is a sense in which the blueprint is a miniaturized 'preformed' building, albeit the miniature may be recoded into fewer dimensions than the building has.

The reason for mentioning the reduction of blueprints to one dimension is, of course, that DNA is a one-dimensional code. Just as it is theoretically possible to transmit a scale model of a building via a one-dimensional telephone line – a digitized set of blueprints – so it is theoretically possible to transmit a scaled-down body via the one-dimensional digital DNA code. This doesn't happen but, if it did, it would be fair to say that modern molecular biology had vindicated the ancient theory of preformationism. Now to consider the other great theory of embryology, epigenesis, the recipe or 'cookery book' theory.

A recipe in a cookery book is not, in any sense, a blueprint for the cake that will finally emerge from the oven. This is not because the recipe is a one-dimensional string of words whereas the cake is a three-dimensional object. As we have seen, it is perfectly possible, by a scanning procedure, to render a scale model into a one-dimensional code. But a recipe is not a scale model, not a description of a finished cake, not in any sense a point-for-point representation. It is a set of *instructions* which, if obeyed in the right order, will result in a cake. A true one-dimensionally coded blueprint of a cake would consist of a series of scans through the cake, as though a skewer were passed through it repeatedly in an orderly sequence across and down the cake. At millimetre intervals the immediate surroundings of the skewer's point would be recorded in code; for instance, the exact coordinates of every currant and crumb would be retrievable from the serial data. There would be strict one-to-one mapping between each bit of the cake and a corresponding bit of the blueprint. Obviously this is nothing like a real recipe. There is no one-to-one mapping between 'bits' of cake and words or letters of the recipe. If the words of the recipe map onto anything, it is not single bits of the finished cake but single steps in the procedure for making a cake.

Now, we don't yet understand everything, or even most things, about how animals develop from fertilized eggs. Nevertheless, the indications are very strong that the genes are much more like a recipe than like a blueprint. Indeed, the recipe analogy is really rather a good one, while the blueprint analogy, although it is often unthinkingly used in elementary textbooks, especially recent ones, is wrong in almost every particular. Embryonic development is a process. It is an orderly sequence of events, like the procedure for making a cake, except that there are millions more steps in the process and different steps are going on simultaneously in many different parts of the 'dish'. Most of the steps involve cell multiplication, generating prodigious numbers of cells, some of which die, others of which join up with each other to form organs, tissues and other many-celled structures. As we saw in an earlier chapter, how a *particular* cell behaves depends not on the genes that it contains – for all the cells in a body contain the same set of genes – but on which subset of the genes is turned on in that cell. In any one place in the developing body, at any one time during development, only a minority of the genes will be switched on. In different parts of the embryo, and at different times during development, other sets of genes will be turned on. Precisely which genes are switched on in any one cell at any one time depends on chemical conditions in that cell. This, in turn, depends upon past conditions in that part of the embryo.

Moreover, the effect that a gene has when it *is* turned on depends upon what there is, in the local part of the embryo, to have an effect on. A gene turned on in cells at the base of the spinal cord in the third week of development will have a totally different effect from the same gene turned on in cells of the shoulder in the sixteenth week of development. So, the effect, if any, that a gene has is *not* a simple property of the gene itself, but is a property of the gene in interaction with the recent history of its local surroundings in the embryo. This makes nonsense of the idea that the genes are anything like a blueprint for a body. The same thing was true, you will remember, of the computer biomorphs.

There is no simple one-to-one mapping, then, between genes and bits of body, any more than there is mapping between words of recipe and crumbs of cake. The genes, taken together, can be seen as a set of instructions for carrying out a process, just as the words of a recipe, taken together, are a set of instructions for carrying out a process. The reader may be left asking how, in that case, it is possible for geneticists to make a living. How is it possible ever to speak of, let alone do research on, a gene 'for' blue eyes, or a gene 'for' colour blindness? Doesn't the very fact that geneticists can study such single-gene effects suggest that there really *is* some sort of one-gene/one-bit-of-body mapping? Doesn't it disprove everything I have been saying about the set of genes being a recipe for developing a body? Well no, it certainly doesn't, and it is important to understand why.

Perhaps the best way to see this is to go back to the recipe analogy. It will be agreed that you can't divide a cake up into its component crumbs and say 'This crumb corresponds to the first word in the recipe, this crumb corresponds to the second word in the recipe', etc. In this sense it will be agreed that the whole recipe maps onto the whole cake. But now suppose we change one word in the recipe; for instance, suppose 'baking-powder' is deleted or is changed to 'yeast'. We bake 100 cakes according to the new version of the recipe, and 100 cakes according to the old version of the recipe. There is a key difference between the two sets of 100 cakes, and this *difference* is due to a one-word difference in the recipes. Although there is no one-to-one mapping from word to crumb of cake, there *is* one-to-one mapping from word *difference* to whole-cake *difference*. 'Baking-powder' does not correspond to any particular part of the cake: its influence affects the rising, and hence the final shape, of the whole cake. If 'baking-powder' is deleted, or replaced by 'flour', the cake will not rise. If it is replaced by 'yeast', the cake will rise but it will taste more like bread. There will be a reliable, identifiable difference between cakes baked according to the original version and the 'mutated' versions of the recipe, even though there is no particular 'bit' of any cake that corresponds to the words in question. This is a good analogy for what happens when a gene mutates.

An even better analogy, because genes exert quantitative effects and mutations change the quantitative magnitude of those effects, would be a change from '350 degrees' to '450 degrees'. Cakes baked according to the 'mutated', higher-temperature version of the recipe will come out different, not just in one part but throughout their substance, from cakes baked according to the original lower-temperature version. But the analogy is still too simple. To simulate the 'baking' of a baby, we should imagine not a single process in a single oven, but a tangle of conveyor belts, passing different parts of the dish through 10 million different miniaturized ovens, in series and in parallel, each oven bringing out a different combination of flavours from 10,000 basic ingredients. The point of the cooking analogy, that the genes are not a blueprint but a recipe for a process, comes over from the complex version of the analogy even more strongly than from the simple one.

It is time to apply this lesson to the question of the inheritance of acquired characteristics. The thing about building something from a blueprint, as opposed to a recipe, is that the process is *reversible*. If you have a house, it is easy to reconstruct its blueprint. Just measure all the dimensions of the house and scale them down. Obviously, if the house were to 'acquire' any characteristics – say an interior wall were knocked down to give an open-plan ground floor – the 'reverse blueprint' would faithfully record the alteration. Just so if the genes were a description of the adult body. If the genes were a blueprint, it would be easy to imagine any characteristic that a body acquired during its lifetime being faithfully transcribed back into the genetic code, and hence passed into the next generation. The blacksmith's son really could inherit the consequences of his father's exercise. It is because the genes are not a blueprint but a recipe that this is not possible. We can no more imagine acquired characteristics being inherited than we can imagine the following. A cake has one slice cut out of it. A description of the alteration is now fed back into the recipe, and the recipe changes in such a way that the next cake baked according the altered recipe comes out of the oven with one slice already neatly missing.

Lamarckians are traditionally fond of calluses, so let us use that example. Our hypothetical bank clerk had soft, pampered hands except for a hard callus on the middle finger of his right hand, his writing finger. If generations of his descendants all write a great deal, the Lamarckian expects that genes controlling the development of skin in that region will be altered in such a way that babies come to be born with the appropriate finger already hardened. If the genes were a blueprint this would be easy. There would be a gene 'for' each square millimetre (or appropriate small unit) of skin. The whole surface of the skin of an adult bank clerk would be 'scanned', the hardness of each square millimetre carefully recorded and fed back into the genes 'for' that particular square millimetre, in particular the appropriate genes in his sperms.

But the genes are not a blueprint. There is no sense in which there is a gene 'for' each square millimetre. There is no sense in which the adult body could be scanned and its description fed back into the genes. The 'coordinates' of a callus could not be 'looked up' in the genetic record and the 'appropriate' genes altered. Embryonic development is a process, in which all working genes participate; a process which, if correctly followed in the forward direction, will result in an adult body; but it is a process that is inherently, by its very nature, irreversible. The inheritance of acquired characteristics not only *doesn't* happen: it *couldn't* happen in any life-form whose embryonic development is epigenetic rather than preformationistic. Any biologist that advocates Lamarckism is, though he may be shocked to hear it, implicitly advocating an atomistic, deterministic, reductionistic embryology. I didn't want to burden the general reader with that little string of pretentious jargon words: I just couldn't resist the irony, for the biologists who come closest to sympathizing with Lamarckism today also happen to be particularly fond of using those same cant words in criticizing others.

This is not to say that, somewhere in the universe, there may not be some alien system of life in which embryology *is* preformationistic; a life-form that really does have 'blueprint genetics', and that really could, therefore, inherit acquired characteristics. All that I have shown so far is that Lamarckism is incompatible with embryology as we know it. My claim at the outset of this chapter was stronger: that, even if acquired characteristics *could* be inherited, the Lamarckian theory would still be incapable of explaining adaptive evolution. This claim is so strong that it is intended to apply to all life-forms, everywhere in the universe. It is based upon two lines of reasoning, one concerned with difficulties over the principle of use and disuse, the other with further problems with the inheritance of acquired characteristics. I shall take these in reverse order.

The problem with acquired characteristics is basically this. It is all very well inheriting acquired characteristics, but not all acquired characteristics are improvements. Indeed, the vast majority of them are injuries. Obviously evolution is not going to proceed in the general direction of adaptive improvement if acquired characteristics are inherited indiscriminately: broken legs and smallpox scars being passed down the generations just as much as hardened feet and suntanned skin. Most of the characteristics that any machine acquires as it gets older tend to be the accumulated ravages of time: it wears out. If they were gathered up by some kind of scanning process and fed into the blueprint for the next generation, successive generations would get more and more decrepit. Instead of starting afresh with a new blueprint, each new generation would begin life encumbered and scarred with the accumulated decay and injuries of previous generations.

This problem is not necessarily insuperable. It is undeniable that some acquired characteristics are improvements, and it is theoretically conceivable that the inheritance mechanism might somehow discriminate the improvements from the injuries. But in wondering how this discrimination might work, we are now led to ask why acquired characteristics sometimes *are* improvements. Why, for instance, do areas of skin that are used, like the soles of a barefoot runner, become thicker and tougher? On the face of it, it would seem more probable that the skin would become thinner: on most machines, parts that are subject to wear and tear become thinner, for the obvious reason that wear removes particles rather than adding them.

The Darwinian, of course, has a ready answer. Skin that is subject to wear and tear gets thicker, because natural selection in the ancestral past has favoured those individuals whose skin happened to respond to wear and tear in this advantageous way. Similarly, natural selection favoured those members of ancestral generations who happened to respond to sunlight by going brown. The Darwinian maintains that the only reason even a minority of acquired characteristics are improvements is that there is an underpinning of past Darwinian selection. In other words, the Lamarckian theory can explain adaptive improvement in evolution only by, as it were, riding on the back of the Darwinian theory. Given that Darwinian selection is there in the background, to ensure that some acquired characteristics are advantageous, and to provide a mechanism for discriminating the advantageous from the disadvantageous acquisitions, the inheritance of acquired characteristics might, conceivably, lead to some evolutionary improvement. But the *improvement*, such as it is, is all due to the Darwinian underpinning. We are forced back to Darwinism to explain the adaptive aspect of evolution.

The same is true of a rather more important class of acquired improvements, those that we lump together under the heading of learning. During the course of its life, an animal becomes more skilled at the business of making its living. The animal learns what is good for it and what is not. Its brain stores a large library of memories about its world, and about which actions tend to lead to desirable consequences and which to undesirable consequences. Much of an animal's behaviour therefore comes under the heading of acquired characteristics, and much of this type of acquisition – 'learning' – really does deserve the title of improvement. If parents could somehow transcribe the wisdom of a lifetime's experience into their genes, so that their offspring were born with a library of vicarious experience built in and ready to be drawn upon, those offspring could begin life one jump ahead. Evolutionary progress might indeed speed up, as learned skills and wisdom would automatically be incorporated into the genes.

But this all presupposes that the changes in behaviour that we call learning are, indeed, improvements. Why *should* they necessarily be improvements? Animals do, as a matter of fact, learn to do what is good for them, rather than what is bad for them, but why? Animals tend to avoid actions that have, in the past, led to pain. But pain is not a substance. Pain is just what the brain treats as pain. It is a fortunate fact that those occurrences that are treated as painful, for instance violent puncturing of the body surface, also happen to be those occurrences that tend to endanger the animal's survival. But we could easily imagine a race of animals that *enjoyed* injury and other occurrences endangering their survival; a race of animals whose brain was so constructed that it took pleasure in injury and felt as painful those stimuli, such as the taste of nutritious food, which augur well for their survival. The reason we do not in fact see such masochistic animals in the world is the Darwinian reason that masochistic ancestors, for obvious reasons, would not have survived to leave descendants that inherited their masochism. We could probably, by artificial selection in padded cages, in pampered conditions where survival is assured by teams of vets and minders, breed a race of hereditary masochists. But in nature such masochists would not survive, and this is the fundamental reason why the changes that we call learning tend to be improvements rather than the reverse. We have again arrived at the conclusion that there must be a Darwinian underpinning to ensure that acquired characteristics are advantageous.

We now turn to the principle of use and disuse. This principle does seem to work rather well for some aspects of acquired improvements. It is a general rule that does not depend upon specific details. The rule says simply, 'Any bit of the body that is frequently used should grow larger; any bit that is not used should become smaller or even wither away altogether'. Since we can expect that useful (and therefore presumably used) bits of body in general will benefit by being enlarged, while useless (and therefore presumably unused) bits might as well not be there at all, the rule does seem to have some general merit. Nevertheless, there is a big problem about the principle of use and disuse. This is that, even if there were no other objection to it, it is much too crude a tool to fashion the exquisitely delicate adaptations that we actually see in animals and plants.

The eye has been a useful example before, so why not again? Think of all the intricately cooperating working parts: the lens with its clear transparency, its colour correction and its correction for spherical distortion; the muscles that can instantly focus the lens on any target from a few inches to infinity; the iris diaphragm or 'stopping down' mechanism, which fine-tunes the aperture of the eye continuously, like a camera with a built-in lightmeter and fast special-purpose computer; the retina with its 125 million colour-coding photocells; the fine network of blood vessels that fuels every part of the machine; the even finer network of nerves – the equivalent of connecting wires and electronic chips. Hold all this fine-chiselled complexity in your mind, and ask yourself whether it could have been put together by the principle of use and disuse. The answer, it seems to me, is an obvious 'no'.

The lens is transparent and corrected against spherical and chromatic aberration. Could this have come about through sheer *use*? Can a lens be washed clear by the volume of photons that pour through it? Will it become a better lens because it is used, because light has passed through it? Of course not. Why on earth should it? Will the cells of the retina sort themselves into three colour-sensitive classes, simply because they are bombarded with light of different colours? Again, why on earth should they? Once the focusing muscles exist, it is true that exercising them will make them grow bigger and stronger; but this will not in itself make images come into sharper focus. The truth is that the principle of use and disuse is incapable of shaping any but the crudest and most unimpressive of adaptations.

Darwinian selection, on the other hand, has no difficulty in explaining every tiny detail. Good eyesight, accurate and true down to pernickity detail, can be a matter of life and death for an animal. A lens, properly focused and corrected against aberration, can make all the difference, for a fast-flying bird like a swift, between catching a fly and smashing into a cliff. A well-modulated iris diaphragm, stopping down rapidly when the sun comes out, can make all the difference between seeing a predator in time to escape and being dazzled for a fatal instant. Any improvement in the effectiveness of an eye, no matter how subtle and no matter how deeply buried in internal tissues, can contribute to the animal's survival and reproductive success, and hence to the propagation of the genes that made the improvement. Therefore Darwinian selection can explain the evolution of the improvement. The Darwinian theory explains the evolution of successful apparatus for survival, as a direct consequence of its very success. The coupling between the explanation, and that which is to be explained, is direct and detailed.

The Lamarckian theory, on the other hand, relies on a loose and crude coupling: the rule that anything that is used a great deal would be better if it were bigger. This amounts to relying on a correlation between the size of an organ and its effectiveness. If there is such a correlation, it is surely an exceedingly weak one. The Darwinian theory in effect relies on a correlation between the *effectiveness* of an organ and its effectiveness: a necessarily perfect correlation! This weakness of the Lamarckian theory does not depend upon detailed facts about the particular forms of life that we see on this planet. It is a general weakness that applies to any kind of adaptive complexity, and I think it must apply to life anywhere in the universe, no matter how alien and strange the details of that life may be.

Our refutation of Lamarckism, then, is a bit devastating. First, its key assumption, that of the inheritance of acquired characteristics, seems to be false in all life-forms that we have studied. Second, it not only is false but it *has* to be false in any life-form that relies upon an epigenetic ('recipe') rather than a preformationistic ('blueprint') kind of embryology, and this includes all life-forms that we have studied. Third, even if the assumptions of the Lamarckian theory were true, the theory is in principle, for two quite separate reasons, incapable of explaining the evolution of serious adaptive complexity, not just on this earth but anywhere in the universe. So, it isn't that Lamarckism is a rival to the Darwinian theory that happens to be wrong. Lamarckism isn't a rival to Darwinism at all. It isn't even a serious *candidate* as an explanation for the evolution of adaptive complexity. It is doomed from the start as a potential rival to Darwinism.

There are a few other theories that have been, and even occasionally still are, advanced as alternatives to Darwinian selection. Once again, I shall show that they are not really serious alternatives at all. I shall show (it is really obvious) that these 'alternatives' – 'neutralism', 'mutationism', and so on – may or may not be responsible for some proportion of observed evolutionary change, but they cannot be responsible for *adaptive* evolutionary change, that is for change in the direction of building up improved devices for survival like eyes, ears, elbow joints, and echo-ranging devices. Of course, large quantities of evolutionary change may be non-adaptive, in which case these alternative theories may well be important in parts of evolution, but only in the boring parts of evolution, not the parts concerned with what is special about life as opposed to non-life. This is especially clear in the case of the neutralist theory of evolution. This has a long history, but it is particularly easy to grasp in its modern, molecular guise in which it has been promoted largely by the great Japanese geneticist Motoo Kimura, whose English prose style, incidentally, would shame many a native speaker.

We have already briefly met the neutralist theory. The idea, you will remember, is that different versions of the same molecule, for instance versions of the haemoglobin molecule differing in their precise amino-acid sequences, are exactly as good as each other. This means that mutations from one alternative version of haemoglobin to another are *neutral* as far as natural selection is concerned. Neutralists believe that the vast majority of evolutionary changes, at the level of molecular genetics, are neutral – *random* with respect to natural selection. The

alternative school of geneticists, called selectionists, believe that
natural selection is a potent force even at the level of detail at every
point along molecular chains.

It is important to distinguish two distinct questions. First is the
question that is relevant to this chapter, whether neutralism is an
alternative to natural selection as an explanation for adaptative
evolution. Second, and quite distinct, is the question whether most of
the evolutionary change that actually occurs is adaptive. Given that
we are talking about an evolutionary change from one form of a
molecule to another, how likely is it that the change came about
through natural selection, and how likely is it that it is a neutral
change which came about through random drift? Over this second
question, a ding-dong battle has raged among molecular geneticists,
first one side gaining the upper hand, then the other. But if we happen
to be focusing our interest on adaptation – the first question – it is all a
storm in a teacup. As far as we are then concerned, a neutral mutation
might as well not exist because neither we, nor natural selection, can
see it. A neutral mutation *isn't* a mutation at all, when we are thinking
about legs and arms and wings and eyes and behaviour! To use the
recipe analogy again, the dish will taste the same even if some words of
the recipe have 'mutated' to a different print font. As far as those of us
who are interested in the final dish are concerned, it is still the same
recipe, whether printed like this or *like this* or **like this.** Molecular
geneticists are like pernickety printers. They care about the actual
form of the words in which recipes are written down. Natural selection
doesn't care, and nor should we when we are talking about the
evolution of adaptation. When we are concerned with other aspects of
evolution, for instance rates of evolution in different lineages, neutral
mutations will be of surpassing interest.

Even the most ardent neutralist is quite happy to agree that natural
selection is responsible for all adaptation. All he is saying is that most
evolutionary change is not adaptation. He may well be right, although
one school of geneticists would not agree. From the sidelines, my own
hope is that the neutralists will win, because this will make it so much
easier to work out evolutionary relationships and rates of evolution.
Everybody on both sides agrees that neutral evolution cannot lead to
adaptive improvement, for the simple reason that neutral evolution is,
by definition, random; and adaptive improvement is, by definition,
non-random. Once again, we have failed to find any alternative to
Darwinian selection, as an explanation for the feature of life that
distinguishes it from non-life, namely adaptive complexity.

We now come to another historical rival to Darwinism – the theory of 'mutationism'. It is hard for us to comprehend now but, in the early years of this century when the phenomenon of mutation was first named, it was regarded not as a necessary part of Darwinian theory but as an *alternative* theory of evolution! There was a school of geneticists called the mutationists, which included such famous names as Hugo de Vries and William Bateson among the early rediscoverers of Mendel's principles of heredity, Wilhelm Johannsen the inventor of the word gene, and Thomas Hunt Morgan the father of the chromosome theory of heredity. De Vries in particular was impressed by the magnitude of the change that mutation can wreak, and he thought that new species always originated from single major mutations. He and Johannsen believed that most variation *within* species was non-genetic. All the mutationists believed that selection had at best a minor weeding-out role to play in evolution. The really creative force was mutation itself. Mendelian genetics was thought of, not as the central plank of Darwinism that it is today, but as anti-thetical to Darwinism.

It is extremely hard for the modern mind to respond to this idea with anything but mirth, but we must beware of repeating the patronizing tone of Bateson himself: 'We go to Darwin for his in-comparable collection of facts [but . . .] for us he speaks no more with philosophical authority. We read his scheme of Evolution as we would those of Lucretius or Lamarck.' And again, 'the transformation of masses of populations by imperceptible steps guided by selection is, as most of us now see, so inapplicable to the fact that we can only marvel both at the want of penetration displayed by the advocates of such a proposition, and at the forensic skill by which it was made to appear acceptable even for a time.' It was above all R. A. Fisher who turned the tables and showed that, far from being antithetical to Darwinism, Mendelian particulate heredity was actually essential to it.

Mutation is necessary for evolution, but how could anybody ever have thought it was sufficient? Evolutionary change is, to a far greater extent than chance alone would expect, *improvement*. The problem with mutation as the sole evolutionary force is simply stated: how on earth is mutation supposed to '*know*' what will be good for the animal and what will not? Of all possible changes that might occur to an existing complex mechanism like an organ, the vast majority will make it worse. Only a tiny minority of changes will make it better. Anybody who wants to argue that mutation, without selection, is the driving force of evolution, must explain how it comes about that mutations tend to be for the better. By what mysterious, built-in wisdom does the body choose to mutate in the direction of getting better, rather than getting worse? You will observe that this is really the same question, in another guise, as we posed for Lamarckism. The mutationists, needless to say, never answered it. The odd thing is that the question hardly seems to have occurred to them.

Nowadays, and unfairly, this seems all the more absurd to us because we are brought up to believe that mutations are 'random'. If mutations are random, then, by definition, they cannot be biased towards improvement. But the mutationist school did not, of course, regard mutations as random. They thought that the body had a built-in tendency to change in certain directions rather than others, though they left open the question of how the body 'knew' what changes would be good for it in the future. While we write this off as mystical nonsense, it is important for us to be clear exactly what we mean when we say that mutation is random. There is randomness and randomness, and many people confuse different meanings of the word. There are, in truth, many respects in which mutation is not random. All I would insist on is that these respects do *not* include anything equivalent to anticipation of what would make life better for the animal. And something equivalent to anticipation would indeed be needed if mutation, without selection, were to be used to explain evolution. It is instructive to look a little further at the senses in which mutation is, and is not, random.

The first respect in which mutation is non-random is this. Mutations are caused by definite physical events; they don't just spontaneously happen. They are induced by so-called 'mutagens' (dangerous because they often start cancers): X-rays, cosmic rays, radioactive substances, various chemicals, and even other genes called 'mutator genes'. Second, not all genes in any species are equally likely to mutate. Every locus on the chromosomes has its own characteristic *mutation rate.* For instance, the rate at which mutation creates the gene for the disease Huntington's chorea (similar to St Vitus's Dance), which kills people in early middle age, is about 1 in 200,000. The corresponding rate for achondroplasia (the familiar dwarf syndrome, characteristic of basset hounds and dachsunds, in which the arms and legs are too short for the body) is about 10 times as high. These rates are measured under normal conditions. If mutagens like X-rays are present, all normal mutation rates are boosted. Some parts of the chromosome are so-called 'hot spots' with a high turnover of genes, a locally very high mutation rate.

Third, at each locus on the chromosomes, whether it is a hot spot or not, mutations in certain directions can be more likely than mutations in the reverse direction. This gives rise to the phenomenon known as 'mutation pressure' which can have evolutionary consequences. Even if, for instance, two forms of the haemoglobin molecule, Form 1 and Form 2, are selectively neutral in the sense that both are equally good at carrying oxygen in the blood, it could still be that mutations from 1 to 2 are commoner than reverse mutations from 2 to 1. In this case, mutation pressure will tend to make Form 2 commoner than Form 1. Mutation pressure is said to be zero at a given chromosomal locus, if the forward mutation rate at that locus is exactly balanced by the backward mutation rate.

We can now see that the question of whether mutation is really random is not a trivial question. Its answer depends on what we understand random to mean. If you take 'random mutation' to mean that mutations are not influenced by external events, then X-rays disprove the contention that mutation is random. If you think 'random mutation' implies that all genes are equally likely to mutate, then hot spots show that mutation is not random. If you think 'random mutation' implies that at all chromosomal loci the mutation pressure is zero, then once again mutation is not random. It is only if you define 'random' as meaning 'no general bias towards bodily improvement' that mutation is truly random. All three of the kinds of real non-randomness we have considered are powerless to move evolution in the direction of adaptive improvement as opposed to any other (functionally) 'random' direction. There is a fourth kind of non-randomness, of which this is also true but slightly less obviously so. It will be necessary to spend a little time on this because it is still muddling even some modern biologists.

There are people for whom 'random' would have the following meaning, in my opinion a rather bizarre meaning. I quote from two opponents (P. Saunders and M-W. Ho) of Darwinism, on their conception of what Darwinians believe about 'random mutation': 'The neo-Darwinian concept of random variation carries with it the major fallacy that everything conceivable is possible'. '*All* changes are held to be possible and all *equally likely*' (my emphasis). Far from holding this belief, I don't see how you would begin to set about making such a belief even *meaningful*! What could it possibly mean to hold that 'all' changes are equally likely? *All* changes? In order for two or more things to be 'equally likely', it is necessary that those things should be definable as discrete events. For instance, we can say 'Heads and Tails are equally likely', because Heads and Tails are two discrete events. But 'all possible' changes to an animal's body are not discrete events of this type. Take the two possible events: 'Cow's tail lengthens by one inch'; and 'Cow's tail lengthens by two inches'. Are these two separate events, and therefore 'equally likely'? Or are they just quantitative variants of the same event?

It is clear that a kind of caricature of a Darwinian has been set up, whose notion of randomness is an absurd, if not actually meaningless, extreme. It took me a while to understand this caricature, for it was so foreign to the way of thinking of the Darwinians that I know. But I think I do now understand it, and I shall try to explain it, as I think it helps us to understand what lies behind quite a lot of alleged opposition to Darwinism.

Variation and selection work together to produce evolution. The Darwinian says that variation is random in the sense that it is not directed towards improvement, and that the tendency towards improvement in evolution comes from selection. We can imagine a kind of continuum of evolutionary doctrines, with Darwinism at one end and Mutationism at the other. The extreme mutationist believes that selection plays no role in evolution. The direction of evolution is determined by the direction of the mutations that are offered. For instance, suppose we take the enlargement of the human brain that has occurred during the last few million years of our evolution. The Darwinian says that the variation that was offered up by mutation for selection included some individuals with smaller brains and some individuals with larger brains; selection favoured the latter. The mutationist says that there was a bias in favour of larger brains in the variation that was offered up by mutation; there was no selection (or no need for selection) after variation was offered up; brains got bigger because mutational change was biased in the direction of bigger brains. To summarize the point: in evolution there was a bias in favour of larger brains; this bias could have come from selection alone (the Darwinian view) or from mutation alone (the mutationist view); we can imagine a continuum between these two points of view, almost a kind of trade-off between the two possible sources of evolutionary bias. A middle view would be that there was *some* bias in mutations towards enlargement of the brain, and that selection increased the bias in the population that survived.

The element of caricature comes in the portrayal of what the Darwinian means when he says that there is no bias in the mutational variation that is offered up for selection. To me, as a real-life Darwinian, it means only that mutation is not systematically biased in the direction of adaptive improvement. But to the larger-than-life caricature of a Darwinian, it means that all conceivable changes are 'equally likely'. Setting aside the logical impossibility of such a belief, already noted, the caricature of a Darwinian is thought to believe that the body is infinitely malleable clay, ready to be shaped by all-powerful selection into any form that selection might favour. It is important to understand the difference between the real-life Darwinian and the caricature. We shall do so in terms of a particular example, the difference between the flight techniques of bats and of angels.

Angels are always portrayed as having wings sprouting from their backs, leaving their arms unencumbered by feathers. Bats, on the other hand, along with birds and pterodactyls, have no independent arms. Their ancestral arms have become incorporated into wings, and cannot be used, or can only be used very clumsily, for other purposes such as picking up food. We shall now listen in on a conversation between a real-life Darwinian and an extreme caricature of a Darwinian.

Real-life.   I wonder why bats didn't evolve wings like angels. You'd think that they could use a free pair of arms. Mice use their arms all the time for picking up food and nibbling it, and bats look terribly clumsy on the ground without arms. I suppose one answer might be that mutation never provided the necessary variation. There just never were any mutant ancestral bats that had wing buds sticking out of the middle of their backs.

Caricature.   Nonsense. Selection is everything. If bats haven't got wings like angels, this can only mean that selection didn't favour wings like angels. There certainly were mutant bats with wing buds sticking out of the middle of their backs, but selection just didn't favour them.

Real-life.   Well, I quite agree that selection might not have favoured them if they *had* sprouted. For one thing they would have increased the weight of the whole animal, and surplus weight is a luxury no aircraft can afford. But surely you don't think that, *whatever* selection might in principle favour, mutation will always come up with the necessary variation?

Caricature.   Certainly I do. Selection is everything. Mutation is random.

Real-life.   Well yes, mutation is random, but this only means that it can't see into the future and plan what would be good for the animal. It doesn't mean that absolutely *anything* is possible. Why do you think that no animal breathes fire out of its nostrils like a dragon, for instance? Wouldn't it be useful for catching and cooking prey?

Caricature.   That's easy. Selection is everything. Animals don't breathe fire, because it wouldn't pay them to do so. Fire-breathing mutants were eliminated by natural selection, perhaps because making fire was too costly in energy.

Real-life.   I don't believe there ever were fire-breathing mutants. And if there had been, presumably they would have been in grave danger of burning themselves!

Caricature.   Nonsense. If that was the only problem, selection would have favoured the evolution of asbestos-lined nostrils.

Real-life.   I don't believe any mutation ever produced asbestos-lined nostrils. I don't believe mutant animals could secrete asbestos, any more than mutant cows could jump over the moon.

Caricature.   Any moon-jumping mutant cow would promptly be eliminated by natural selection. There's no oxygen up there you know.

Real-life.   I'm surprised you don't postulate mutant cows with genetically determined space-suits and oxygen masks.

Caricature.   Good point! Well, I suppose the real explanation must be that it just wouldn't pay cows to jump over the moon. And we mustn't forget the energetic cost of reaching escape velocity.

Real-life.   This is absurd.

Caricature.   You are obviously not a true Darwinian. What are you, some kind of crypto-mutationist deviationist?

Real-life.   If you think that, you should meet a real mutationist.

Mutationist. Is this a Darwinian in-group argument, or can anyone join in? The trouble with both of you is that you give far too much prominence to selection. All that selection can do is weed out gross deformities and freaks. It can't produce really constructive evolution. Go back to the evolution of bat wings. What really happened is that in an ancient population of ground-dwelling animals, mutations started turning up with elongated fingers and webs of skin between. As the generations went by, these mutations became more and more frequent until, eventually, the whole population had wings. It had nothing to do with selection. There was just this built-in tendency in the ancestral bat constitution to evolve wings.

Real-life &   Rank mysticism! Get back in the last century where you
Caricature   belong.
(in unison).

I hope I am not being presumptuous when I take it that the reader's sympathies are with neither the Mutationist nor with the caricature of a Darwinian. I assume that the reader agrees with the real-life Darwinian, as, of course, do I. The caricature does not really exist.

Unfortunately some people *think* he exists, and think that, since they disagree with him, they are disagreeing with Darwinism itself. There is a school of biologists who have taken to saying something like the following. The trouble with Darwinism is that it neglects the constraints imposed by embryology. Darwinians (this is where the caricature comes in) think that, if selection would favour some conceivable evolutionary change, then the necessary mutational variation will turn out to be available. Mutational change in any direction is equally likely: selection provides the only bias.

But any real-life Darwinian would acknowledge that, although any gene on any chromosome may mutate at any time, the consequences of mutation on *bodies* are severely limited by the processes of embryology. If I ever doubted this (I didn't), my doubts would have been dispelled by my biomorph computer simulations. You can't just postulate a mutation 'for' sprouting wings in the middle of the back. Wings, or anything else, can only evolve if the process of development allows them to. Nothing magically 'sprouts'. It has to be made by the processes of embryonic development. Only a minority of the things that conceivably could evolve are actually permitted by the status quo of existing developmental processes. Because of the way arms develop, it is possible for mutations to increase the length of fingers and cause webs of skin to grow between them. But there may not be anything in the embryology of backs that lends itself to 'sprouting' angel wings. Genes can mutate till they are blue in the face, but no mammal will ever sprout wings like an angel unless mammalian embryological processes are susceptible to this kind of change.

Now as long as we don't know all the ins and outs of how embryos develop, there is room for disagreement over how likely it is that particular imagined mutations have or have not ever existed. It might turn out, for instance, that there is nothing in mammalian embryology to forbid angel wings, and that the caricature Darwinian was right, in this *particular* case, to suggest that angel wing-buds arose but were not favoured by selection. Or it might turn out that when we know more about embryology we shall see that angel wings were always a nonstarter, and that therefore selection never had a chance to favour them. There is a third possibility, which we should list for completeness, that embryology never allowed the possibility of angel wings and that selection would never have favoured them even if it had. But what we must insist on is that we can't afford to ignore the constraints on evolution that embryology imposes. All serious Darwinians would agree about this, yet some people portray Darwinians as denying it. It turns out that people who make a lot of noise about 'developmental constraints' as an alleged anti-Darwinian force are confusing Darwinism with the caricature of Darwinism that I parodied above.

This all began with a discussion over what is meant when we say that mutation is 'random'. I listed three respects in which mutation is not random: it is induced by X-rays, etc.; mutation rates are different for different genes; and forward mutation rates do not have to equal backward mutation rates. To this, we have now added a fourth respect in which mutation is not random. Mutation is non-random in the sense that it can only make alterations to _existing_ processes of embryonic development. It cannot conjure, out of thin air, any conceivable change that selection might favour. The variation that is available for selection is constrained by the processes of embryology, as they actually exist.

There is a fifth respect in which mutation _might_ have been non-random. We can imagine (just) a form of mutation that was systematically biased in the direction of improving the animal's adaptedness to its life. But although we can imagine it, nobody has ever come close to suggesting any means by which this bias could come about. It is only in this fifth respect, the 'mutationist' respect, that the true, real-life Darwinian insists that mutation is random. Mutation is not systematically biased in the direction of adaptive improvement, and no mechanism is known (to put the point mildly) that could guide mutation in directions that are non-random in this fifth sense. Mutation is random with respect to adaptive advantage, although it is non-random in all sorts of other respects. It is selection, and only selection, that directs evolution in directions that are non-random with respect to advantage. Mutationism is not just wrong in fact. It never could have been right. It is not in principle capable of explaining the evolution of improvement. Mutationism belongs with Lamarckism, not as a disproved rival to Darwinism but as no rival at all.

The same is true of my next alleged rival to Darwinian selection, championed by the Cambridge geneticist Gabriel Dover under the odd name 'molecular drive' (since everything is made of molecules it is not obvious why Dover's hypothetical process should deserve the name _molecular_ drive any more than any other evolutionary process; it reminds me of a man I knew who complained of a gastric stomach, and worked things out using his mental brain). Motoo Kimura and the other proponents of the neutralist theory of evolution do not, as we saw, make any false claims for their theory. They have no illusions about random drift being a rival to natural selection as an explanation for adaptive evolution. They recognize that only natural selection can drive evolution in adaptive directions. Their claim is simply that a lot of evolutionary change (as a molecular geneticist sees evolutionary change) is not adaptive. Dover makes no such modest claims for his theory. He thinks that he can explain *all* of evolution without natural selection, although he generously concedes that there may be *some* truth in natural selection as well!

Throughout this book, our first recourse when considering such matters has been to the example of the eye, although it has, of course, been only a representative of the large set of organs that are too complex and well designed to have come about by chance. Only natural selection, I have repeatedly argued, even comes close to offering a plausible explanation for the human eye and comparable organs of extreme perfection and complexity. Fortunately, Dover has explicitly risen to the challenge, and has offered his own explanation of the evolution of the eye. Assume, he says, that 1,000 steps of evolution are needed to evolve the eye from nothing. This means that a sequence of 1,000 genetic changes were needed to transform a bare patch of skin into an eye. This seems to me to be an acceptable assumption for the sake of argument. In the terms of Biomorph Land, it means that the bare-skin animal is 1,000 genetic steps distant from the eyed animal.

Now, how do we account for the fact that just the right set of 1,000 steps were taken to result in the eye as we know it? Natural selection's explanation is well known. Reducing it to its simplest form, at each one of the 1,000 steps, mutation offered a number of alternatives, only one of which was favoured because it aided survival. The 1,000 steps of evolution represent 1,000 successive choice points, at each of which most of the alternatives led to death. The adaptive complexity of the modern eye is the end-product of 1,000 successful unconscious 'choices'. The species has followed a particular path through the labyrinth of all possibilities. There were 1,000 branch-points along the path, and at each one the survivors were the ones that happened to take the turning that led to improved eyesight. The wayside is littered with the dead bodies of the failures who took the wrong turning at each one of the 1,000 successive choice points. The eye that we know is the end-product of a sequence of 1,000 successful selective 'choices'.

That was (one way of expressing) natural selection's explanation of the evolution of the eye in 1,000 steps. Now, what of Dover's explanation? Basically, he argues that it wouldn't have mattered which choice the lineage took at each step: it would retrospectively have found a use for the organ that resulted. Each step that the lineage took, according to him, was a random step. At Step 1, for example, a random mutation spread through the species. Since the newly evolved characteristic was functionally random, it didn't aid the animals' survival. So the species searched the world for a new place or new way of life in which they could use this new random feature that had been imposed upon their bodies. Having found a piece of environment that suited the random part of their bodies, they lived there for a while, until a new random mutation arose and spread through the species. Now the species had to scour the world for a new place or way of life where they could live with their new random bit. When they found it, Step 2 was completed. Now the Step 3 random mutation spread through the species, and so on for 1,000 steps, at the end of which the eye as we know it had been formed. Dover points out that the human eye happens to use what we call 'visible' light rather than infrared. But if random processes had happened to impose an infrared sensitive eye upon us, we would, doubtless, have made the most of it, and found a way of life that exploited infrared rays to the full.

At first glance this idea has a certain seductive plausibility, but only at a very brief first glance. The seductiveness comes from the neatly symmetrical way in which natural selection is turned on its head. Natural selection, in its most simple form, assumes that the environment is imposed upon the species, and those genetic variants best fitted to that environment survive. The environment is imposed, and the species evolves to fit it. Dover's theory turns this on its head. It is the nature of the species that is 'imposed', in this case by the vicissitudes of mutation, and other internal genetic forces in which he has a special interest. The species then locates that member of the set of all environments that best fits its imposed nature.

But the seductiveness of the symmetry is superficial indeed. The wondrous cloud-cuckooism of Dover's idea is displayed in all its glory the moment we begin to think in terms of numbers. The essence of his scheme is that, at each of the 1,000 steps, it didn't matter which way the species turned. Each new innovation that the species came up with was functionally random, and the species then found an environment to suit it. The implication is that the species *would have found* a suitable environment, no matter which branch it had taken at every fork in the way. Now just think how many possible environments this lets us in for postulating. There were 1,000 branch points. If each branch point was a mere bifurcation (as opposed to a 3-way or 18-way branch, a conservative assumption), the total number of livable environments that must, in principle, exist, in order to allow Dover's scheme to work, is 2 to the power 1,000 (the first branch gives two pathways; then each of those branches into two, making four in all; then each of these branches, giving 8; then 16, 32, 64, . . . all the way to $2^{1,000}$). This number may be written as a 1 with 301 noughts after it. It is far far greater than the total number of atoms in the entire universe.

Dover's alleged rival to natural selection could never work, not just never in a million years but never in a million times longer than the universe has existed, never in a million universes each lasting a million times as long again. Notice that this conclusion is not materially affected if we change Dover's initial assumption that 1,000 steps would be needed to make an eye. If we reduce it to only 100 steps, which is probably an underestimate, we still conclude that the set of possible livable environments that must be waiting in the wings, as it were, to cope with whatever random steps the lineage might take, is more than a million million million million million. This is a smaller number than the previous one, but it still means that the vast majority of Dover's 'environments' waiting in the wings would each have to be made of less than a single atom.

It is worth explaining why the theory of natural selection is not susceptible to a symmetrical destruction by a version of the large-numbers argument'. In Chapter 3 we thought of all real and conceivable animals as sitting in a gigantic hyperspace. We are doing a similar thing here, but simplifying it by considering evolutionary branch points as 2-way, rather than 18-way branches. So the set of all possible animals that might have evolved in 1,000 evolutionary steps are perched on a gigantic tree, which branches and branches so that the total number of final twigs is 1 followed by 301 noughts. Any actual evolutionary history can be represented as a particular pathway through this hypothetical tree. Of all conceivable evolutionary pathways, only a minority actually ever happened. We can think of most of this 'tree of all possible animals' as hidden in the darkness of non-existence. Here and there, a few trajectories through the darkened tree are illuminated. These are the evolutionary pathways that actually happened, and, numerous as these illuminated branches are, they are still an infinitesimal minority of the set of all branches. Natural selection is a process that is capable of picking its way through the tree of all conceivable animals, and finding just that minority of pathways that are viable. The theory of natural selection cannot be attacked by the kind of large-numbers argument with which I attacked Dover's theory, because it is of the essence of the theory of natural selection that it is continually cutting down most of the branches of the tree. That is precisely what natural selection does. It picks its way, step by step, through the tree of all conceivable animals, avoiding the almost infinitely large majority of sterile branches – animals with eyes in the soles of their feet, etc. – which the Dover theory is obliged, by the nature of its peculiar inverted logic, to countenance.

We have dealt with all the alleged alternatives to the theory of natural selection except the oldest one. This is the theory that life was created, or its evolution master-minded, by a conscious designer. It would obviously be unfairly easy to demolish some particular version of this theory such as the one (or it may be two) spelled out in Genesis. Nearly all peoples have developed their own creation myth, and the Genesis story is just the one that happened to have been adopted by one particular tribe of Middle Eastern herders. It has no more special status than the belief of a particular West African tribe that the world was created from the excrement of ants. All these myths have in common that they depend upon the deliberate intentions of some kind of supernatural being.

At first sight there is an important distinction to be made between what might be called 'instantaneous creation' and 'guided evolution'. Modern theologians of any sophistication have given up believing in instantaneous creation. The evidence for some sort of evolution has become too overwhelming. But many theologians who call themselves evolutionists, for instance the Bishop of Birmingham quoted in Chapter 2, smuggle God in by the back door: they allow him some sort of supervisory role over the course that evolution has taken, either influencing key moments in evolutionary history (especially, of course, *human* evolutionary history), or even meddling more comprehensively in the day-to-day events that add up to evolutionary change.

We cannot disprove beliefs like these, especially if it is assumed that God took care that his interventions always closely mimicked what would be expected from evolution by natural selection. All that we can say about such beliefs is, firstly, that they are superfluous and, secondly, that they *assume* the existence of the main thing we want to *explain*, namely organized complexity. The one thing that makes evolution such a neat theory is that it explains how organized complexity can arise out of primeval simplicity.

If we want to postulate a deity capable of engineering all the organized complexity in the world, either instantaneously or by guiding evolution, that deity must already have been vastly complex in the first place. The creationist, whether a naive Bible-thumper or an educated bishop, simply *postulates* an already existing being of prodigious intelligence and complexity. If we are going to allow ourselves the luxury of postulating organized complexity without offering an explanation, we might as well make a job of it and simply postulate the existence of life as we know it! In short, divine creation, whether instantaneous or in the form of guided evolution, joins the list of other theories we have considered in this chapter. All give some superficial appearance of being alternatives to Darwinism, whose merits might be tested by an appeal to evidence. All turn out, on closer inspection, not to be rivals of Darwinism at all. The theory of evolution by cumulative natural selection is the only theory we know of that is in principle *capable* of explaining the existence of organized complexity. Even if the evidence did not favour it, it would *still* be the best theory available! In fact the evidence does favour it. But that is another story.

Let us hear the conclusion of the whole matter. The essence of life is statistical improbability on a colossal scale. Whatever is the explanation for life, therefore, it cannot be chance. The true explanation for the existence of life must embody the very antithesis of chance. The antithesis of chance is nonrandom survival, properly understood. Nonrandom survival, improperly understood, is not the antithesis of chance, it is chance itself. There is a continuum connecting these two extremes, and it is the continuum from single-step selection to cumulative selection. Single-step selection is just another way of saying pure chance. This is what I mean by nonrandom survival improperly understood. *Cumulative selection*, by slow and gradual degrees, is the explanation, the only workable explanation that has ever been proposed, for the existence of life's complex design.

The whole book has been dominated by the idea of chance, by the astronomically long odds against the spontaneous arising of order, complexity and apparent design. We have sought a way of taming chance, of drawing its fangs. 'Untamed chance', pure, naked chance, means ordered design springing into existence from nothing, in a single leap. It would be untamed chance if once there was no eye, and then, suddenly, in the twinkling of a generation, an eye appeared, fully fashioned, perfect and whole. This is possible, but the odds against it will keep us busy writing noughts till the end of time. The same applies to the odds against the spontaneous existence of any fully fashioned, perfect and whole beings, including – I see no way of avoiding the conclusion – deities.

To 'tame' chance means to break down the very improbable into less improbable small components arranged in series. No matter how improbable it is that an X could have arisen from a Y in a single step, it is always possible to conceive of a series of infinitesimally graded intermediates between them. However improbable a large-scale change may be, smaller changes are less improbable. And provided we postulate a sufficiently large series of sufficiently finely graded intermediates, we shall be able to derive anything from anything else, without invoking astronomical improbabilities. We are allowed to do this only if there has been sufficient time to fit all the intermediates in. And also only if there is a mechanism for guiding each step in some particular direction, otherwise the sequence of steps will career off in an endless random walk.

It is the contention of the Darwinian world-view that both these provisos are met, and that slow, gradual, cumulative natural selection is the ultimate explanation for our existence. If there are versions of the evolution theory that deny slow gradualism, and deny the central role of natural selection, they may be true in particular cases. But they cannot be the whole truth, for they deny the very heart of the evolution theory, which gives it the power to dissolve astronomical improbabilities and explain prodigies of apparent miracle.

BIBLIOGRAPHY

1. Alberts, B., Bray, D., Lewis, J., Raff, M., Roberts, K. & Watson, J. D. (1983) *Molecular Biology of the Cell*. New York: Garland.

2. Anderson, D. M. (1981) Role of interfacial water and water in thin films in the origin of life. In J. Billingham (ed.) *Life in the Universe*. Cambridge, Mass: MIT Press.

3. Andersson, M. (1982) Female choice selects for extreme tail length in a widow bird. *Nature*, 299: 818–20.

4. Arnold, S. J. (1983) Sexual selection: the interface of theory and empiricism. In P. P. G. Bateson (ed.), *Mate Choice*, pp. 67–107. Cambridge: Cambridge University Press.

5. Asimov, I. (1957) *Only a Trillion*. London: Abelard-Schuman.

6. Asimov, I. (1980) *Extraterrestrial Civilizations*. London: Pan.

7. Asimov, I. (1981) *In the Beginning*. London: New English Library.

8. Atkins, P. W. (1981) *The Creation*. Oxford: W. H. Freeman.

9. Attenborough, D. (1980) *Life on Earth*. London: Reader's Digest, Collins & BBC.

10. Barker, E. (1985) Let there be light: scientific creationism in the twentieth century. In J. R. Durant (ed.) *Darwinism and Divinity*, pp. 189–204. Oxford: Basil Blackwell.

11. Bowler, P. J. (1984) *Evolution: the history of an idea*. Berkeley: University of California Press.

12. Bowles, K. L. (1977) *Problem-Solving using Pascal*. Berlin: Springer-Verlag.

13. Cairns-Smith, A. G. (1982) *Genetic Takeover.* Cambridge: Cambridge University Press.

14. Cairns-Smith, A. G. (1985) *Seven Clues to the Origin of Life.* Cambridge: Cambridge University Press.

15. Cavalli-Sforza, L. & Feldman, M. (1981) *Cultural Transmission and Evolution.* Princeton, N. J.: Princeton University Press.

16. Cott, H. B. (1940) *Adaptive Coloration in Animals.* London: Methuen.

17. Crick, F. (1981) *Life Itself.* London: Macdonald.

18. Darwin, C. (1859) *The Origin of Species.* Reprinted. London: Penguin.

19. Dawkins, M. S. (1986) *Unravelling Animal Behaviour.* London: Longman.

20. Dawkins, R. (1976) *The Selfish Gene.* Oxford: Oxford University Press.

21. Dawkins, R. (1982) *The Extended Phenotype.* Oxford: Oxford University Press.

22. Dawkins, R. (1982) Universal Darwinism. In D. S. Bendall (ed.) *Evolution from Molecules to Men*, pp. 403–25. Cambridge: Cambridge University Press.

23. Dawkins, R. & Krebs, J. R. (1979) Arms races between and within species. *Proceedings of the Royal Society of London*, B, 205: 489–511.

24. Douglas, A. M. (1986) Tigers in Western Australia. *New Scientist*, 110 (1505): 44–7.

25. Dover, G. A. (1984) Improbable adaptations and Maynard Smith's dilemma. Unpublished manuscript, and two public lectures, Oxford, 1984.

26. Dyson, F. (1985) *Origins of Life.* Cambridge: Cambridge University Press.

27. Eigen, M., Gardiner, W., Schuster, P., & Winkler-Oswatitsch. (1981) The origin of genetic information. *Scientific American*, 244 (4): 88–118.

28. Eisner, T. (1982) Spray aiming in bombardier beetles: jet deflection by the Coander Effect. *Science*, 215: 83–5.

29. Eldredge, N. (1985) *Time Frames: the rethinking of Darwinian evolution and the theory of punctuated equilibria.* New York: Simon & Schuster (includes reprinting of original Eldredge & Gould paper).

30. Eldredge, N. (1985) *Unfinished Synthesis: biological hierarchies and modern evolutionary thought.* New York: Oxford University Press.

31. Fisher, R. A. (1930) *The Genetical Theory of Natural Selection.* Oxford: Clarendon Press. 2nd edn paperback. New York: Dover Publications.

32. Gillespie, N. C. (1979) *Charles Darwin and the Problem of Creation.* Chicago: University of Chicago Press.

33. Goldschmidt, R. B. (1945) Mimetic polymorphism, a controversial chapter of Darwinism. *Quarterly Review of Biology*, 20: 147–64 and 205–30.

34. Gould, S. J. (1980) *The Panda's Thumb.* New York: W. W. Norton.

35. Gould, S. J. (1980) Is a new and general theory of evolution emerging? *Paleobiology*, 6: 119–30.

36. Gould, S. J. (1982) The meaning of punctuated equilibrium, and its role in validating a hierarchical approach to macroevolution. In R. Milkman (ed.) *Perspectives on Evolution*, pp. 83–104. Sunderland, Mass: Sinauer.

37. Gribbin, J. & Cherfas, J. (1982) *The Monkey Puzzle.* London: Bodley Head.

38. Griffin, D. R. (1958) *Listening in the Dark.* New Haven: Yale University Press.

39. Hallam, A. (1973) *A Revolution in the Earth Sciences.* Oxford: Oxford University Press.

40. Hamilton, W. D. & Zuk, M. (1982) Heritable true fitness and bright birds: a role for parasites? *Science*, 218: 384–7.

41. Hitching, F. (1982) *The Neck of the Giraffe, or Where Darwin Went Wrong.* London: Pan.

42. Ho, M–W. & Saunders, P. (1984) *Beyond Neo-Darwinism.* London: Academic Press.

43. Hoyle, F. & Wickramasinghe, N. C. (1981) *Evolution from Space.* London: J. M. Dent.

44. Hull, D. L. (1973) *Darwin and his Critics.* Chicago: Chicago University Press.

45. Jacob, F. (1982) *The Possible and the Actual.* New York: Pantheon.

46. Jerison, H. J. (1985) Issues in brain evolution. In R. Dawkins & M. Ridley (eds) *Oxford Surveys in Evolutionary Biology*, 2: 102–34.

47. Kimura, M. (1982) *The Neutral Theory of Molecular Evolution.* Cambridge: Cambridge University Press.

48. Kitcher. P. (1983) *Abusing Science: the case against creationism.* Milton Keynes: Open University Press.

49. Land, M. F. (1980) Optics and vision in invertebrates. In H. Autrum (ed.) *Handbook of Sensory Physiology*, pp. 471–592. Berlin: Springer.

50. Lande, R. (1980) Sexual dimorphism, sexual selection, and adaptation in polygenic characters. *Evolution*, 34: 292–305.

51. Lande, R. (1981) Models of speciation by sexual selection of polygenic traits. *Proceedings of the National Academy of Sciences*, 78: 3721–5.

52. Leigh, E. G. (1977) How does selection reconcile individual advantage with the good of the group? *Proceedings of the National Academy of Sciences*, 74: 4542–6.

53. Lewontin, R. C. & Levins, R. (1976) The Problem of Lysenkoism. In H. & S. Rose (eds) *The Radicalization of Science*. London: Macmillan.

54. Mackie, J. L. (1982) *The Miracle of Theism*. Oxford: Clarendon Press.

55. Margulis, L. (1981) *Symbiosis in Cell Evolution*. San Francisco: W. H. Freeman.

56. Maynard Smith, J. (1983) Current controversies in evolutionary biology. In M. Grene (ed.) *Dimensions of Darwinism*, pp. 273–86. Cambridge: Cambridge University Press.

57. Maynard Smith, J. (1986) *The Problems of Biology*. Oxford: Oxford University Press.

58. Maynard Smith, J. *et al.* (1985) Developmental constraints and evolution. *Quarterly Review of Biology*, 60: 265–87.

59. Mayr, E. (1963) *Animal Species and Evolution*. Cambridge, Mass: Harvard University Press.

60. Mayr, E. (1969) *Principles of Systematic Zoology*. New York: McGraw-Hill.

61. Mayr, E. (1982) *The Growth of Biological Thought*. Cambridge, Mass: Harvard University Press.

62. Monod, J. (1972) *Chance and Necessity*. London: Fontana.

63. Montefiore, H. (1985) *The Probability of God*. London: SCM Press.

64. Morrison, P., Morrison, P., Eames, C. & Eames, R. (1982) *Powers of Ten*. New York: Scientific American.

65. Nagel, T. (1974) What is it like to be a bat? *Philosophical Review*, reprinted in D. R. Hofstadter & D. C. Dennett (eds). *The Mind's I*, pp. 391–403, Brighton: Harvester Press.

66. Nelkin, D. (1976) The science textbook controversies. *Scientific American* 234 (4): 33–9.

67. Nelson, G. & Platnick, N. I. (1984) Systematics and evolution. In M–W Ho & P. Saunders (eds), *Beyond Neo-Darwinism*. London: Academic Press.

68. O'Donald, P. (1983) Sexual selection by female choice. In P. P. G. Bateson (ed.) *Mate Choice*, pp. 53–66. Cambridge: Cambridge University Press.

69. Orgel, L. E. (1973) *The Origins of Life*. New York: Wiley.

70. Orgel, L. E. (1979) Selection in vitro. *Proceedings of the Royal Society of London*, B, 205: 435–42.

71. Paley, W. (1828) *Natural Theology*, 2nd edn. Oxford: J. Vincent.

72. Penney, D., Foulds, L. R. & Hendy, M. D. (1982) Testing the theory of evolution by comparing phylogenetic trees constructed from five different protein sequences. *Nature*, 297: 197–200.

73. Ridley, M. (1982) Coadaptation and the inadequacy of natural selection. *British Journal for the History of Science*, 15: 45–68.

74. Ridley, M. (1986) *The Problems of Evolution*. Oxford: Oxford University Press.

75. Ridley, M. (1986) *Evolution and Classification: the reformation of cladism*. London: Longman.

76. Ruse, M. (1982) *Darwinism Defended*. London: Addison-Wesley.

77. Sales, G. & Pye, D. (1974) *Ultrasonic Communication by Animals*. London: Chapman & Hall.

78. Simpson, G. G. (1980) *Splendid Isolation*. New Haven: Yale University Press.

79. Singer, P. (1976) *Animal Liberation*. London: Cape.

80. Smith, J. L. B. (1956) *Old Fourlegs: the story of the Coelacanth*. London: Longmans, Green.

81. Sneath, P. H. A. & Sokal, R. R. (1973) *Numerical Taxonomy*. San Francisco: W. H. Freeman.

82. Spiegelman, S. (1967) An *in vitro* analysis of a replicating molecule. *American Scientist*, 55: 63–8.

83. Stebbins, G. L. (1982) *Darwin to DNA, Molecules to Humanity*. San Francisco: W. H. Freeman.

84. Thompson, S. P. (1910) *Calculus Made Easy*. London: Macmillan.

85. Trivers, R. L. (1985) *Social Evolution*. Menlo Park: Benjamin-Cummings.

86. Turner, J. R. G. (1983) 'The hypothesis that explains mimetic resemblance explains evolution': the gradualist-saltationist schism. In M. Grene (ed.) *Dimensions of Darwinism*, pp. 129–69. Cambridge: Cambridge University Press.

87.   Van Valen, L. (1973) A new evolutionary law. *Evolutionary Theory*, 1: 1–30.

88.   Watson, J. D. (1976) *Molecular Biology of the Gene*. Menlo Park: Benjamin-Cummings.

89.   Williams, G. C. (1966) *Adaptation and Natural Selection*. New Jersey: Princeton University Press.

90.   Wilson, E. O. (1971) *The Insect Societies*. Cambridge, Mass: Harvard University Press.

91.   Wilson, E. O. (1984) *Biophilia*. Cambridge, Mass: Harvard University Press

92.   Young, J. Z. (1950) *The Life of Vertebrates*. Oxford: Clarendon Press.

COMPUTER PROGRAMS AND 'THE EVOLUTION OF EVOLVABILITY'

The biomorph computer program described in Chapter 3 is now available for Apple Macintosh, RM Nimbus and IBM-compatible computers (see advertisement on p. 341). All three programs have the basic nine 'genes' necessary to produce the biomorphs illustrated in Chapter 3 and trillions more like them – or not so like them. The Macintosh version of the program also possesses a range of additional genes, producing 'segmented' biomorphs (with segmentation 'gradients') and biomorphic images reflected in various planes of symmetry. These enhancements of the biomorphic chromosome, together with a new colour version of the program now being developed for the Macintosh II and not yet released, led me to reflect on 'the evolution of evolvability'. This new reprinting of *The Blind Watchmaker* provides an opportunity to share some of these thoughts.

Natural selection can act only on the range of variation thrown up by mutation. Mutation is described as 'random' but this means only that it is not systematically directed towards improvement. It is a highly non-random subset of all the variation that we can conceive of. Mutation has to act by altering the processes of existing embryology. You can't make an elephant by mutation if the existing embryology is octopus embryology. That is obvious enough. What was less obvious to me until I started playing with the expanded Blind Watchmaker program is that not all embryologies are equally 'fertile' when it comes to fostering future evolution.

Imagine that a wide-open space of evolutionary opportunity has suddenly opened up – say a deserted continent has suddenly become available through natural catastrophe. What kinds of animals will fill the evolutionary vacuum? They will surely have to be descendants of individuals good at surviving in the post-catastrophe conditions. But more interestingly, some kinds of embryology might be especially good not just at surviving but at *evolving*. Perhaps the reason the mammals took over after the extinction of the dinosaurs is not just that mammals were good at individual *survival* in the post-dionsaur world. It may be that the mammalian way of growing a new body is also 'good' at throwing up a great variety of types – carnivores, herbivores, anteaters, tree-climbers, burrowers, swimmers and so on – and the mammals can therefore be said to be good at *evolving*.

What has this to do with computer biomorphs? Shortly after developing the Blind Watchmaker program, I experimented with other computer programs that were the same except that they employed a different basic embryology – a different fundamental body-drawing rule upon which mutation and selection could act. These other programs, although superficially similar to Blind Watchmaker, turned out to be sadly impoverished in the range of evolutionary possibility that they offered. Evolution continually became stuck up sterile blind alleys. Degeneration seemed to be the commonest outcome of even the most carefully guided evolution. In contrast, the branching-tree embryology at the heart of the Blind Watchmaker program seemed ever-pregnant with renewable evolutionary resources; there was no tendency towards automatic degeneration as evolution proceeded – the richness, versatility, even beauty, seemed to be indefinitely refreshed as the generations flashed by.

Nevertheless, prolific and varied as the biomorphic fauna produced by the original Blind Watchmaker program was, I continually found myself coming up against apparent barriers to further evolution. If Blind Watchmaker's embryology was so evolutionarily superior to those alternative programs, might there not be modifications, extensions to the embryological drawing rule that could make Blind Watchmaker itself even more luxuriant with evolutionary diversity? Or – another way of putting the same question – could the basic chromosome of nine genes be expanded in fruitful directions?

In designing the original Blind Watchmaker program, I deliberately tried to avoid deploying my biological knowledge. My purpose was to exhibit the power of non-random selection of random variation. I wanted the biology, the design, the beauty, to *emerge* as a result of selection. I didn't want to be able to accuse myself, later, of having built it in when I wrote the program in the first place. The branching-tree embryology of Blind Watchmaker was the very first embryology that I tried. That I had in fact been lucky was suggested by my subsequent disappointing experience with alternative embryologies. At all events, in thinking of ways to expand the basic 'chromosome', I did allow myself the luxury of using some of my biological knowledge and intuition. Among the most evolutionarily successful animal groups are those that have a *segmented* body plan. And among the most fundamental features of animal body plans are their plans of *symmetry*. Accordingly, the new genes that I added to the biomorphic chromosome controlled variations in segmentation and symmetry.

We, and all vertebrates, are segmented. This is clear in our ribs and our vertebral columns, whose repetitive nature is seen not just in the bones themselves but in the associated muscles, nerves and blood-vessels. Even our heads are fundamentally segmented, but in the adult head the segmental structure has become obscured to all but professionals schooled in embryonic anatomy. Fish are more obviously segmented than we are (think of the battery of muscles lying along the backbone of a kippered herring). In crustaceans, insects, centipedes and millipedes the segmentation is even manifest on the outside. The difference, in this respect, between a centipede and a lobster is one of homogeneity. The centipede is like a long goods-train with all the trucks almost identical to one another. The lobster is like a train with a motley variety of carriages and trucks, all basically the same and with the same jointed appendages sticking out of each. But in some cases the trucks are welded together in groups and the appendages have become large legs or pincers. In the tail region the trucks are smaller and more uniform, and their clawed side-apparatuses have become small, feathery 'swimmerets'.

To make biomorphs segmented I did the obvious thing: I invented a new gene controlling 'Number of Segments', and another new gene controlling 'Distance between Segments'. One complete old-style biomorph became a single segment of a new-style biomorph.

Above are seven biomorphs differing only in their 'Number of Segments' gene or their 'Distance between Segments' gene. The left-hand biomorph is the old, familiar branching tree, and the others are just serially arranged repetitive trains of the same basic tree. The simple tree, like all the original Blind Watchmaker biomorphs, is the special case of a 'one-segment animal'.

So far, I have talked only about uniform, centipede-like segmentation. Lobster segments differ from one another in complicated ways. A simpler way in which segments can vary is through 'gradients'. A woodlouse's segments are more like one another than a lobster's, but they are not as uniform as a typical millipede's (actually some apparent 'woodlice' or 'pillbugs' are technically millipedes). A woodlouse is narrow at the front and back, and broadens in the middle. As you work your way from the front to the back of the train, the segments have a size *gradient* which peaks in the middle. Other segmented animals, such as the extinct trilobites, are widest at the front and taper to the rear. They have a simpler size gradient which peaks at one end. It was this simpler kind of gradient that I sought to imitate in my segmented biomorphs. I did it by adding a constant number (which could be a negative number) to the expressed *value* of a particular gene going from front to back. Of the following three biomorphs, the left-hand one has no gradients, the middle one has a gradient on Gene 1, and the right-hand one on Gene 4.

Having expanded the basic biomorphic chromosome by these two genes and the associated gradient genes, I was ready to unleash the new-style biomorphic embryology into the computer and see what it could do by way of evolution. Compare the following picture with Chapter 3's Figure 5, all of whose biomorphs lack segmentation.

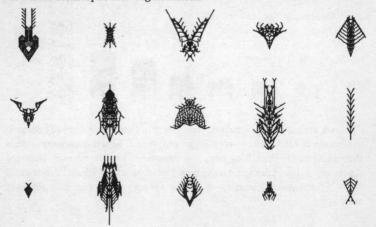

I think you will agree that a more 'biologically interesting' range of evolutionary versatility has now become available. The 'invention' of segmentation, as a new breakthrough in embryology, has opened floodgates of evolutionary potential in the land of computer biomorphs. My conjecture is that something like that happened in the origin of the vertebrates, and in the origin of the first segmented ancestors of insects, lobsters and millipedes. The invention of segmentation was a watershed event in evolution.

Symmetry was the other obvious innovation. The original Blind Watchmaker biomorphs were all constrained to be symmetrical about the midline. I introduced a new gene to make this optional. This new gene determined whether a biomorph with its original nine gene values set to those of the basic tree looked like (a) or like (b). Other genes determined whether there was symmetric reflection in the updown plane, (c), or full four-way symmetry, (d). These new genes could vary in all combinations, as in (e) and (f). When segmented animals were asymmetrical in the midline plane, I introduced a botanically inspired constraint: alternate segments should be asymmetrical in opposite directions, as in (g). Armed with these further genes, I again set about a vigorous breeding program, to see whether the new embryology could foster a more exuberant evolution than the old. Here is a portfolio of segmented biomorphs with midline asymmetry:

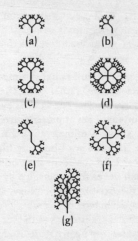

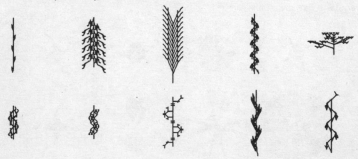

And here are some radially symmetrical biomorphs whose segmentation, if any, may be as cryptic as that of the adult human head:

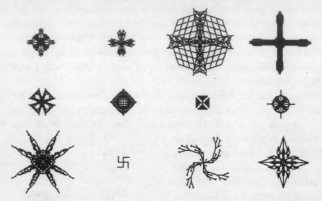

The gene for full radial symmetry tempts the selector to breed pleasing abstract designs rather than the biologically realistic ones that I had previously sought. The same is even more true of the colour version of the program that I am at present developing.

One group of animals, the echinoderms (including starfish, sea urchins, brittle stars and sea lilies), are highly unusual in being five-sidedly symmetrical. I am confident that, no matter how hard I or anybody else tries, we shall never find five-sided symmetry emerging by random mutation from the existing embryology. This would require a new 'watershed' innovation in biomorph embryology, and I have not attempted it. But freak starfish and sea-urchins with four or six arms rather than the usual five do sometimes turn up in nature. And in the course of exploring biomorph land I have encountered superficially starfish-like or urchin-like forms that have encouraged me to select for an increased resemblance. Here is a collection of echinoderm-like biomorphs, although none of them has the requisite five arms:

As a final test of the versatility of my new biomorphic embryology, I set myself the task of breeding a biomorphic alphabet good enough to sign my own name. Every time I encountered a biomorph that resembled, however slightly, a letter of the alphabet, I bred and bred to enhance the resemblance. The verdict on this ambitious endeavour is mixed, to say the least. 'I' and 'N' are well-nigh perfect. 'A' and 'H' are respectable if slightly ungainly. 'D' is poor, while to breed a proper 'K' is, I suspect, downright impossible – I had to cheat by borrowing the upright stroke of the 'W'. Yet another gene would have to be added, I suspect, before a plausible 'K' could be evolved.

ΓICHAΓ▷　▷AWKINᶂ

After my somewhat illiterate attempt to sign my own name, I had more luck with evolving that of the inspired artefact with which all this work was done:

MACINTOᶂH

It is my strong impression, borne out, I hope, by the illustrations here, that the introduction of a few radical changes in the fundamental embryology of the biomorphs has opened up new vistas of evolutionary possibility which simply were not available to the original program described in Chapter 3. And, as I said earlier, I believe that something similar happened at various junctures in the evolution of some prominent groups of animals and plants. The invention of segmentation by our own ancestors, and separately by the ancestors of insects and crustaceans, is probably only one of several examples of 'watershed' events in our evolutionary history. These watershed events are, at least when seen with the wisdom of hindsight, different in kind from ordinary evolutionary changes. Our first segmented ancestors, and the first segmented ancestor of earthworms and insects, may not have been particularly good at surviving as individuals – though obviously they *did* survive as individuals, or we, their descendants, would not be here. My point now is that the invention of segmentation by these ancestors was more significant than just a new technique for surviving, like sharper teeth or keener eyes. When segmentation was added to the embryonic procedures of our ancestors, whether or not the individual animals concerned became better at surviving, the lineages to which they belonged suddenly became *better at evolving*.

Modern animals, we vertebrates and all our fellow-travellers on this planet, inherit the genes of an unbroken line of ancestors that were good at individual survival. That much I tried to make clear in *The Blind Watchmaker*. But we also inherit the embryological procedures of ancestral lineages that were good at *evolving*. There has been a kind of higher-order selection among lineages, not for their ability to survive but for their longer-term ability to evolve. We bear the accumulated improvements of a number of watershed events, of which the invention of segmentation is just one example. It is not just bodies and behaviour that have evolved in improved directions. We could even say that evolution itself has evolved. There has been a progressive evolution of evolvability.

The Macintosh version of the Blind Watchmaker program has menu options to turn on or off each of the main categories of mutation. By turning off all the new types of mutation one reverts to the earlier version of the program (or the present IBM version). Breed for a while under these conditions, and you get some feel for the enormous range of faunas permitted by the earlier program, but also for its limitations. If you then switch on, say, segmentation mutations, or symmetry mutations (or if you switch from IBM to a Macintosh!), you can exult in something of the feeling of liberation that may have attended evolution's great watershed events.

Reference   Dawkins, R. (1989) The evolution of evolvability. In C. Langton (ed.) *Artificial Life*. New York: Addison-Wesley.

INDEX AND KEY TO BIBLIOGRAPHY

This book is meant to be read from cover to cover. It is not a work of reference. Many items in the index will mean something only to people that have already read the book and want to find a particular place again. In such a book, footnotes are an irritating distraction. The following index, in addition to performing the normal function of an index, is intended to replace footnotes by acting as a key to the bibliography. The numbers in brackets refer to the numbered books or articles in the bibliography. Other numbers refer to pages in the book. Where an indexed word recurs on a consecutive series of pages, normally only the first page, or the page where a definition will be found, is given.

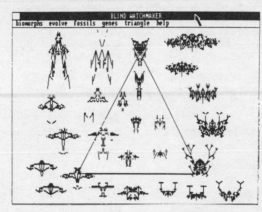

Visit Penguin on the Internet
and browse at your leisure

- preview sample extracts of our forthcoming books
- read about your favourite authors
- investigate over 10,000 titles
- enter one of our literary quizzes
- win some fantastic prizes in our competitions
- e-mail us with your comments and book reviews
- instantly order any Penguin book

and masses more!

'To be recommended without reservation ... a rich and rewarding on-line experience' – Internet Magazine

www.penguin.co.uk

READ MORE IN PENGUIN

In every corner of the world, on every subject under the sun, Penguin represents quality and variety – the very best in publishing today.

For complete information about books available from Penguin – including Puffins, Penguin Classics and Arkana – and how to order them, write to us at the appropriate address below. Please note that for copyright reasons the selection of books varies from country to country.

In the United Kingdom: Please write to *Dept. EP, Penguin Books Ltd, Bath Road, Harmondsworth, West Drayton, Middlesex UB7 ODA*

In the United States: Please write to *Consumer Sales, Penguin Putnam Inc., P.O. Box 12289 Dept. B, Newark, New Jersey 07101-5289.* VISA and MasterCard holders call 1-800-788-6262 to order Penguin titles

In Canada: Please write to *Penguin Books Canada Ltd, 10 Alcorn Avenue, Suite 300, Toronto, Ontario M4V 3B2*

In Australia: Please write to *Penguin Books Australia Ltd, P.O. Box 257, Ringwood, Victoria 3134*

In New Zealand: Please write to *Penguin Books (NZ) Ltd, Private Bag 102902, North Shore Mail Centre, Auckland 10*

In India: Please write to *Penguin Books India Pvt Ltd, 11 Community Centre, Panchsheel Park, New Delhi 110017*

In the Netherlands: Please write to *Penguin Books Netherlands bv, Postbus 3507, NL-1001 AH Amsterdam*

In Germany: Please write to *Penguin Books Deutschland GmbH, Metzlerstrasse 26, 60594 Frankfurt am Main*

In Spain: Please write to *Penguin Books S. A., Bravo Murillo 19, 1° B, 28015 Madrid*

In Italy: Please write to *Penguin Italia s.r.l., Via Benedetto Croce 2, 20094 Corsico, Milano*

In France: Please write to *Penguin France, Le Carré Wilson, 62 rue Benjamin Baillaud, 31500 Toulouse*

In Japan: Please write to *Penguin Books Japan Ltd, Kaneko Building, 2-3-25 Koraku, Bunkyo-Ku, Tokyo 112*

In South Africa: Please write to *Penguin Books South Africa (Pty) Ltd, Private Bag X14, Parkview, 2122 Johannesburg*

READ MORE IN PENGUIN

SCIENCE AND MATHEMATICS

Six Easy Pieces Richard P. Feynman

Drawn from his celebrated and landmark text *Lectures on Physics*, this collection of essays introduces the essentials of physics to the general reader. 'If one book was all that could be passed on to the next generation of scientists it would undoubtedly have to be *Six Easy Pieces*' John Gribbin, *New Scientist*

A Mathematician Reads the Newspapers John Allen Paulos

In this book, John Allen Paulos continues his liberating campaign against mathematical illiteracy. 'Mathematics is all around you. And it's a great defence against the sharks, cowboys and liars who want your vote, your money or your life' Ian Stewart

Dinosaur in a Haystack Stephen Jay Gould

'Today we have many outstanding science writers ... but, whether he is writing about pandas or Jurassic Park, none grabs you so powerfully and personally as Stephen Jay Gould ... he is not merely a pleasure but an education and a chronicler of the times' *Observer*

Does God Play Dice? Ian Stewart

As Ian Stewart shows in this stimulating and accessible account, the key to this unpredictable world can be found in the concept of chaos, one of the most exciting breakthroughs in recent decades. 'A fine introduction to a complex subject' *Daily Telegraph*

About Time Paul Davies

'With his usual clarity and flair, Davies argues that time in the twentieth century is Einstein's time and sets out on a fascinating discussion of why Einstein's can't be the last word on the subject' *Independent on Sunday*

READ MORE IN PENGUIN

SCIENCE AND MATHEMATICS

In Search of SUSY   John Gribbin

Many physicists believe that we are on the verge of developing a complete 'theory of everything' which can reduce the four basic forces of nature – gravity, electromagnetism, the strong and weak nuclear forces – to a single superforce. At its heart is the principle of super-symmetry (SUSY).

Fermat's Last Theorem   Amir D. Aczel

Here, weaving together history and science, Amir D. Aczel offers a thrilling, step-by-step account of the search for the mathematicians' Holy Grail. 'Mr Aczel has written a tale of buried treasure . . . This is a captivating volume' *The New York Times*

Insanely Great   Steven Levy

It was Apple's co-founder Steve Jobs who referred to the Mac as 'insanely great'. He was absolutely right: the machine that revolu-tionized the world of personal computing was and is great – yet the machinations behind its inception were nothing short of insane. 'A delightful and timely book' *The New York Times Book Review*

The Artful Universe   John D. Barrow

This thought-provoking investigation illustrates some unexpected links between art and science. 'Full of good things . . . In what is probably the most novel part of the book, Barrow analyses music from a mathematical perspective . . . an excellent writer' *New Scientist*

The Jungles of Randomness   Ivars Peterson

Taking us on a fascinating journey into the ambiguities and un-certainties of randomness, Ivars Peterson explores the complex interplay of order and disorder, giving us a new understanding of nature and human activity.

READ MORE IN PENGUIN

POPULAR SCIENCE

In Search of Nature   Edward O. Wilson

A collection of essays of 'elegance, lucidity and breadth' *Independent*.
'A graceful, eloquent, playful and wise introduction to many of the
subjects he has studied during his long and distinguished career in
science' *The New York Times*

Clone   Gina Kolata

'A thoughtful, engaging, interpretive and intelligent account ... I
highly recommend it to all those with an interest in ... the new
developments in cloning' *New Scientist*. 'Superb but unsettling' J. G.
Ballard, *Sunday Times*

The Feminization of Nature   Deborah Cadbury

Scientists around the world are uncovering alarming facts. There is
strong evidence that sperm counts have fallen dramatically. Testicular
and prostate cancer are on the increase. Different species are showing
signs of 'feminization' or even 'changing sex'. 'Grips you from page
one ... it reads like a Michael Crichton thriller' John Gribbin

Richard Feynman: A Life in Science   John Gribbin and Mary Gribbin

'Richard Feynman (1918–88) was to the second half of the century
what Einstein was to the first: the perfect example of scientific genius'
Independent. 'One of the most influential and best-loved physicists of
his generation ... This biography is both compelling and highly
readable' *Mail on Sunday*

***T. rex* and the Crater of Doom**   Walter Alvarez

Walter Alvarez unfolds the quest for the answer to one of science's
greatest mysteries – the cataclysmic impact on Earth which brought
about the extinction of the dinosaurs. 'A scientific detective story par
excellence, told with charm and candour' Niles Eldredge

READ MORE IN PENGUIN

POPULAR SCIENCE

How the Mind Works Steven Pinker

'Presented with extraordinary lucidity, cogency and panache ...
Powerful and gripping ... To have read [the book] is to have consulted
a first draft of the structural plan of the human psyche ... a glittering
tour de force' *Spectator*. 'Witty, lucid and ultimately enthralling'
Observer

At Home in the Universe Stuart Kauffman

Stuart Kauffman brilliantly weaves together the excitement of
intellectual discovery and a fertile mix of insights to give the general
reader a fascinating look at this new science – the science of complexity
– and at the forces for order that lie at the edge of chaos. 'Kauffman
shares his discovery with us, with lucidity, wit and cogent argument,
and we see his vision ... He is a pioneer' Roger Lewin

Stephen Hawking: A Life in Science
Michael White and John Gribbin

'A gripping account of a physicist whose speculations could prove as
revolutionary as those of Albert Einstein ... Its combination of
erudition, warmth, robustness and wit is entirely appropriate to their
subject' *New Statesman & Society*. 'Well-nigh unputdownable' *The
Times Educational Supplement*

Voyage of the *Beagle* Charles Darwin

The five-year voyage of the *Beagle* set in motion the intellectual
currents that culminated in the publication of *The Origin of Species*.
His journal, reprinted here in a shortened version, is vivid and
immediate, showing us a naturalist making patient observations,
above all in geology. The editors have provided an excellent
introduction and notes for this edition, which also contains maps and
appendices.

BY THE SAME AUTHOR

Climbing Mount Improbable

'Dazzling' David Attenborough

'A beautiful, barnstorming thunderclap of a book' Michael White, *Mail on Sunday*

'Mount Improbable ... is Dawkins's metaphor for natural selection: its peaks standing for evolution's most complex achievements: the human brain, the squid's eye, and the albatross's aeronautical prowess ... exhilarating – a perfect, elegant riposte to a great deal of fuzzy thinking about natural selection and evolution' Robin McKie, *Observer*

'One of the most gifted storytellers of our generation ... He is a missionary who writes like an angel. He is to Darwinism what Saint Paul is to Christianity' Mike Maran, *Scotland on Sunday*

'The parables – riveting biological narratives, enthralling as the *Arabian Nights* tales – continue to ring the changes. Yet the central message – that DNA transcends the significance of the organism – remains the same ... organisms are merely vehicles for genes ... This is vintage Dawkins' John Cornwell, *New Scientist*

'An elegant series of lectures on Darwinian selection ... Dawkins continues a tradition of scientific writing from Galileo to Darwin' Ian Thomson, *Daily Telegraph*

'Dawkins has done more than anyone else now writing to make evolutionary biology comprehensible and acceptable' John Maynard Smith, *Sunday Times*